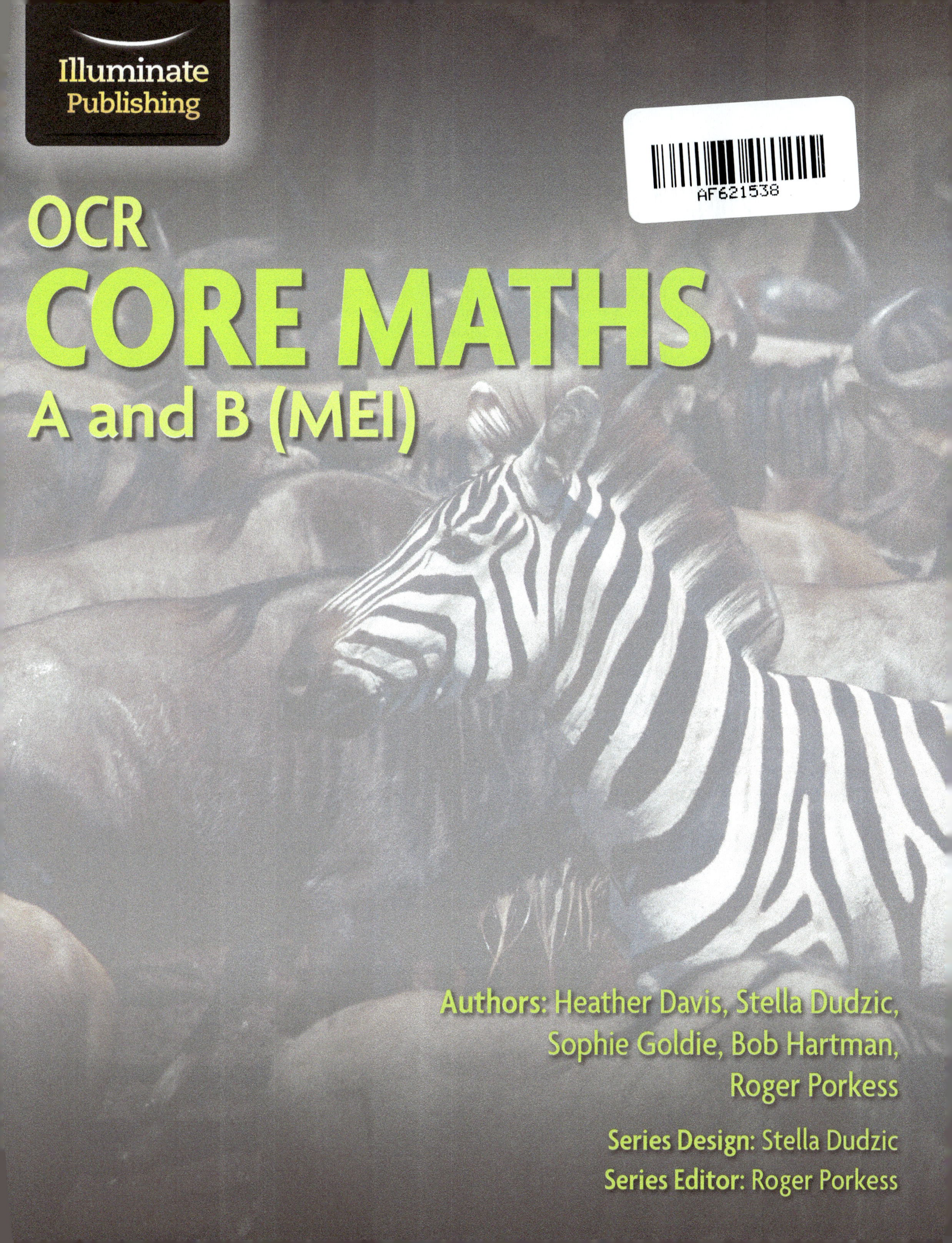

Illuminate Publishing

OCR CORE MATHS A and B (MEI)

Authors: Heather Davis, Stella Dudzic, Sophie Goldie, Bob Hartman, Roger Porkess

Series Design: Stella Dudzic

Series Editor: Roger Porkess

Published in 2021 by Illuminate Publishing, an imprint of
Hodder Education (a trading division of Hodder & Stoughton Limited),
an Hachette UK Company, Carmelite House,
50 Victoria Embankment, London EC4Y 0DZ

Orders: please contact Hachette UK Distribution, Hely Hutchinson Centre, Milton Road, Didcot, Oxfordshire, OX11 7HH. Telephone: +44 (0)1235 827827. Email: education@hachette.co.uk. Lines are open from 9 a.m. to 5 p.m., Monday to Friday. You can also order through our website: www.hoddereducation.com

British Library Cataloguing-in-Publication Data

A catalogue record for this book is available from the British Library

ISBN 978-1-913963-00-2

Printed by: Ashford Colour Ltd, UK

The authorised representative in the EEA is Hachette Ireland,
8 Castlecourt Centre, Dublin 15, D15 XTP3, Ireland (email: info@hbgi.ie)

Impression 4
Year 2026

Hachette UK's policy is to use papers that are natural, renewable and recyclable products and made from wood grown in well-managed forests and other controlled sources. The logging and manufacturing processes are expected to conform to the environment regulations of the country of origin.

Editor: Dawn Booth
Design and layout: Kamae Design
Cover design: Kamae Design

The teaching content of this resource is endorsed by OCR for use with specification Core Maths A (MEI) Level 3 Certificate (H868) and Core Maths B (MEI) Level 3 Certificate (H869). In order to gain OCR endorsement, this resource has been reviewed against OCR's endorsement criteria.

This resource was designed using the most up to date information from the specification. Specifications are updated over time which means there may be contradictions between the resource and the specification, therefore please use the information on the latest specification and Sample Assessment Materials at all times when ensuring students are fully prepared for their assessments.

Any references to assessment and/or assessment preparation are the publisher's interpretation of the specification requirements and are not endorsed by OCR. OCR recommends that teachers consider using a range of teaching and learning resources in preparing learners for assessment, based on their own professional judgement for their students' needs. OCR has not paid for the production of this resource, nor does OCR receive any royalties from its sale. For more information about the endorsement process, please visit the OCR website, www.ocr.org.uk.

MIX
Paper | Supporting responsible forestry
FSC™ C104740
www.fsc.org

Contents

* Chapter 17 covers content that is in both Part 2: Critical Maths and Part 3: Statistical Problem Solving.

Introduction

Core Maths is designed to meet the needs of the very large number of post-16 students who are not taking AS or A Level mathematics but who use maths in their other subjects, everyday life and in the workplace. Typically, they have obtained at least a standard pass at GCSE.

This book meets the particular needs of students following the OCR Core Maths A and B (MEI) course but it will also be useful to those who are not taking it but might have done. The book is designed to motivate students whose interest in mathematics is as a means to an end. So the content is introduced through contexts that readers should find interesting and relevant. It encourages a way of thinking in which applying mathematics to any situation is an entirely natural thing to do.

The structure of the book is designed to help readers use it successfully. The first ten chapters cover the first part of the OCR Core Maths A and B (MEI) course, Introduction to Quantitative Reasoning. There are then two alternative parts, corresponding to the A and B qualifications. Chapters 11 to 17 cover Critical Maths and Chapters 17 to 20 cover Statistical Problem Solving. (Chapter 17 is common to both of these parts.)

Each chapter consists of a number of sections. Each section has essentially the same structure with an opening Activity, a Worked example and a short Exercise of typically two questions. Between these main features there are short passages of text, advice and discussion. Readers should work through each section, including the exercise questions, checking the answers against those in the back of the book. Fuller information is available in the accompanying online resource.

Towards the end of each chapter there is a summary exercise with the title 'Test yourself'; these questions give readers the opportunity to check that they have mastered the skills covered in the chapter. At the very end there is a Practice question, allowing readers to test their knowledge and understanding at the expected level, followed by a summary of what has been covered in the chapter.

On a final note, the authors would like to extend a personal message to anyone who is using this book. We wish you success. This will involve more than gaining good marks in the Core Maths exam. You will also be able to use maths to make more sense of your other subjects, and so do better in them. You will also be able to use maths successfully in the rest of your lives, in further study, at work and in having a better understanding of the world around you.

How to use this book

The pages use a variety of features to support your learning.

- Each chapter is divided into **sections**, which cover the important points.

1B Using percentages

Percentages can be used to express risk but risk can also be expressed as 1 in a number. You need to understand both ways.

- Each section starts with an **Activity**. This allows you to remind yourself of existing knowledge relevant to a new topic, or to do some exploratory work or thinking before getting started on it. An Activity always includes something for you to do. Answers to these are given towards the end of the book.

Activity 2F Think of a number

Rishi is thinking of a number.

He multiplies it by 2 and adds 11.

He then takes the square root and his final answer is 5.

What number does he start with?

- **Worked examples** and **solutions** cover the key concepts that you need to know in context.

Worked example 2F(2)

Solve the equation: $t^2 - 15t - 50 = 0$.

$t^2 - 15t - 50 = 0$ is a quadratic equation.

Solution

$t^2 - 15t - 50 = 0$

$t^2 - 5t - 10t + 50 = 0$ ← Split the term in t: $-5 + -10 = -15$, the coefficient of t is -15. $-5 \times -10 = 50$, the constant term is 50.

$t(t - 5) - 10(t - 5) = 0$ ← Factorise each pair of terms.

$(t - 5)(t - 10) = 0$ ← Factorise, extracting $(t - 5)$ as a common factor.

Either $t - 5 = 0 \Rightarrow t = 5$.

Or $t - 10 = 0 \Rightarrow t = 10$. ← One of the factors must be zero if their product is zero.

So, $t = 5$ or $t = 10$.

- **Exercises** give lots of opportunities for you to apply the skills you have just encountered in the Worked example, with answers provided towards the end of the book. The colour bar down the side of each question indicates whether these are starter level (green), general level (orange) or extension level (purple).

Exercise 7B

1 Which of the following is the gradient of the graph on the right?
(a) 2 (b) −2 (c) $\frac{1}{2}$ (d) $-\frac{1}{2}$

2 Calculate the gradients of the lines joining:
(a) (0, 1) and (6, 3) (b) (1, 1) and (4, 9) (c) (1, 7) and (5, 1).

3 State what the gradient represents for the graphs of:
(a) distance travelled in kilometres against time taken in hours
(b) volume of water in a leaking tank in litres against time in minutes.

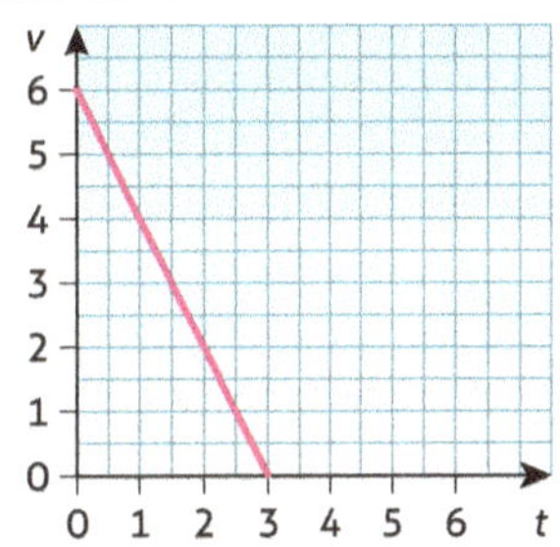

- **Call-outs** are provided to give guidance or explanations. They may also give notes that you will find helpful.

You would expect that most female students are around average height and that there are few very short or very tall students.

- **Notice that** boxes draw attention to specific points that you need to be aware of.

> The flood risk is expressed as a probability for each year. For the risks you looked at earlier, they were over a lifetime.

- **Discussion points** are used to help you think about a particular issue.

> **The average person?**
>
> Risks are usually given for an average person.
>
> What do we mean by an average person?
>
> Is this the best way to communicate risk?

- A **remember** box highlights an important point that you will find useful to know.

> Per cent (%) means out of every hundred.

- **Cautions** highlight common mistakes that you may make.

> If you are asked to give an explanation or a reason, there may be more than one possible answer.

- **Test yourself questions** provide a summary test at the end of each chapter, giving you the opportunity to check that you have mastered the skills covered. These questions also use the colour bar system to show what level they are. Answers to these are also given towards the end of the book.

Test yourself 9H

1 A particular virus is spherical with diameter 80 nanometres.

(a) Write the radius of the virus in metres, as a decimal.

(b) Work out the volume of the virus, giving your answer in cubic metres.

> 1 nanometre = 10^{-9} metres.

> The volume of a sphere of radius r is given by $V = \frac{4}{3}\pi r^3$.

- A **Practice question** at the end of each Test yourself section provides a wider question that enables you to test application of their knowledge and understanding at the expected level.

Practice question

A typical adult human contains 39 trillion bacteria.

The average length of one of these bacteria is 1 micrometre.

(a) Write 39 trillion in standard form.

(b) Write 1 micrometre in metres, in standard form.

The distance from the Earth to the Moon is 400 thousand kilometres.

(c) If all the bacteria in an adult human were laid end to end, would the line reach the Moon? Give suitable calculations to justify your answer.

> A trillion is a million million. There are 1,000 micrometres in 1 millimetre.

- **Now that you have completed this chapter:** a series of bullet points at the end of each chapter providing a handy summary of the key learning points.

Now that you have completed this chapter, you should know:

- How you conduct a hypothesis test.
- The vocabulary associated with hypothesis testing.
- How to carry out a χ^2 test on a contingency table.

CHAPTER 1

Probability and risk

We don't know what will happen in the future but probability allows us to measure how likely events are to happen. This is especially useful when planning for the possibility of adverse events (risks).

1A Risk

Risk assessments are used to look at what might go wrong in order to help planning in businesses, schools and hospitals.

Activity 1A Risk assessment

The table below shows part of a risk assessment for a small business.

Risk	Likelihood	Impact	Risk rating
Severe weather disrupts travel	low	low	green
Office unusable due to fire or flood damage	low	medium	yellow
Computer system failure	low	high	yellow
Loss of staff	high	low	yellow
Customers find alternative supplier	medium	medium	amber
Loss of reputation	medium	high	amber
Cashflow problems	high	high	red

(a) What do you think 'likelihood' and 'impact' mean?

(b) How do you think the risk rating in the final column relates to the previous two columns?

(c) Why do you think it would be useful for a business to do a risk assessment?

Quantifying risk

To give an idea of how likely the risks are to happen, the table on the previous page uses the terms low, medium and high. This gives an idea of which are more likely to happen but not how much more likely.

The term '*risk*' is used in different ways. It can mean any of the following:

- An *event* someone might worry about.
- *How likely* such an event is to happen.
- A *combined measure* of how likely and how bad the event is.

Worked example 1A

For the events in Activity 1A, the table below attaches numbers to the likelihood and impact. Complete the table to get an overall measure of risk for each event. 1 is low, 2 is medium and 3 is high for both likelihood and impact.

What values of likelihood × impact are indicated by each of the colours in the risk rating column?

Solution

Risk	Likelihood	Impact	Likelihood × impact	Risk rating
Severe weather disrupts travel	1	1	1	green
Office unusable due to fire or flood damage	1	2	2	yellow
Computer system failure	1	3	**3**	yellow
Loss of staff	3	1	**3**	yellow
Customers find alternative supplier	2	2	**4**	amber
Loss of reputation	2	3	**6**	amber
Cashflow problems	3	3	**9**	red

The bold numbers in the dark green squares have been completed by multiplying likelihood and impact.

The values indicated by the colours are shown in the table below.

Risk rating	Green	Yellow	Amber	Red
Values of likelihood × impact	1	2, 3	4, 6	9

Combining likelihood and impact

Multiplying measures of likelihood and impact to give an overall measure is one way to order risks. An alternative way of doing this is to add measures of likelihood and impact.

Risks with the highest ratings must be managed by seeking to reduce the likelihood or impact.

Exercise 1A

1 Risks can be shown in a chart like the one on the right.

Which of the four regions, A, B, C or D, do you think each of the following risks for a business belongs in?

(a) Unhappy customer sues the business.

(b) Member of staff gets locked in the office at the end of the day.

(c) Visitor slips on ice when entering the office.

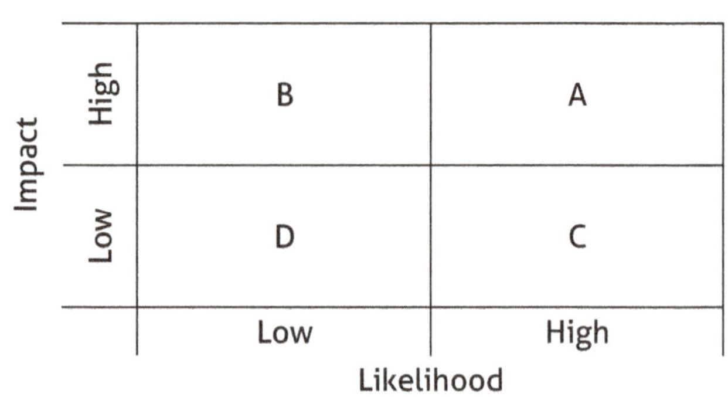

2 Here are four risks that people might be concerned about. Imagine an average person in the UK. Put the risks in order of the chance of them happening at some time in the life of an average person in the UK, from least likely to most likely. An estimate of the probability of each risk has been given.

Getting cancer	Developing high blood pressure	Suffering from depression	Going into a care home in old age
1 in 2	90%	1 in 5	24%

3 How different would the risks given in Q2 above be for different individuals? What kinds of people might have higher than average risk? Lower than average risk?

The average person?

Risks are usually given for an average person.

What do we mean by an average person?

Is this the best way to communicate risk?

There are different ways of measuring how likely a risk is to happen. Two common ways are:

- as a percentage
- as 1 in *n*, where *n* is a whole number.

You need to be able to move between these to help you understand and compare risks.

1B Using percentages

Percentages can be used to express risk but risk can also be expressed as 1 in a number. You need to understand both ways.

Activity 1B Visualising percentages

Here is a diagram of 100 people.

(a) What percentage is 1 in 7?

(b) What percentage is 1 in 12?

Per cent (%) means out of every hundred.

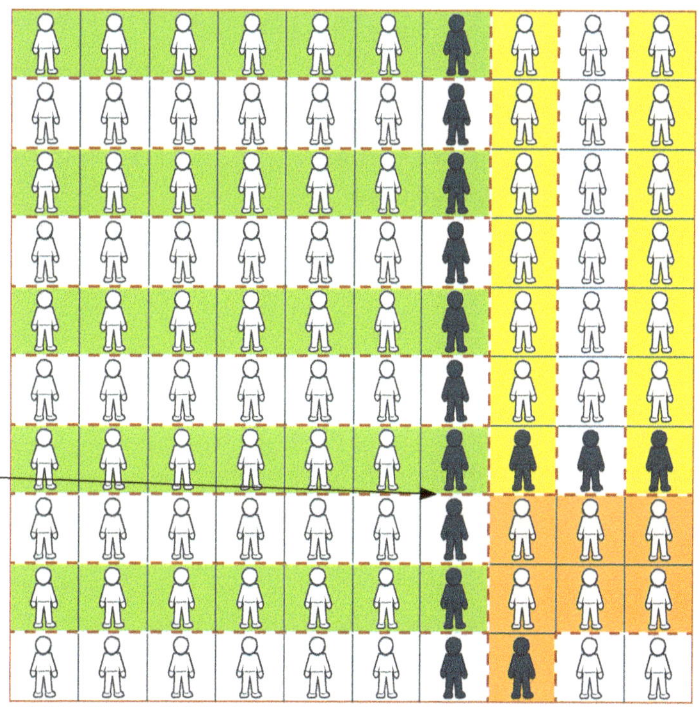

1 in 7 means one out of every seven people. There are just over 14 groups of 7 in 100, so 1 in 7 is just over 14%.

Worked example 1B

Which is the more likely risk? One with a probability of 14% or one with a probability of 1 in 9?

Solution

1 in 9 means 1 out of every 9.

There are 100 ÷ 9 ≈ 11.1 groups of 9 in 100.

So, 1 in 9 is 11.1%.

14% is the higher probability.

Percentage means out of 100.

To compare the probabilities, they need to be in the same form. It is usually easiest to have them both as percentages.

Exercise 1B

1 For each part of this question, four risks have the given probabilities of happening. Put each group of four in order, from least likely to most likely.

(a)	1 in 5	1 in 1000	1 in 2	1 in 100
(b)	10%	55%	25%	77%
(c)	1 in 30	76%	1 in 6	23%

1C Risk and time

This section introduces time, which is an important factor when considering risk.

Activity 1C Flood risk: the time factor

The map shows the flood risk for an area south of Worcester.

Why is time important when expressing risk?

The flood risk is expressed as a probability for each year. For the risks you looked at earlier, they were over a lifetime.

Risk of flooding from rivers and seas

- High (greater than 1 in 30 every year)
- Medium (between 1 in 30 and 1 in 100 every year)
- Low (between 1 in 100 and 1 in 1000 each year)
- Very low (less than 1 in 1000 every year)

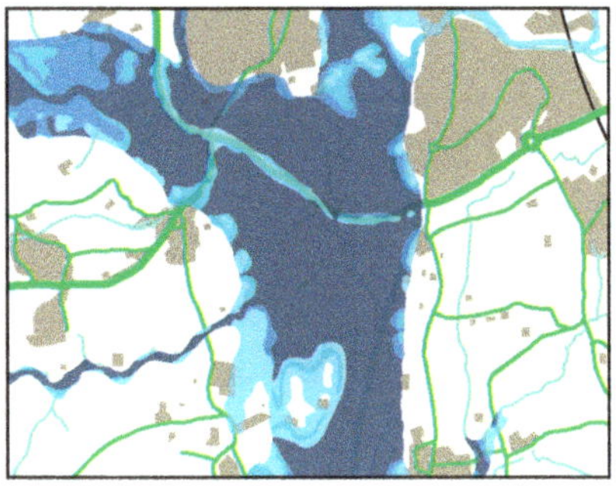

Worked example 1C

A 50-year-old woman has a medical check-up. She is told that her 10-year risk of suffering from heart disease is 2% and her lifetime risk is 40%.

Give a reason why the lifetime risk is higher than the 10-year risk.

Solution

Her likely lifetime is longer than 10 years, so the risk over the longer period of time will be higher.

Another possible factor is that risk increases with age, so it will get higher as she gets older.

If you are asked to give an explanation or a reason, there may be more than one possible answer.

Exercise 1C

1 Which of the following statements are true for an area with a flood risk of 1 in 30 every year?

(a) On average, there will be a flood every 30 years.

(b) When there is a flood, the area will not flood for another 30 years.

(c) If there hasn't been a flood for a long time, the risk of flooding goes up.

(d) There is a 50–50 chance of there being at least one flood in an interval of 15 years.

1D Calculating probability

Probabilities are used to measure how likely events are when assessing risk or analysing games. They are also often used in medicine, law and insurance.

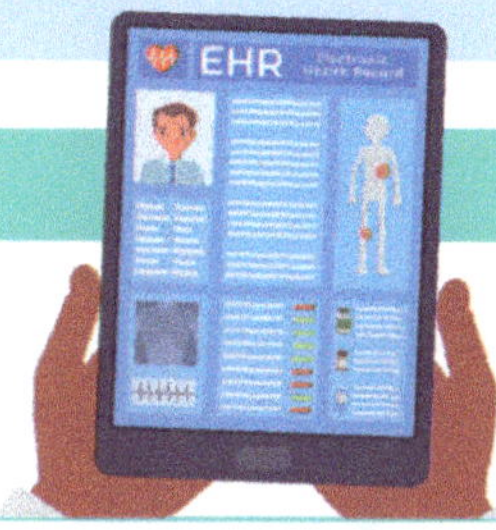

Activity 1D Calculating risk

In the section on risk, the lifetime risk of being diagnosed with cancer for someone in the UK born after 1960 was given as 1 in 2. How do you think this probability has been worked out?

People have been calculating probabilities for hundreds of years. Early calculations were often about dice, coins or cards because people playing games wanted to know their chances of winning (or losing). **Equally likely outcomes** can be identified for these situations if the coins, dice, etc. are **fair**. This allows probabilities to be worked out.

Equally likely outcomes are possible outcomes that have equal probabilities of happening.

Worked example 1D

A game at a funfair is played as follows. Contestants choose a folded ticket from a bag. There is a number on each ticket. If the number is even, the contestant wins a prize.

(a) Explain why a contestant might not have a 50–50 chance of winning.

(b) Explain how you could estimate the probability of a contestant winning.

Solution

(a) If there is an equal number of odd and even numbers to choose from then the contestant has a 50–50 chance of winning, but you don't know whether this is the case. There may be more odd numbers in the bag.

(b) You could estimate the probability of a contestant winning by watching the game played a large number of times. The estimate of probability would be $\frac{\text{number of wins}}{\text{number of plays}}$.

You only have a 50–50 chance if there are two equally likely outcomes. Games of chance might be described in a way that makes you think you have a better chance of winning than you actually do.

Experiments, outcomes and events

Doing something that has a result where uncertainty is involved is sometimes called an ***experiment***. Each basic result of the experiment is an ***outcome***. An ***event*** can be a single outcome or a combination of outcomes.

For example, in (d) in Exercise 1D (next page), choosing a person at random is the experiment; the outcomes are all the individual people in the town but there are only two events of interest: male or female. In this case, each event is made up of many possible outcomes.

Mutually exclusive and exhaustive events

Mutually exclusive events cannot both happen at once.

Exhaustive events cover all possibilities.

Exercise 1D

1 For each of the following situations, decide whether the events listed are equally likely. Explain your reasoning.

	Situation	Events
(a)	Toss one fair coin	• Heads • Tails
(b)	Toss two fair coins	• No heads • One head • Two heads
(c)	Plant a seed	• It grows • It does not grow
(d)	Choose a person at random from your town	• Male • Female
(e)	Choose a person at random from your town	• Under 16 years old • Aged 16 or over

1E Using relative frequency to estimate probability

Relative frequency is sometimes referred to as experimental probability; it is the proportion of experiments that result in the event you are interested in.

Activity 1E Increasing the sample size

A new, quicker test has been developed for a particular illness. The new test was tried on people who have already been tested for the illness using a test known to be reliable. The percentage of new tests that are correct is calculated as new people join the trial. The results are shown in the table and graph on the right.

How accurate is the new test?

Number of people tested so far	1	2	3	4	5	10	15	20	25	30
Number of tests correct	0	0	1	2	3	7	12	16	20	25

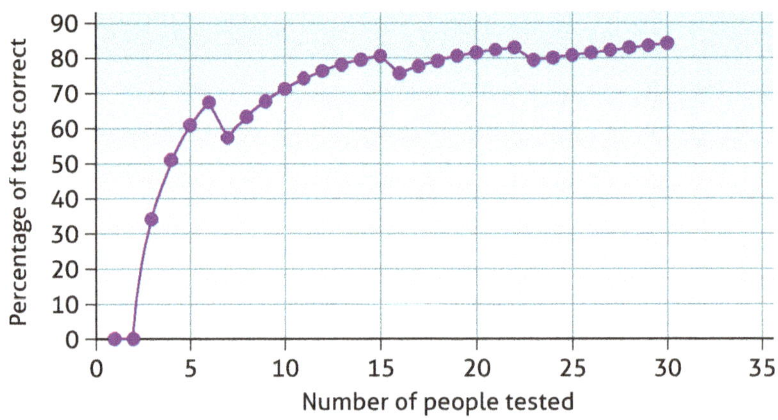

For probabilities related to medicine, weather or other events where we cannot find equally likely outcomes, data from a sample (often from the past) are used to estimate probability.

Worked example 1E

The table below shows the usual adult population of York (not including university students) in 2011.

	Age								
Sex	**16–24**	**25–34**	**35–44**	**45–54**	**55–64**	**65–74**	**75–84**	**85 and over**	**Total**
All persons	23 937	26 393	26 231	25 705	22 368	16 926	11 630	4 856	158 046
Males	12 039	13 321	13 004	12 680	10 901	7 842	4 927	1 639	76 353
Females	11 898	13 072	13 227	13 025	11 467	9 084	6 703	3 217	81 693

Data: Crown Copyright ONS Census 2011

Use the data to estimate the probability that a person chosen at random from the population of York is over the age of 65.

Solution

Based on the Census 2011 data, the **relative frequency** of people aged 65 or over in York is

(16 926 + 11 630 + 4 856) out of 158 046. This is 33 412 out of 158 046.

As a decimal this is 33 412 ÷ 158 046 ≈ 0.211. As a percentage, this is 21.1%.

As time moves on from 2011, the census data become less representative but they can still be used to **estimate** probabilities.

Exercise 1E

The first two questions refer to the table giving the adult population of York in Worked example 1E.

1 A person is chosen at random from this adult population of York to take part in a survey. What is the probability of choosing each of the following? Give your answers as decimals.

(a) A male.

(b) A female.

(c) Someone aged between 16 and 24 (inclusive).

(d) A male below the age of 65.

(e) Check that answers (a) and (b) add up to 1. Explain why this should happen.

76 353 out of 158 046 are male. To find this as a decimal, divide 76 353 by 158 046.

2 A female is chosen at random from the adult population of York represented in the table. What is the probability of choosing each of the following?

(a) Someone aged between 16 and 24 (inclusive).

(b) Someone below the age of 65.

3 The following table gives the number of 17-year olds who had progressed to Higher Education (HE) in 2011/12 according to the type of school they attended.

School/college type	Independent	Selective state	Other state	Total state	Total
Total age 17 in 2009/10	29 875	25 115	259 535	284 650	314 525
Progressed to HE by age 19 in 2011/12	25 535	22 525	168 365	190 890	216 425

Data Crown Copyright 2014, BIS

Use the data in the table to estimate the probabilities of the following:

(a) A 17-year-old student going on to HE.

(b) A 17-year-old student in a selective state school going on to HE.

(c) Someone in HE having been to a selective state school at age 17.

1F Dependent and independent events

In this section you will consider whether additional information changes the probability.

Activity 1F Does it make a difference?

Does going to a selective state school change the probability of progressing to Higher Education?
Use the data from question 3 in Exercise 1E, on the previous page, to help you decide.

Dependent and independent events and conditional probability

The probability that a 17-year old in a selective state school goes on to HE is a ***conditional probability***. It is sometimes written as 'the probability that a 17-year old goes on to HE ***given that*** he/she is at a selective state school'. The extra information that the 17-year old goes to a selective state school might affect the probability that he/she goes on to HE. If it does affect the probability, the events 'goes to selective school' and 'goes on to HE' are said to be ***dependent***.

Worked example 1F

? A random sample of people are surveyed about playing computer games and sleeping.

The results are shown in the following two-way table:

	Sleeps well	Difficulty sleeping	Total
Plays computer games often	82	18	100
Rarely or never plays computer games	123	27	150
Total	205	45	250

Based on the information in this survey, is sleeping well independent of frequency of playing computer games?

Solution

For those who play computer games often, the probability of sleeping well is:
$\frac{82}{100} = 0.82.$

For those who play computer games rarely or never, the probability of sleeping well is: $\frac{123}{150} = 0.82.$

These probabilities are the same, so sleeping well is independent of frequency of playing computer games.

If the events are independent then the probability of sleeping well will be the same for both groups.

Exercise 1F

1 Decide whether these pairs of events are **dependent** or **independent**.

	Situation	Event one	Event two
(a)	Toss two fair coins	Outcome on first coin	Outcome on second coin
(b)	Choose a person at random from your town	Sex of person	Whether person ends up in a care home
(c)	Going on two non-connecting flights	Suitcase goes missing on first flight	Suitcase goes missing on second flight

2 A clinic collects data about the blood pressure of a random sample of adults. The results are shown in the table below:

	Normal blood pressure	High blood pressure	Total
Smoker	10	20	30
Non-smoker	80	40	120
Total	90	60	150

Based on the information in the table, is high blood pressure independent of smoking?

It is sometimes difficult to be sure whether events are independent or dependent and the only way to find out is by collecting data. For example, some illnesses are more common in men than in women (or vice versa) and some aren't.

1G Probability simulations

The kinds of probability questions you will learn to deal with here come up in medicine and law.

Activity 1G Quincunx or Galton grid

The quincunx or Galton grid is an arrangement of rows of pegs. Balls drop through the grid, landing on a peg on each row. The balls have equal probability of going either side of each peg they land on. It is possible to have virtual grids where the probabilities of dropping either side of the peg are different.

The Galton grid in the diagram is being used to simulate tosses of 6 fair coins. The number of heads each time is a number between 0 and 6.

Find an online Galton grid and see how it works.

Worked example 1G

An online quincunx (or Galton grid) is used to simulate 600 tosses of 5 fair coins.

The results of the simulation are shown in the table below:

No. of heads	0	1	2	3	4	5	Total
Frequency	23	83	180	185	102	27	600

Use the simulation results to estimate the probability of getting 3 heads when 5 fair coins are tossed.

Solution

The probability is $\frac{185}{600} = 0.308$

Number of outcomes with 3 heads.

Total number of outcomes.

Exercise 1G

1 An online quincunx (or Galton grid) is used to simulate 200 families with 3 children each. The simulation assumes that each child is equally likely to be a boy or a girl.

The results of the simulation are shown in the table below:

No. of boys	0	1	2	3	Total
Frequency	23	89	68	20	200

Use the simulation to estimate the probabilities of each of the following numbers of boys.

(a) 0 (b) 1 (c) 2 (d) 3

2 Use a Galton grid to do another simulation. Use 3 rows and a probability of 0.5. Simulate 500 or 1000 families using the quincunx to estimate the probabilities of each of the following numbers of boys in families of 3 children.

(a) 0 (b) 1 (c) 2 (d) 3

3 Which simulation do you think gives more reliable results? Why?

4 What are the theoretical probabilities?

1H Frequency trees

Frequency trees are a way of representing what would happen in an ideal or average simulation; they can be used to find answers to probability questions.

Activity 1H A question from the news

A business is worried about whether their workers take drugs.

The Crime Survey for England and Wales says that 19% of 16–24 year olds have taken illegal drugs at least once in the past year. For 25–59 year olds, it is 6%.

10% of the business's employees are aged 16–24 with the other 90% aged 25–59.

(a) What is the probability that a randomly chosen worker has taken drugs in the last year?

(b) Would it be easier to work out if you knew the total number of workers at the company?

Worked example 1H

? Use a frequency tree to work out the probability in Activity 1H.

Solution

Imagine a large number of people who could be workers in the business and use a frequency tree to decide how many are in each category.

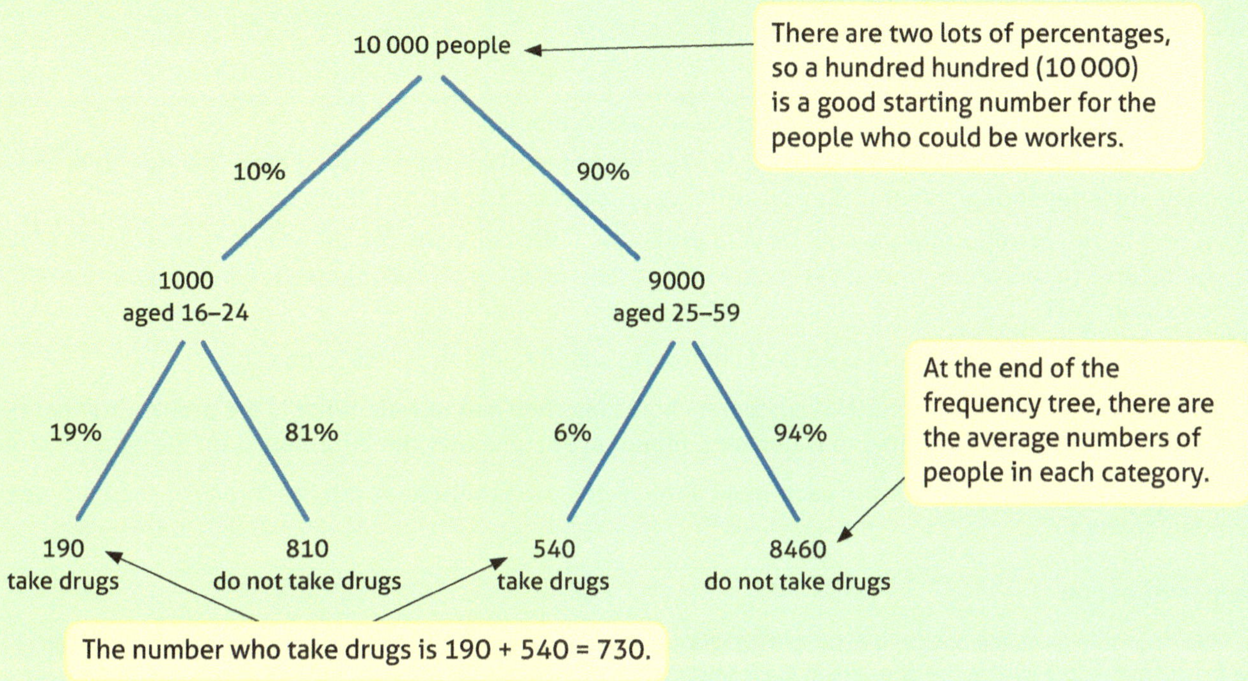

The probability is 730 ÷ 10 000 = 0.073 or 7.3%.

Exercise 1H

1 Alison and Ben are 21 years old. They share a birthday but are not related. They want to plan a special holiday for their 60th birthdays. The probability of a 21-year-old man living to at least 60 is 92%. For a woman, the corresponding probability is 95%. What is the probability that both Alison and Ben reach their 60th birthday?

2 A street has a 1 in 30 chance of being flooded in any given year. What is the probability of at least one flood in two consecutive years?

3 57% of taxpayers are men. 19% of male taxpayers earn enough to pay tax at the higher rate. 10% of female taxpayers earn enough to pay tax at the higher rate.

What is the probability that a randomly chosen taxpayer pays tax at the higher rate?

Test yourself 1I

1 Use the following statistics to compare road safety in Great Britain in 2000 and 2013. Decide what measure(s) to use in your answer.

Year	Deaths of car occupants	Billions of km driven	Number of motor vehicles (1000s)	Injuries from road accidents (1000s)	Population (millions)
2000	1655	466.2	28 818	317	57.2
2013	785	488.8	34 981	182	62.3

Department for Transport statistics

2 A company uses the following risk matrix.

		Likelihood			
		Unlikely (1)	**Possible (2)**	**Likely (3)**	**Very likely (4)**
Impact	Minor (1)	1	2	3	4
	Moderate (2)	2	4	6	8
	Major (3)	3	6	9	12
	Huge (4)	4	8	12	16

The company decides that a risk score of over 10 is unacceptable.

(a) There are two likely risks in the company's risk assessment. The probability of each is about 60% and they are independent. What is the approximate probability that both of them happen?

(b) A new business opportunity would be very profitable if the company can deliver on time. The risk of not being able to deliver on time is assessed as being very likely, with major impact. What could the company do about this?

3 The sign at the post office says that 90% of customers wait 5 minutes or less to be served.

A researcher asks a random sample of customers how long they had to wait. What is the probability that the first three people she asks all had to wait over 5 minutes? Assume that the information on the sign is correct.

4 An athlete has a 1 in 3 risk of injury each week. What is the probability that she has three consecutive weeks without an injury?

Practice question

The following table gives the numbers of people living alone in 2012 for the UK. The total population was 63.7 million. Use these figures to estimate the answers to the following questions for the present population.

People living alone by age and sex (thousands)						
	16–24	**25–44**	**45–64**	**65–74**	**75 and over**	**Total**
Male	111	1065	1281	516	553	3526
Female	97	536	1140	861	1472	4106
Total	208	1601	2421	1377	2025	7632

ONS statistics

(a) Estimate the probability that a randomly chosen person from the UK lives alone.

(b) Estimate the probability that a randomly chosen male who lives alone is aged over 75.

(c) Estimate the probability that a female is 16–24 years old given that she lives alone.

(d) There are more people living alone aged 45–64 than aged 65–74.

Explain why this does not mean that the probability of living alone is higher for 45–64 year olds than for 65–74 year olds.

(e) John says: 'For people in the age group 25–44, men are about twice as likely as women to live alone.'

Either show that John is correct or explain why John could be wrong.

Now that you have completed this chapter, you should know:

- That risk can involve impact as well as probability.
- How to express risks as probabilities or in the form 1 in *n*.
- What outcomes and events are.
- How to use relative frequency to estimate probability.
- How to interpret two-way tables.
- The difference between independent and dependent events.
- How to find probabilities of combined events.
- The meaning of conditional probabilities.
- How to draw tree diagrams.
- How to use tree diagrams to find probabilities of combined events.

CHAPTER 2

Working with formulae and spreadsheets

This chapter is about using algebra. Usually this will involve working with formulae: substituting values for the variables or turning them into equations that you will need to solve. The work can be done by hand or on a calculator, but it is often easier to use a spreadsheet.

2A Using formulae

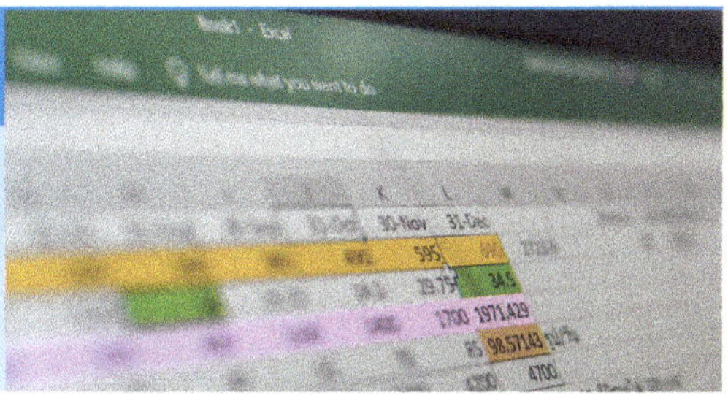

A formula allows you to calculate the value of a quantity, given the values of the variables on which it depends. There are many ways of describing formulae, including using words, flowcharts and algebra. You will often meet them through using spreadsheets.

Activity 2A Taxi fares

A taxi firm charges a basic fee of £2.60, plus £1.25 per km.

(a) Work out the cost of a journey of 6 km.

(b) Another journey, using the same taxi firm, cost £15.10. How far was the journey?

The cost of the journey, £C, could be written using algebra as $C = 1.25d + 2.60$, where d is the distance in kilometres.

1.25 is the rate per kilometre in £.

2.60 is the charge for zero kilometres in £.

Worked example 2A

The height h metres of a cricket ball t seconds after it has been hit is given by the formula: $h = 25t - 5t^2$. Assume the flight path is over level ground.

(a) Work out the height of the cricket ball after (i) 2 seconds and (ii) 3 seconds.

(b) At what time does the cricket ball land?

Solution

(a) $h = 25t - 5t^2$ (i) When $t = 2$, $h = 25 \times 2 - 5 \times 2^2$; the height is 30 m.

(ii) When $t = 3$, $h = 25 \times 3 - 5 \times 3^2$; the height is 30 m.

You will notice that the height is the same at time 3 s as it was at 2 s. This tells you that the ball is at its maximum height sometime between 2 s and 3 s. In fact, it is at 2.5 s and is 31.25 m.

(b) $0 = 25t - 5t^2$, so $25t = 5t^2$ ← Replace h by 0 to form an equation.

$5t = t^2$ ← Divide both sides by 5.

$t = 0$, $t = 5$; both satisfy the equation. ← $t = 0$ is when it is hit.

The cricket ball lands 5 seconds after it is hit.

Exercise 2A

1 A taxi firm charges a basic fee of £2.20 plus £1.35 per kilometre.

(a) How much does a ride of 8 km cost?

(b) Write the cost, £C, in terms of d, the distance travelled, using algebra.

2 The monthly cost of a landline, £C, is given by $C = 20.20 + 0.15m$ where m is the number of minutes of calls made.

(a) What is the cost per minute for calls?

(b) In August Yvonne made 320 minutes of calls. What was her bill for the month?

(c) In October Yvonne's bill was £49.60. How many minutes of calls did she make in October?

3 The height, h m, of a javelin, t seconds after it is thrown, is given by the formula $h = 14t - 4.9t^2$.

(a) Find the height of the javelin after 2 seconds.

(b) How long is the javelin in the air?

2B Interpreting spreadsheets

This section shows how information can be stored in a spreadsheet.

Activity 2B Sandwiches

The total sales for 10 days were £878.80.

How can you obtain that information from the spreadsheet?

The figures in column G are in Accounting mode, so in pounds and pence.

	A	B	C	D	E	F	G
1	Sandwiches						
2		£2.50	£2.15	£2.90	£2.40		
3	Day	tuna	cheese	ham	tomato	total	
4	1	16	13	9	5	43	£106.05
5	2	16	14	0	6	36	£84.50
6	3	12	5	13	12	42	£107.25
7	4	12	15	8	9	44	£107.05
8	5	9	11	13	2	35	£88.65
9	6	6	1	5	14	26	£65.25
10	7	9	11	8	14	42	£102.95
11	8	18	6	10	7	41	£103.70
12	9	2	1	2	13	18	£44.15
13	10	3	9	8	8	28	£69.25

Worked example 2B

? Look at the spreadsheet above.

(a) What number is in cell C7?

(b) What does the number in cell F10 represent?

(c) What does £2.15 represent?

(d) How has £107.05 in cell G7 been calculated?

Solution

(a) 15 ← Look for column C, row 7.

(b) The total number of sandwiches sold on Day 7. ← Column F contains the total numbers of sandwiches sold, row 10 has Day 7's data.

(c) £2.15 is the cost of a cheese sandwich. ← The price is above the cheese sandwich column.

(d) 12 × £2.50 + 15 × £2.15 + 8 × £2.90 + 9 × £2.40

The total income from all the sandwiches sold on Day 4.

Exercise 2B

1 The spreadsheet on the right shows what percentage of students passed English and Maths at GCSE at grade C/4 or higher. FSM stands for Free School Meals.

(a) What is in cell A8?

(b) What cell contains the percentage of students who gained a C or better in English and Maths in 2013?

(c) What calculation has been done to get 26.4 in cell E6?

	A	B	C	D	E
1	**English and Maths C/4 or higher**				
2		**All**	**Non-FSM**	**FSM**	**Gap**
3	2011	58.7	62.5	35.1	27.4
4	2012	59.3	63	36.8	26.2
5	2013	61.3	65.3	38.7	26.6
6	2014	58.9	62.7	36.3	26.4
7	2015	59.2	63	35.7	27.3
8	2016	63	66.7	39.1	27.6
9	2017	63.9	67.4	40.3	27.1
10	2018	70	77	51	26

2C Spreadsheet formulae

This section shows how formulae are used in spreadsheets.

Activity 2C From algebra to cell references

Match the algebraic expressions to the spreadsheet formulae.

(a) What does * mean?

(b) What do you think A2 refers to?

(c) What does / mean?

Algebraic expressions: x^2, $2x - 2$, $\frac{x}{2}$, $\sqrt{x}$

Spreadsheet formulae: A2/2, A2^2, A2*2-2, SQRT (A2)

Worked example 2C

A taxi has a basic charge of £3.40 and a rate of £1.43 per kilometre.

(a) Set up a spreadsheet to find the cost of fares with this taxi and use it to find the cost of a ride of 38 km.

(b) Change the spreadsheet to calculate the cost of a journey of 100 km.

Solution

(a) The number to use in the calculation is stored in cell A1.

38 is entered in cell A1.

	A	B
1	38	57.74
2		

The formula is entered in cell B1.

The command in cell B1 is =A1*1.43+3.4 and the answer of 57.74 appears there.

The journey of 38 km costs £57.74.

'=' tells the spreadsheet to calculate the formula that follows.

(b)

	A	B
1	100	146.40
2		

The distance changes, so the value in cell A1 also changes.

Remember to include = at the start of the formula.

The journey of 100 km costs £146.40.

Exercise 2C

1 Cell A1 contains the length of a room in metres. What formula goes into cell A2 to calculate the length of the room in centimetres?

2 Cell A1 contains the radius of a circle. What formula in B1 calculates the area of the circle?

3 The lengths of the two shorter sides of a right-angled triangle are in cells A1 and B1. What formula in C1 calculates the length of the longest side?

4 What does the formula =(A1+A2+A3+A4+A5)/5 calculate?

5 What does the formula =AVERAGE(B1:B10) calculate?

2D The power of a spreadsheet

This section is about copying the contents of the cells of a spreadsheet.

Activity 2D Sequences

Open a new spreadsheet page.

Enter 1 in cell A1 and =A1+4 in cell A2.

	A
1	1
2	5

Click on cell A2 and hover over the bottom right-hand corner until a plus sign appears, drag downwards to produce the sequence shown in the spreadsheet on the right.

(a) What if cell A2 is changed to =A1+5?

(b) What if A1 is changed to 2 instead of 1 but the command in A1 remains =A1+4 ?

	A
1	1
2	5
3	9
4	13
5	17
6	21
7	25

Try different values in A1 with =A1+4 in cell A2. Try different numbers added or subtracted from A1 in the formula in A2. The cell references increase as you drag the formula down or across. Dragging down =A1*2 produces =A2*2 in cell A3 then =A3*2 in cell A4. Dragging across =A1*2 produces =B1*2 and =C1*2.

Exercise 2D

1 Write down the values or formulae needed in A1 and A2, so that dragging down produces each of the sequences:

(a)	(b)	(c)	(d)
3	1	2187	65536
5	2	729	256
7	4	243	16
9	8	81	4
11	16	27	2

2 Write down the values in A1, and the formula in B1, so that dragging across produces each of these sequences:

(a) 3, 7, 15, 31, 63

(b) 2, 5, 26, 677, 458330

2E Using the $ symbol

The work in Section D is now taken further by allowing cell references to be fixed or changed as you need.

Activity 2E What's happening?

Look at the two sequences on the right.

	A	B	C	D	E	F	G	H
1	4	7	11	18	29	47	76	123
2	4	7	11	15	19	23	27	31

Each one is generated using the numbers 4 and 7 with A1 and A2 both containing 4, and B1 and B2 both containing 7.

(a) Describe how to obtain the next term in each case.

Cell C1 contains `=A1+B1`.

(b) What is the formula in F1?

In the second sequence each term is 4 more than the previous one and 4 is in A2. So, C2 = A2 + B2, and D2 = A2 + C2.

The command for cell C2 is written `=$A2+B2`.

(c) What happens when you copy this across?

(d) What is the effect of the $ sign here?

(e) What formula do you think is in D2?

Sometimes you want the cell reference to stay the same. A $ sign before the letter (or number) in the cell reference stops that letter (or number) changing.

Worked example 2E

? The spreadsheet below shows the total cost of buying fence panels at £21 and posts at £8 each.

	A	B	C	D	E	F	G	H
1		**post**	**£8**		**panel**	**£21**		
2					**Panels**			
3			**1**	**2**	**3**	**4**	**5**	**6**
4		**1**	£29	£50	£71	£92	£113	£134
5		**2**	£37	£58	£79	£100	£121	£142
6	**Posts**	**3**	£45	£66	£87	£108	£129	£150
7		**4**	£53	£74	£95	£116	£137	£158
8		**5**	£61	£82	£103	£124	£145	£166
9		**6**	£69	£90	£111	£132	£153	£174

The rows increase by 21. The columns increase by 8.

To change the cost of a fence post only cell C1 is changed and the numbers automatically update.

Write down the formula in cell C4 and explain how it works.

Solution

The formula in cell C4 is `=$B4*$C$1+C$3*$F$1`.

The diagram below shows how each part of the formula contributes to the calculation.

C1 is the cost of one post. Both row number and column reference must stay the same.

F1 is the cost of one panel. Both row number and column refence must stay the same.

`=$B4*$C$1+C$3*$F$1`

$B4 is the number of posts. Only the row number changes.

The $ precedes the letters and/or numbers that are to remain unchanged.

C$3 is the number of panels. Only the column number changes.

Exercise 2E

1 (a) Create the spreadsheet on the right, which calculates the daily takings for a sandwich shop, using \$ to enable you to update the sandwich prices quickly.

(b) Enter the formula =SUM(B4:B9) into cell B10. What does this give you?

(c) Copy the formula in B10 across to F10. What information is in E10?

(d) What information is in F10?

	A	B	C	D	E	F
1	Sandwiches					
2		tuna	cheese	ham	tomato	
3		£2.50	£2.15	£2.90	£2.40	
4	Monday	15	7	12	14	£120.95
5	Tuesday	5	14	11	6	£88.90
6	Wednesday	11	10	10	14	£111.60
7	Thursday	13	8	7	13	£101.20
8	Friday	12	15	9	12	£117.15
9	Saturday	11	9	9	5	£84.95
10						

2 In a new spreadsheet the number 2 is entered into cell A1 and the number 7 is entered into D1. The formula =D1+A\$1 is placed in cell D2 and copied down to cell D9. Empty cells are interpreted as containing the value zero. So, B1, C1, E1, F1, G1 and so on contain the value zero.

(a) What formula is in D9?

(b) The formula in cell D2 is copied across to cell J2. What number is in cell J2?

(c) Row 2 contains two non-zero entries. Which cells are they and what numbers are in them?

2F Solving equations

This section is about solving equations algebraically.

Activity 2F Think of a number

Rishi is thinking of a number.

He multiplies it by 2 and adds 11.

He then takes the square root and his final answer is 5.

What number does he start with?

Working backwards to find the original number by 'undoing' each step is a common algebraic method for solving equations. Factorisation is used to solve quadratic equations.

Another method for solving equations is trial and improvement. This is shown in the third worked example.

Worked example 2F(1)

Solve the equation: $4\left(\frac{1}{2}x - 1\right) = 20$.

Solution

$4\left(\frac{1}{2}x - 1\right) = 20$ ← Divide both sides by 4.

$\frac{1}{2}x - 1 = 5$ ← Add 1 to both sides.

$\frac{1}{2}x = 6$ ← Multiply both sides by 2.

$x = 12$

Worked example 2F(2)

? Solve the equation: $t^2 - 15t - 50 = 0$.

$t^2 - 15t - 50 = 0$ is a quadratic equation.

Solution

$t^2 - 15t - 50 = 0$

$t^2 - 5t - 10t + 50 = 0$ ← Split the term in t: $-5 + -10 = -15$, the coefficient of t is -15. $-5 \times -10 = 50$, the constant term is 50.

$t(t-5) - 10(t-5) = 0$ ← Factorise each pair of terms.

$(t-5)(t-10) = 0$ ← Factorise, extracting $(t-5)$ as a common factor.

Either $t - 5 = 0 \Rightarrow t = 5$.

Or $t - 10 = 0 \Rightarrow t = 10$. ← One of the factors must be zero if their product is zero.

So, $t = 5$ or $t = 10$.

Worked example 2F(3)

? Use trial and improvement to find a solution to the equation $15t - t^2 = 49$, correct to 1 decimal place.

Solution

$15t - t^2 = 49$

Try $t = 1$, $15 \times 1 - 1^2 = 14$, too small.

Try $t = 4$, $15 \times 4 - 4^2 = 44$, too small but much closer.

Try $t = 5$, $15 \times 5 - 5^2 = 50$, too big but closer.

Each trial value should be chosen to get closer to the correct answer.

Try $t = 4.9$, $15 \times 4.9 - 4.9^2 = 49.49$, too big.

Try $t = 4.8$, $15 \times 4.8 - 4.8^2 = 48.96$, too small but very close.

The solution lies between 4.9 and 5.0.

Try $t = 4.85$, $15 \times 4.85 - 4.85^2 = 49.2275$, too big.

The solution lies between 4.8 and 4.85 and so rounds to 4.8 to 1 decimal place.

Notice that in the previous example, the solution to $15t - t^2 = 50$ is exact, whereas in this case however many decimal places you take, you still have to round the answer.

Exercise 2F

1 Solve:

(a) $10 + 2t = 17$ (b) $5t - 11 = 7$ (c) $10 - 2x = 7$

2 Solve:

(a) $4(x + 5) = 30$ (b) $\frac{1}{3}(5x - 4) = 1$ (c) $\frac{1}{4}(5 - 2x) = 2$

3 Solve:

(a) $5(x-1)^2 - 7 = 118$ (b) $x^2 - 5x + 4 = 0$ (c) $7t - t^2 = 12$

4 Use trial and improvement to find a solution, correct to 1 dp, for:

(a) $x^2 - 7x = 10$ (b) $36t - 5t^2 = 64$ (c) $x^3 = 100$

2G Solving equations using technology

Trial and improvement is laborious to do 'by hand'.

Using a calculator or spreadsheet with the shortcuts available makes it more efficient.

Activity 2G Using your calculator

(a) Find the cube root of 10 000 to 1 decimal place, using trial and improvement.

(b) What are the key presses on your calculator to work it out in one step?

(c) In what ways can you do the trial and improvement method more efficiently on your calculator?

Worked example 2G

Use a spreadsheet to solve $x^3 + x = 100$ correct to 1 decimal place.

Solution

First find the whole number part of the solution.

	A	B
1	x	x^3+x
2	1	2
3	2	10
4	3	30
5	4	68
6	5	130
7	6	222
8	7	350
9	8	520
10	9	738
11	10	1010

Enter 1 in A2 and 2 in A3.

Select both cells A2 and A3 and drag down to fill more cells.

Enter =A2^3+A2 into cell B2.

Select cell B2 and drag down to fill more cells until 100 is passed.

The solution is between 4 and 5.

Next, determine the number in the first decimal place.

	A	B
1	x	x^3+x
2	4.1	73.021
3	4.2	78.288
4	4.3	83.807
5	4.4	89.584
6	4.5	95.625
7	4.6	101.936
8	4.7	108.523
9	4.8	115.392
10	4.9	122.549
11	5.0	130.000

Enter 4.1 in A2 and 4.2 in A3.

Select both cells A2 and A3 and drag down to fill more cells.

=A2^3+A2 in cell B2 should still be there.

The other cells in column B should automatically update.

The solution is between 4.5 and 4.6.

The numbers suggest that it is nearer 4.6 but this should be checked by calculating the mid-value of the interval, 4.55.

	A	B
1	x	x^3+x
2	4.55	98.74638

4.55 gives a value below 100, so 4.55 is too small.

The solution, correct to 1 decimal place, is 4.6.

Exercise 2G

1 Use a calculator to work out, by trial and improvement or otherwise:

(a) $\sqrt[3]{750}$ (b) The positive solution to $x^4 - x = 4$.

Give your answers correct to 2 dp.

2 Use trial and improvement on a spreadsheet to solve:

(a) $x^4 - 3x = 20$ (b) $\sqrt[3]{x} + x = 15$

Give your answers correct to 2 dp.

2H Rearranging formulae

Formulae are usually presented with one of the variables as the subject, but often you need to change which of them it is.

Activity 2H Match the formulae

(a) Match the formulae that are equivalent.

$A = BC + D$ $BC = A$ $A = B(C + D)$ $A = \frac{B}{C}$

$C = \frac{A}{B}$ $C = \frac{A}{B} - D$ $C = \frac{A - D}{B}$ $C = \frac{B}{A}$

(b) Explain why they are the same formula.

Worked example 2H

? For an object moving with constant acceleration a, $v^2 = u^2 + 2as$, where u is the initial velocity, v is the final velocity, a is the acceleration and s is the displacement.

v^2 is the subject of this formula.

(a) Make u the subject of the formula.

(b) Express a in terms of u, v and s.

Taking the positive square root of each side makes v the subject of the formula.

Solution

(a) $v^2 = u^2 + 2as$ ← Subtract $2as$ from both sides.

$v^2 - 2as = u^2$

$u^2 = v^2 - 2as$ ← The term with the new subject is on the left-hand side.

$u = \sqrt{v^2 - 2as}$ ← Take the positive square root of both sides.

u is now the subject of the formula.

(b) $v^2 = u^2 + 2as$ ← Subtract u^2 from both sides.

$v^2 - u^2 = 2as$

$2as = v^2 - u^2$ ← The term with the new subject is on the left-hand side.

$a = \frac{v^2 - u^2}{2s}$ ← Divide both sides by $2s$.

a is now the subject of the formula and is expressed in terms of u, v and s.

Exercise 2H

1 Rearrange $C = 2.35 + 0.25n$ to make n the subject of the formula.

2 $v = u + at$

(a) Make u the subject of the formula.

(b) Rearrange to express t in terms of u, v and a.

3 Rearrange $d = 5t^2$ to express t in terms of d.

4 $y = mx + c$

(a) Express c in terms of m, x and y.

(b) Make x the subject of the formula.

Test yourself 2I

1 A rough guide for the time to cook a roast chicken is 45 minutes per kilogram, plus 20 minutes.

(a) How long does it take to cook a chicken weighing 2 kg?

(b) A spreadsheet is set up with the weight of the chicken in kilograms in cell A2. What formula should cell B2 contain if it is to calculate the cooking time in minutes?

2 The maximum speed of a hailstone, in m s^{-1}, of diameter d mm, is given by $v = 1.4d^{0.8}$.

(a) Find the maximum speed of a hailstone with diameter 1 cm.

(b) Use trial and improvement to find the diameter of a hailstone that has a maximum speed of 5 m s^{-1}. Give your answer correct to 2 dp.

You will meet different styles for the units for speed. In this question m s^{-1} is used but it could have been ms^{-1} or m/s or written in words as 'metres per second'. Different styles are used in this book to help you get used to them. Equivalent styles are used for other compound units, for example those for density and acceleration.

3 The age of the rock exposed when a glacier has melted can be estimated by examining the lichen growing on it. The diameter, d mm, of the patch of lichen is given by:

$d = 7\sqrt{t - 12}$, where t is the number of years since the ice disappeared.

(a) Use the formula to estimate the diameter of a patch of lichen on a rock exposed 2000 years ago. Give your answer correct to 2 significant figures.

(b) Rearrange the formula to make t the subject.

(c) A spreadsheet is set up with the value for t in cell A2. What formula should go in cell B2 to calculate d?

4 The height, h cm, that a tennis ball bounces after being dropped from a height, H cm, is given by the formula:

$h = H \times 0.754^n$, where n is the number of bounces.

Roger enters the information into a spreadsheet, dragging down the formula in cell B3.

	A	B
1	*H*	100.0
2	*n*	*h*
3	0	100.0
4	1	75.4
5	2	56.9
6	3	42.9
7	4	32.3
8	5	24.4
9	6	18.4

(a) How high does the ball bounce after the 4th bounce?

(b) What cell is the dropping height, H, stored in?

(c) What formula is in cell B3?

(d) Find the number of bounces until the height of the bounce is less than 10 cm, for an initial drop height of 100 cm.

5 A formula for calculating the child's dosage, in millilitres (ml), for a medicine is $C = \frac{Y}{Y+12} \times A$, where Y is the age in years and A is the adult dose in ml.

Suja enters the information into a spreadsheet, dragging down the formula that is entered in B3.

	A	B	C
1		Adult dose	20
2	Age of child		
3	1	1.5	
4	2	2.9	
5	3	4.0	
6	4	5.0	
7	5	5.9	
8	6	6.7	
9	7	7.4	
10	8	8.0	
11	9	8.6	
12	10	9.1	
13	11	9.6	
14	12	10.0	
15			

(a) What does the 20 in cell C1 represent?

(b) What dosage should you give a child who is 11 years old?

(c) A child is given a dose of 8.6 ml. How old is the child?

(d) Show how the value in cell B3 is calculated.

(e) Explain why the entries in cells B5 and B6 are whole numbers.

(f) Write down the formula in cell B3.

(g) What is the effect of the $ in the formula in cell B3?

(h) The spreadsheet results go as far as age 12. Medical personnel use their judgement to decide whether to use this method for calculating doses, or other available methods.

Give a reason why this formula would be inappropriate for an 18-year old.

Practice question

A formula for rough conversion between degrees Celsius, C, and degrees Fahrenheit, F, is $F = 2C + 30$.

(a) Calculate the temperature in °F (degrees Fahrenheit) equivalent to 12°C.

(b) Rearrange the formula to give C in terms of F.

(c) What temperature is the same in both °F and °C, according to this formula?

The correct formula is $F = 1.8C + 32$.

(d) What temperature is the same in both °F and °C, according to the correct formula?

(e) What is the percentage error in using the rough conversion for room temperature (20°C)? Give your answer to the nearest 0.1%.

Now that you have completed this chapter, you should know:

- How to use and rearrange algebraic formulae.
- How to solve equations, both algebraically and using ICT.
- How to work with spreadsheets.

CHAPTER 3
Percentages

You need to be able to use percentages in a variety of different contexts, for example working out a sale price, calculating interest earned on an investment or comparing different proportions.

3A Writing one amount as a percentage of another

The picture shows the 100 steps of Campo di Giove, Italy; each one you climb means you are 1% nearer the top. A common situation involves writing one quantity as a percentage of another.

Activity 3A Recycling

This percentage bar chart shows the proportion of household waste that has been recycled in England from 2016 to 2018.

(a) Work out the missing percentage for 2016.

(b) Show that in neither 2017 nor 2018 the percentages add up to exactly 100.

Why does this happen?

(c) The unrounded percentages have a total of 100%.

Give an example of what the percentages to 2 decimal places for each year could have been.

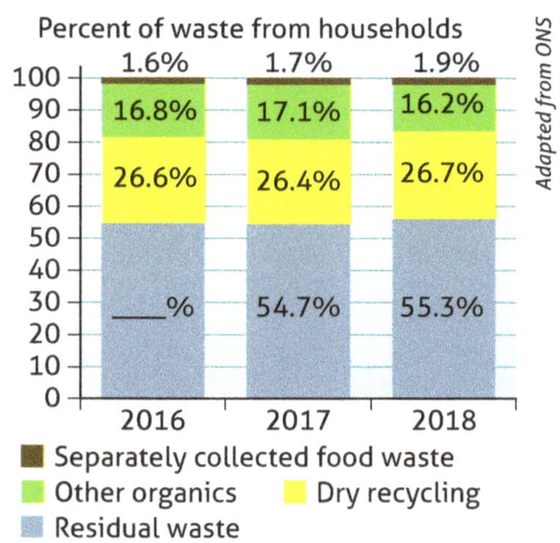

Worked example 3A

The table below shows the composition breakdown and recycling rate of waste from households in England, 2016 to 2018. The units are thousand tonnes.

Waste type	2016	2017	2018	% change from 2017 to 2018
Total recycling *of which*:	10 217	10 139	9 840	−2.9%
Dry recycling	6 042	5 917	5 866	___%
Separately collected food waste	355	386	414	___%
Other organics	3 820	3 836	3 560	___%
Total residual	12 535	12 266	____	___%
Total waste from households	22 752	____	21 991	___%
Waste from households recycling rate	44.9%	___%	___%	

Data adapted from ONS

Show that (a) the recycling rate for 2016 is 44.9%

(b) the percentage change from 2017 to 2018 is −2.9%.

Solution

(a) 10 217 thousand tonnes out of 22 752 thousand tonnes are recycled.

The words 'out of' tell you that the number that follows goes on the bottom line of a fraction.

As a fraction, this is $\frac{10\,217}{22\,752}$.

To convert a fraction to a decimal you divide the numerator (top) by the denominator (bottom).

As a decimal, it is 0.449...

As a percentage, it is $\frac{10\,217}{22\,752} \times 100\% = 44.9\%$.

To convert a decimal to a percentage you multiply by 100.

(b) 2017: 10 139 thousand tonnes were recycled. 2018: 9840 thousand tonnes were recycled.

So, the change (or **difference**) is:

9840 − 10139 = −299 thousand tonnes.

Percentage change = $\frac{-299}{10\,139} \times 100\% = -2.94...\%$

Percentage change is $\frac{\text{Difference}}{\text{Original amount}} \times 100\%$.

So, the amount recycled decreased by 2.9%, to 2 significant figures.

Exercise 3A

1 Find the missing values from the table in Worked example 3A.

2 Mia weighs her household's general waste and recycling for one fortnight.

Here are her results.

Tins	Bottles	Paper	Plastic	Compost	General waste
0.94 kg	0.74 kg	1.6 kg	3.6 kg	6.86 kg	7.63 kg

(a) Find the percentage of Mia's rubbish that is

(i) tins (ii) compost (iii) bottles and plastic.

(b) Compare Mia's recycling rate with that of the average household in England.

3B Finding a percentage of an amount

This section covers the common situation where you need to work out a certain percentage of a given amount.

Activity 3B Buying a house

Buying a house is the biggest investment that many people will make. It involves a lot of different calculations.

Discuss which of the calculations involve percentages.

Worked example 3B

? When you buy a property, you have to pay **stamp duty**. This is a tax paid to the government.

You pay stamp duty on the value of your house over £125,000

Value of house	Stamp duty
Up to £125,000	0%
On the next £125,000	2%
On the next £675,000	5%

Work out how much stamp duty is due on a house that costs £700,000.

When a large number is written, it is usually split into groups of 3 digits. There may be commas between them or there may be gaps. So, you can have 125,000 (as here) or 125 000. Because you will come across both styles in everyday life, they are used interchangeably in this book.

In Currency and Accounting modes a spreadsheet uses a comma.

Solution

Break £700,000 up into the three different bands of stamp duty.

£125,000 £0 stamp duty
£125,000 2% stamp duty
£450,000 5% stamp duty

2% of £125,000 $= \frac{2}{100} \times$ £125,000 ←
= 0.02 × £125,000
= £2500

0.02 is called a **multiplier** – you can go straight from the percentage to the multiplier by dividing by 100.

5% of £450,000 = 0.05 × £450,000
= £22,500

Total stamp duty = £2500 + £22,500 = £25,000

Exercise 3B

1 Calculate these.

(a) 15% of £360,000 (b) 67% of £950 (c) 0.5% of £1,000,000

2 (a) Work out how much stamp duty should be paid on each of these properties.

A house costing £200,000	A flat costing £120,000	A house costing £750,000

(b) An estate agent charges 1.5% of the value of a property to advertise and sell it.
How much would the estate agent charge the owners of the above properties?

3C Percentage increase and decrease

Most quantities increase or decrease. This section covers the associated percentage calculations.

Activity 3C Extra effort

After winning her competition, an athlete says 'I gave it 110%.'

Can you ever have more than 100%?

Worked example 3C

? Ella, Meena and Tamsin win the three medals in a long jump competition.

The table shows their previous personal best jumps.

Ella	6.40 m
Meena	6.28 m
Tamsin	6.93 m

Ella jumps 5% further than her previous best.
Meena jumps 7.5% further than her previous best.
Tamsin's jump is 3% less than her previous best.

Who wins the gold medal? How many centimetres does she win by?

Solution

Ella: 100% + 5% = 105%
105% of 6.40 m = 1.05 × 6.40 = 6.72 m

Meena: 100% + 7.5% = 107.5%
107.5% of 6.28 = 6.751 m = 6.75 m to nearest cm

Tamsin: 100% − 3% = 97%
0.97 × 6.93 m = 6.722 m = 6.72 m to nearest cm

6.75m – 6.72 m = 0.03 m
= 3 cm

So Meena wins by 3 cm. Ella and Tasmin are 2nd equal.

Exercise 3C

1 (a) Which of these multipliers should you use to decrease an amount by 38%?

1.38   0.038   0.62   0.38   1.62

(b) Write down the percentage increase or decrease that each of the other multipliers gives.

2 Find the final amounts when you:

(a) increase £60 by 5.6%   (b) decrease £45 by 45%.

3 After running 100 m, an athlete's heart rate increases by 220% from a resting rate of 55 beats per minute.

(a) Calculate the athlete's heart rate immediately after the race.

One minute after the race, the athlete's heart rate has dropped by 55%.

(b) Calculate the athlete's heart rate 1 minute after the race.

After 5 minutes the athlete's heart rate is back to its usual resting rate.

(c) Work out the percentage decrease in the athlete's heart rate from 1 minute after exercise to 5 minutes after exercise.

3D Appreciation and depreciation

Appreciation occurs when the value of an item increases over time.

Depreciation occurs when the value of an item decreases over time.

Activity 3D The changing value of a car

A new car is often said to lose value as soon as you drive it off the forecourt and can lose around 40% of its value in the first year.

(a) What type of cars will lose more of their value?

(b) Will a car's value depreciate at the same rate each year?

Worked example 3D

? The value of a car depreciates by 25% each year.

The car cost £13,000 when new.

Calculate the value of the car when it is 3 years old.

Give your answer to a suitable degree of accuracy.

Solution

After 1 year the car is worth 75% of £13,000 = 0.75 × £13,000 = £9750.

After 2 years the car is worth 75% of £9750 = 0.75 × £9750 = £7312.50.

After 3 years the car is worth 75% of £7312.50 = 0.75 × £7312.50 = £5484.38.

The car is worth £5500 (to nearest £100) after 3 years.

There is a faster way to answer this. You have multiplied by 0.75 three times (once for each year) so, the calculation is: 0.75^3 × £13,000 = £5484.

Jake buys a new sports car. He is told it will depreciate by 25% a year. He says, 'That means that after 4 years it will be valueless and after 5 years its value will be negative, so I will have to pay to have it taken away.'

What is wrong with Jake's statement?

Exercise 3D

1 The value of a vintage car is currently appreciating by 5% a year.

The car cost £35,000 last year.

(a) Work out the value of the car (i) this year and (ii) in 2 years' time.

(b) Is the car's value likely to keep appreciating at the same rate?

2 A car is worth £16,500 when new. It depreciates by 30% per year. These steps show how a spreadsheet can be used to find how the value of the car changes over time.

Step 1: Enter the following into a spreadsheet.

	A	B
1	Year	Value of car (£)
2	0	16500

Right click in cell A2 to format the cell. Select:
- Format cells
- Category: Number
- Set the number of decimal places to 0.

Repeat in cell B2, but set the number of decimal places to 2.

(a) Why do you sometimes need to format the cells in a spreadsheet?

Step 2: The formula =A2+1 is entered in cell A3 and the formula =B2*0.7 is entered in B3.

	A	B
1	Year	Value of car (£)
2	0	16500
3		

A formula must always start with an =.

(b) Explain these formulae. What numbers appear in cells A3 and B3?

Step 3: Click in cell A2 and select copy.

Drag down from the corner of A2 to copy the formula down the rows.

Step 4: Copy the formula in cell B3 down the rows.

	A	B
1	Year	Value of car (£)
2	0	16500
3	1	11550
4	2	8085
5	3	5659.5
6	4	3961.65

To toggle between seeing the formulae and seeing the values in a spreadsheet:
- Select the 'Formulas' tab in the ribbon.
- Click on 'Show Formulas' in the 'Formula Auditing' group.

(c) Write down the formula that is used in (i) cell A6 and (ii) cell B6.

(d) When is the value of the car less than £1000? Is this a reasonable model?

3E Inflation and index numbers

Inflation is the rate at which consumer prices rise or fall. change. If prices rise, inflation is positive and if they fall it is negative.

Activity 3E A basket of shopping

The price of a huge 'shopping basket' of goods and services is compared with the cost of the same basket for the previous year.

(a) Why isn't it possible to consider all goods and services when working out inflation?

(b) What goods and services would you put in the basket?

Economists use **index numbers** to show how the value of goods change over time.

An index starts in a given year, called the **base year**.

The value in the base year is given an index number of 100.

Each year after the base year is given an index number to show the percentage change.

An index number of 105 means a 5% rise from the base year, and an index number of 98 means a 2% fall.

Worked example 3E

The table below shows the average house price in thousands of pounds for the UK from 2018 to 2020.

Year	Average house price	Index number
2018	281	100
2019	276	
2020	291	

ONS statistics

Take the base year as 2018.

Calculate the index numbers for 2019 and 2020.

Solution

2019: Percentage change from 2018 to 2019 = $\frac{276 - 281}{281} \times 100\% = -1.78\%$.

Index number is $100 - 1.78 = 98.22$

$= 98.2$ (to 1 dp).

2020: Percentage change from 2018 to 2020 = $\frac{291 - 281}{281} \times 100\% = 3.56\%$.

Index number is $100 + 3.56 = 103.56$

$= 103.6$ (to 1 dp).

Exercise 3E

1 The Bank of England has a target rate of inflation of 2.5%.

Here are the costs of some goods and services last year.

1 litre milk 71 pence	Pizza £4.50	Average fuel bill £1254 per year

How much does each of these goods and services cost this year if its price matches the Bank of England's target?

2 The table shows the price of a typical desktop computer from 2018 to 2020.

Year	2018	2019	2020
Price	£750	£825	£690

Take the base year as 2018 and the index number for the computer as 100.

(a) Calculate the index number for 2019 and 2020.

(b) Why is the index number taken as 100?

3 The table shows the index numbers for the cost of a package holiday to a particular resort from 2018 to 2020.

Year	2018	2019	2020
Index	100	105	88

The holiday cost £850 in 2018.

Work out the cost of the holiday in 2019 and 2020.

3F Annual percentage growth

You will often encounter situations where, rather than just considering a percentage increase on one occasion, you need to calculate how a quantity increases repeatedly, for example population size or the cost of living.

Activity 3F Rabbits

The number of rabbits on a small island increases at a rate of 6% each year for 5 years.

Initially, there are 300 rabbits on the island.

(a) How many rabbits are there after 5 years?

The population of another group of rabbits increases by r% each year. Initially, there are N_0 rabbits.

The formula for N, the number of rabbits after n years is

$N = N_0 \times \left(1 + \frac{r}{100}\right)^n$

(b) Explain the formula.

Worked example 3F

In 2015, a town had a population of 14 436 people. In 2020, the town has a population of 20 591 people. Calculate the annual percentage growth. Comment on your answer.

Solution

Number of years = 2020 – 2015 = 5 years

Using the formula $N = N_0 \times \left(1 + \frac{r}{100}\right)^n$ gives the population in 2020 to be

$14\,436 \times \left(1 + \frac{Growth\ rate}{100}\right)^5 = 20\,591$

Store the answer to each step in your calculator.

Divide both sides by 14 386: $\left(1 + \frac{Growth\ rate}{100}\right)^5 = 1.426...$

To undo a square, you find the square root. To undo a cube, you find the cube root. To undo a power of 5, you find the 5th root.

Take the 5th root of both sides $1 + \frac{Growth\ rate}{100} = \sqrt[5]{1.426...}$

$= 1.0736...$

Subtract 1 from both sides $\frac{Growth\ rate}{100} = 0.0736...$

Multiply both sides by 100 $Growth\ rate = 7.4\%$ to 2 sf.

Comment: The growth rate of 7.4% is quite high. The town has probably had a few new housing developments added.

Exercise 3F

1 Solve these equations.

(a) $1 + \frac{x}{100} = 1.4$

(b) $\left(1 + \frac{x}{100}\right)^2 = 1.0609$

(c) $\left(1 - \frac{x}{100}\right)^4 = 0.4096$

(d) $200\left(1 + \frac{x}{100}\right)^5 = 322.102$

2 In 2010 there were 6.956 billion people in the world.
In 2020 there were 7.822 billion people in the world.

(a) Calculate the annual population growth rate.

(b) Estimate the number of people there will be in 2035.

3 The table below shows the number of clients and the profit made by a small IT company in 2015 and 2020.

Date	Number of clients	Profit
1st January 2015	156	£80,000
1st January 2020	45	£145,000

Calculate the annual reduction rate in number of clients and the annual growth rate in profit.

Comment on your answers.

3G Reverse percentages

Sometimes you need to work out what a quantity was before it was subject to a percentage change. Such calculations are called reverse percentages.

Activity 3G Saving money

A clothes shop reduces the price of all its goods by 15%.

You spend £255 at the shop.

How can you work out how much money you have saved?

Worked example 3G

A clothes shop advertises a sale as '15% off marked price'.

A jacket costs £68 in the sale.

Calculate the normal price of the jacket.

Solution

Normal price = 100%

Sale price = 100% − 15% = 85%

85% of normal price = £68

0.85 × normal price = £68

So, normal price = $\frac{£68}{0.85}$ = £80.

You can't just add on 15% to £68 because 15% of £68 is less than 15% of £80.

Exercise 3G

1 A clothes shop advertises a sale as '15% off marked price'.

The table below shows the normal selling price of some items and their sale price.

Item	Normal selling price	Sale price
Jacket	£80	£68
Jeans	£40	
Shirt	£20	
Trainers		£42.50
Hoodie		£29.75

(a) Calculate the missing prices.

(b) How can you check your answers?

2 Jamal is given a pay increase of 4% for his Saturday job.

His new pay is £8.84 per hour.

What was Jamal's pay before the pay rise?

Test yourself 3H

1 The table below shows how the population of seals and penguins in neighbouring colonies have changed over 4 years.

End of year	Population of seals	Population of penguins
2016	6500	11 400
2020	5600	13 200

Work out the annual population growth or reduction rates for each colony.

Comment on your answers.

2 A location with a sensitive marine environment requires anyone who wants to dive there to pass a test. There are two parts: Part 1 is practical and marked out of 80; Part 2 is theoretical and is marked out of 140. The marks for the two parts are added together and an overall percentage is calculated.

The table shows the grading requirements for the three different levels at which it is possible to pass.

Instructing others	90%
Unaccompanied diving	75%
Accompanied diving	60%

Kim scores 72 out of 80 on Part 1 and 91 out of 140 on Part 2.

(a) What grading does she get on each paper?

(b) What grading is Kim awarded?

Amir gets 95% on Part 1 and overall is passed for Instructing others.

(c) What is the lowest mark Amir could have scored on Part 2?

3 An estate agent tells Gemma that the value of her flat has risen by 12% since she bought the property.

The estate agent values the flat at £287,000.

How much had Gemma paid for her flat?

4 **(a)** The value of a car depreciates by 35% each year.

The car is worth £16,000 new.

Work out the value of the car after 4 years.

(b) Another car depreciates at the same rate.

When the car is 2 years old it is worth £8450.

How much did the car cost when it was new?

5 There is a population of 25 400 sea birds on a cliff on 1st January one year.

The table below shows how the population changes over three consecutive years.

End of year	Percentage change
1	5%
2	2%
3	-4%

Calculate the overall percentage change in the number of sea birds in these three years.

Practice question

The table below shows the average rainfall (in mm) in the UK.

Year	Jan	Feb	Mar	Apr	May	June	July	Aug	Sept	Oct	Nov	Dec	Total
2016	178.7	110.2	83.8	77.3	61.4	95.9	78.7	85.3	96.5	47.0	105.1	77.5	**1097.5**
2017	76.1	93.4	97.8	33.4	57.1	110.4	101.3	102.5	117.3	101.7	107.9	120.7	**1119.5**
2018	132.6	68.8	103.8	83.9	47.8	34.7	54.3	83.1	101.7	104.5	121.7	117.7	**1054.7**
2019	64.8	71.3		48.5	63.7	108.9		132.7	126.0	139.8	118.9	139.7	

Met Office

(a) The monthly figures given for 2016 add up to 1097.4 and not the 1097.5 given by the Met Office. Explain how this difference has occurred.

(b) March 2019 had 81% more rain than February 2019.

Calculate the amount of rainfall in March 2019.

(c) July 2019 had 22.9% less rain than June 2019.

Calculate the amount of rainfall in July 2019.

(d) Work out the mean percentage change in the rainfall from June to July.

What does this tell you about the rainfall in June and July?

(e) Take 2016 as the base year and give the total amount of rainfall as an index number of 100.

Calculate the index number for the total amount of rainfall each year from 2016 to 2019.

What do the index numbers tell you about the total rainfall each year?

Now that you have completed this chapter, you should know:

- How to use percentages to show important features of situations in everyday life.
- How to use percentages in financial problems, including inflation, appreciation and depreciation.
- How to find an original quantity, given the final value and the percentage change.
- How to use an index number to compare a number or value to that in a base year.

CHAPTER 4

Roughly speaking

There are many situations where you need to know the rough size of a quantity. This will often involve estimation. This chapter looks at some of the calculations that might be involved and the level of accuracy that would be reasonable, given the situation.

4A Currency exchange

When abroad, you need to know whether you are getting good value for money when buying food and drink, as well as souvenirs. Being able to convert quickly between the local currency and British pounds (£) helps you decide whether to buy something.

Activity 4A Shopping in Indonesia

The Indonesia rupiah (Rp), has an exchange rate of:

£1 = 18,090.91 Rp or 1 Rp = £0.000,055,276,4 (at time of writing).

Here are the prices of some items at a craft market:

Item	Price
Fridge magnet	40,000 Rp
Earrings	50,000 Rp
Necklace	70,000 Rp
Key tray	100,000 Rp

You want to estimate the cost for each item quickly but reasonably accurately.

(a) Would you use 18,000, 18,100 or 20,000 Rp to the £ to estimate the prices?

(b) What are the advantages and disadvantages of using a less accurate number for the estimate?

When a long number is written it is usually split into groups of 3 digits. There may be commas between them or there may be spaces. So you can have 40,000 (as here) or 40 000. Because you will come across both styles in everyday life, they are used interchangeably in this book.

Worked example 4A

? Estimate the cost in £ of the fridge magnet in Activity 4A.

Solution

Cost = 40,000 ÷ 20,000 = 2 ← 18,090.91 is 20,000 to the nearest 10,000.

So, the fridge magnet costs about £2.

This is likely to be an underestimate because the exchange rate has been rounded up. ← The rough exchange rate is larger than the true one. Dividing by a larger number gives a smaller answer.

Exercise 4A

Assume the exchange rate is £1 = 18,090.91 Rp.

1 Estimate the cost in £ of the following:

(a) a bracelet costing 60,000 Rp

(b) a scarf costing 90,000 Rp

(c) a keyring costing 25,000 Rp.

2 The cost of a different key tray is roughly £8. What do you think its price is in Rp?

3 Are your answers underestimates or overestimates?

4B Costing a journey

Needing to estimate the cost of a journey is a common situation.

Activity 4B Travel expenses

Sales reps have to travel to visit their customers.

(a) What costs do they incur that you can think of?

(b) How accurately is it realistic to present these costs – to the nearest penny, £1, £10 ...?

(c) What other considerations are there?

Worked example 4B

? Jan is a sales rep living in Bude. She plans a round trip to visit retailers in Barnstaple, Bristol, Westbury and Exeter on a regular basis.

Her car averages 48 mpg (miles per gallon) on normal roads and the fuel costs, on average, £1.20 per litre.

Estimate the cost of the round trip.

From	To	Distance in miles
Bude	Barnstaple	34
Barnstaple	Bristol	100
Bristol	Westbury	27
Westbury	Exeter	84
Exeter	Bude	53

1 gallon ≈ 4.5 litres

The symbol ≈ means approximately equals.

Solution

Miles per litre ≈ 50 ÷ 4.5 ≈ 10

Distance travelled ≈ 30 + 100 + 30 + 80 + 50 = 290 miles

Fuel ≈ 290 ÷ 10 = 29 litres

Cost is ≈ 30 × £1.20 = £36

The measures, and interim answers, have mostly been rounded to the nearest 10.

The level of accuracy of the rounding is chosen to make the calculation easier, as you assume no calculator is available. There is no single 'right' answer to an estimation question.

Exercise 4B

Assume that your car averages 38.6 miles per gallon, with a cost of £1.28 per litre.

1 Estimate the fuel cost of a return journey from Bude to Barnstaple (34 miles each way).

2 Estimate the fuel cost of a round trip of 164 miles.

3 The sales representative makes a journey of 156 miles twice a month. Estimate the yearly fuel cost.

4C Rounding

When you estimate a quantity you obtain a rough answer. So, there is no point in giving it too accurately; indeed, doing so can be misleading.

Activity 4C What is sensible?

Jamil has been weighed to the nearest thousandth of a kilogram (kg).

Jamil's weight has been rounded.

(a) Which statement do you think is the most sensible? Why?

(b) Why might you want to round to the nearest 10 kg?

(c) Why might you want to round to 67.9 kg?

A thousandth of a kilogram is a gram.

Jamil's rounded weight	
Nearest 0.001 kg	67.936 kg
Nearest 0.01 kg	67.94 kg
Nearest 0.1 kg	67.9 kg
Nearest kg	68 kg
Nearest 10 kg	70 kg
Nearest 100 kg	100 kg

Worked example 4C

?

(a) Round 28.07 to 1 decimal place (1 dp).

(b) Round 0.003748 to 2 significant figures (2 sf).

(c) Round 27.35 to 1 significant figure (1 sf).

(d) Round 1.5 to 1 significant figure (1 sf).

Solution

(a) 28.1, as 28.07 is nearer 28.1 than 28.0.

(b) 0.0037, as the first non-zero digit is 3 so the first two significant figures are 3 and 7 and the 4 indicates rounding down to 0.0037.

(c) 30, as 2 is the first non-zero digit the 7 indicates rounding up to 30.

(d) 2, as 1.5 is halfway between 1 and 2 and the convention is to round up in this situation.

Exercise 4C

1 (a) Round 4156 to the nearest 100.
(b) Round 0.0341 to the nearest 0.01.
(c) Round 199 to the nearest 10.

2 Round these numbers to the accuracy indicated:
(a) 6.13 to 1 decimal place
(b) 0.079 to 2 decimal places
(c) 10.05 to 1 decimal place.

3 Round these numbers to 1 significant figure:
(a) 24 809 (b) 389 (c) 0.715 (d) 0.006 801.

4 Round these numbers to 2 significant figures:
(a) 73 488 (b) 3999 (c) 0.704 (d) 0.070 707.

4D Checking calculations

Sometimes you need an answer quickly, so use the available figures to give a rough estimate. If it is important the answer needs to be correct, you will check it later.

Activity 4D Response rate

Sam is a psychology student. She gives out 243 questionnaires and gets 47 of them back.

She has lost her calculator, so has to estimate the percentage response rate. She says, '$\frac{47}{243} \times 100$ is approximately 25%.'

(a) How would you check Sam's calculation?

(b) Do you think the answer of 25 is sufficiently close for Sam's purposes?

Worked example 4D

? Sam has also done some face-to-face interviews as part of her research. She plans to transcribe them all as evidence for a case study. The first interview, lasting 27 minutes, takes an hour and 28 minutes to transcribe. She has 526 minutes of interviews recorded.

Her father points out that at this rate it will take her over 28½ hours to transcribe them all.

Use rounding to check whether his figure is about right.

Solution

The accurate calculation is $526 \times \frac{88}{27}$ minutes.

Sam rounds the numbers to 1 sf to get:

$500 \times \frac{90}{30}$ or 1500 minutes.

Dividing by 60 gives 25 hours.

This is quite close to $28\frac{1}{2}$ hours. It is the same order of magnitude.

So, Sam's father is probably correct.

When making a rough estimate, rounding to 1 significant figure is often a good strategy, as you only have one digit to use in the calculation. Sometimes you may want to use 2 significant figures if it means the number divides exactly.

Exercise 4D

1 Estimate by rounding to check:

(a) $4673 \times 28.3 \times 0.028 = 3703$

(b) $\frac{6.78}{0.18} \times 0.076 = 2.863$

(c) $3.141\,59 \times 0.653^2 \times \frac{57}{360} = 0.2121$

(answers are rounded to 4 significant figures).

2 A Sociology researcher has distributed 340 surveys and 127 are returned. Estimate the percentage of surveys that are returned.

4E What must the answer lie between?

It is often useful to know the accuracy of an estimate or rounded answer.

Activity 4E Rough cost

The key tray in Activity 4A on p.41 costs 100,000 Rp.

(a) You could use the approximate conversion of £0.000,06 to 1 Rp to estimate the cost.

(b) What about £0.000,05 to 1 Rp?

(c) Are these overestimates or underestimates?

(d) Which do you think would give a better estimate?

(e) What amount would you use as the cost in £ of the key tray?

The accurate conversion rate is: 1 Rp = £0.000,055,276,4.

Worked example 4E

? A pair of earrings cost 50,000 Rp.

The exchange rate on that day is 1 Rp = £0.000,055,276,4.

Use both of the rough conversions of £0.000,05 to 1 Rp and £0.000,06 to 1 Rp to estimate the cost in £s.

What approximate conversion for the cost of the earrings would you recommend?

Solution

Low estimate 50,000 × 0.000,05 = £2.50.

High estimate 50,000 × 0.000,06 = £3.

£2.50 is below the true cost. £3 is above the true cost.

You could use £2.75 as the cost of the earrings, as it is the average of the two costs, but £3 is a 'round' number that is easier to work with.

In the previous chapter you saw how a large number is written with a gap or a comma every 3 digits. The same thing is done for very small numbers with many digits to the right of the decimal point, like 0.000 06. In this example it is written as 0.000,06, with a comma.

Exercise 4E

1 (a) Use the approximate conversions 20,000 Rp to £1 and 18,000 Rp to £1 to work out what amounts the cost of the earrings (50,000 Rp) lies between.

(b) The cost of a meal is 182,490 Rp. Work out the amounts that the cost of the meal will lie between, in £, using the approximate conversions £0.000,05 to 1 Rp and £0.000,06 to 1 Rp.

(c) Use the approximate conversions 20,000 Rp to £1 and 18,000 Rp to £1 to work out maximum and minimum limits for the cost of the meal.

(d) What amount would you quote for the cost of the meal in GBP?

4F Lower and upper bounds

When you are estimating a quantity, it is often helpful to be able to say, 'it is more than this' and 'it is less than that'.

Activity 4F Raised beds

(a) A rectangular vegetable bed measures 3 m by 4 m, to the nearest metre. You have been asked to create a wooden surround to hold the soil in place. It is cheaper to buy one length of wood and cut it to the sizes needed. What length of wood would you buy?

(b) What might change if the bed is measured to the nearest centimetre?

Worked example 4F

The bed is measured to the nearest tenth of a metre, as 3.0 m by 4.0 m.

What are the minimum and maximum perimeters?

Solution

Width is 3.0 ± 0.05, so $2.95 \le w \le 3.05$.

Length is 4.0 ± 0.05, so $3.95 \le l \le 4.05$.

The minimum possible perimeter is:

$2 \times (2.95 + 3.95) = 13.8$ m.

The maximum possible perimeter is:

$2 \times (3.05 + 4.05) = 14.2$ m.

3 m implies accuracy to the nearest metre.

3.0 m implies accuracy to the nearest 0.1 m.

3.00 m implies accuracy to the nearest 0.01 m, or centimetre.

The smaller value is the **lower bound** for the perimeter, the larger value is the **upper bound** for the perimeter.

This can be written as 14 ± 0.2 m or $13.8 \le p \le 14.2$ m.

The interval between the lower and upper bound is the tolerance for the measurement.

Tolerance is the amount by which a measurement or calculation might change and still be acceptable.

Exercise 4F

1 The dimensions of a vegetable bed are 3 m by 4 m and are accurate to the nearest metre. Work out the lower and upper bounds for the area of the bed. Express your answer using inequality notation.

2 An athlete runs a lap of 400 m in 63 seconds. The distance is correct to the nearest metre and the time is correct to the nearest second. What are the lower and upper bounds for her speed? Write your answer using inequality notation.

4G Writing tolerances

This section looks at the different ways of expressing the accuracy of an estimate.

Activity 4G Nuts

The tolerance for the number of nuts in a snack pack is 30 ± 10%.

(a) What do you think that means?

(b) Do you think that it is reasonable?

(c) Think of some other products. Would ± 10% be an acceptable tolerance for those?

There are three main ways to express tolerances: 15 ± 1.5, 120 ± 5%, $11.5 \le D \le 12.5$.

Worked example 4G

? Write 30 ± 10% as values, using the plus/minus notation (±) and the inequality notation.

10% is 10 out of every hundred. That is, $\frac{10}{100} = \frac{1}{10}$.

Solution

Work out 10% of 30 = 3.

So, 30 ± 10% = 30 ± 3.

Using inequality notation, $27 \le N \le 33$.

Exercise 4G

1 Rewrite these tolerances using inequality signs:

(a) 12 ± 0.5 (b) 12 ± 10% (c) 120 ± 5% (d) 0.1 ± 0.01.

2 Rewrite these tolerances using percentages:

(a) 50 ± 5 (b) $54 \le d \le 66$ (c) 250 ± 5 (d) $19.5 \le w \le 20.5$.

3 Rewrite these tolerances using the plus/minus sign (±):

(a) $45 \le h \le 65$ (b) 150 ± 2% (c) $3.7 \le w \le 4.3$ (d) 40 ± 5%.

4H Using an appropriate degree of accuracy

A useful rule of thumb is that the answer is given to the same accuracy as any measurements in a calculation.

Activity 4H Rounding the area

A vegetable bed is a rectangle of dimensions 3.0 m by 4.0 m.

The minimum area = $2.95 \times 3.95 = 11.6525$ m^2.

The maximum area = $3.05 \times 4.05 = 12.3525$ m^2.

(a) What happens when you round each answer to 2 sf?

(b) Explain why rounding to 2 significant figures would be consistent for the accuracy of the area in this case.

Worked example 4H

Write the answers to the following, to a suitable degree of accuracy:

(a) A circle of radius 2.7 cm has an area = $\pi \times 2.7^2 = 22.8906$ cm^2.

(b) The volume of a cylinder, base radius 2.7 cm and height 8 cm, is = $\pi \times 2.7^2 \times 8 = 183.1248$ cm^3.

($\pi = 3.14$)

Where there is more than one accuracy used, you should give the answer to the lesser accuracy.

Solution

(a) The measurement 2.7 cm is given to 2 significant figures, so the answer is given as 23 cm^2. ($\pi = 3.14$ was used.)

(b) The calculation contains measurements to both 1 and 2 significant figure accuracy, so the answer should be given to 1 significant figure as 200 cm^3.

It is good practice to write down a more accurate answer from your calculator before rounding.

What would be the effect of using different accuracies in finding areas and circumferences of circles?

Exercise 4H

1 The area of a rectangular field measuring 27.7 m by 34.2 m is 947.34 m^2. Write the area to an appropriate degree of accuracy.

2 The average speed of a car travelling 382 km in 7 hours and 12 minutes is 53.0555 km/h. Write the average speed to an appropriate degree of accuracy.

4I The modelling cycle

The diagram on the right shows the steps involved in solving a real-world problem in a type of flow chart called the modelling cycle. It can be appied to many different situations and emphasises that a model should be reviewed in the light of evidence and improved where possible. In the next two chapters you will will meet special versions of this cycle for finance and statistics.

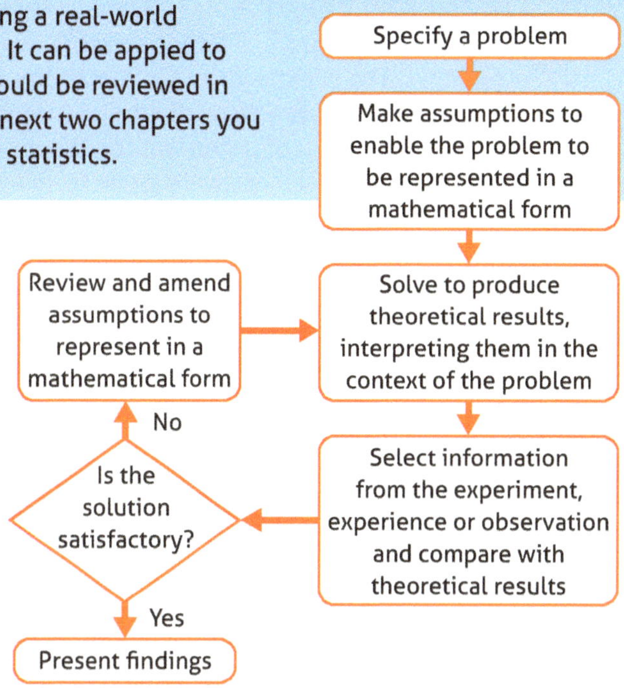

Activity 4I Twins

(a) Using the modelling cycle, estimate how many twins (people born on the same day) you have in the world?

(b) How accurate do you think your answer is?

Worked example 4I

Use the modelling cycle to estimate the cost of the round trip made by Jan, the sales representative in Worked example 4B.

Solution

Specify the problem. → Estimate the cost of Jan's journey starting from Bude, visiting Barnstaple, Bristol, Westbury, Exeter and back to Bude.

Bude and back again

From	To	Distance in miles
Bude	Barnstaple	34
Barnstaple	Bristol	100
Bristol	Westbury	27
Westbury	Exeter	84
Exeter	Bude	53

← Make assumptions to enable the problem to be represented in a mathematical form.

Fuel = 48 mpg at £1.20 per litre.

Solve the problem to produce a theoretical result (36) and interpret it in the context of the problem (£36). →

Add the distances, 300 (rounded) or 298 (not rounded).
Round 48 to 50.
$50 \div 4.5 \approx 10$
$300 \div 10 = 30$
$30 \times 1.20 = 36$
Each round trip costs approximately £36.

After several months Jan is dismayed to discover that the actual cost is nearly £50 for her round trip. This is much more than the £36 she had budgeted. ← Select information from experience and compare with the theoretical result.

Review and amend assumptions. → This is not satisfactory, so Jan looks at the assumptions she made.

- The manufacturer's information suggested 48 mpg but its methods for determining this have since been discredited, so 40 mpg is nearer the truth.
- The cost of petrol was low when Jan made her estimate and has since risen to £1.35 per litre.
- The distances were taken from a reliable source so she does not amend those.
- She decides to work with the more accurate figures this time.

298 ÷ (40 ÷ 4.5) × 1.35 = 45.26 giving an average amount of £45.26. ← Solve the problem to produce a theoretical result and interpret it in the context of the problem.

Compare with experience. Decide whether to further review assumptions. → At this point Jan may decide her solution is satisfactory and present her findings by including them in her budget forecast.

Or, she may still think it underestimates her costs and further revise her assumptions. For example, she may regularly meet heavy traffic and this would affect the fuel use and then the figure of 40 mpg could be reduced further.

She decides that the figure of £45.26 is satisfactory and includes it in her budget for the year. ← Present findings.

Exercise 4I

1 Alan needs to spread moss killer on his lawn. Its shape is roughly a semicircle of radius 11 metres. The moss killer is sold in 2 kg boxes and it must be spread at a rate of 35 g per square metre.

State the problem that Alan needs to solve and use the modelling cycle to do so.

2 Preeyesh is a farmer. He wants to estimate the amount of water in a lake on his land. He models the lake as a cylinder with radius 15 m and height 6 m.

(a) What problem is Preeyesh aiming to solve?

(b) What assumptions has he made?

(c) What is the mathematical problem that he solves?

(d) Solve the mathematical problem.

(e) What is the solution to the actual problem?

(f) During a particularly long dry spell of weather the lake dries up and Preeyesh observes that a cone is a better model for the lake than a cylinder. What part of the modelling cycle does this correspond to?

(g) Review Preeyesh's assumptions and amend them appropriately.

(h) Solve the revised problem.

For a cylinder of base radius r and height h volume $= \pi r^2 h$.

For a cone of base radius r and height h volume $= \frac{1}{3}\pi r^2 h$.

For a sphere of radius r volume $= \frac{4}{3}\pi r^3$.

Test yourself 4J

1 A meteorologist measures the rainfall for July. Each day she measures the amount of rain to the nearest mm. Her results are shown in the table below.

mm	0	1	2	3	4	5	6	7	8	9–18	19	20	21
Frequency	22	1	2	2	0	1	0	0	1	0	1	0	1

(a) Estimate the total rain for July.

(b) State the lower bound and the upper bound for your estimate.

2 A can of drink contains 330 ± 5% ml of liquid. What is the least amount that it could contain?

3 A farmer estimates the dimensions of a rectangular field as 50 ± 5 m by 35 ± 5 m. Calculate upper and lower bounds for the area of the field.

4 A biologist estimates the speed of a snail by timing how long it takes to travel a short distance. He notes that it travels 8.5 mm in 6.2 seconds. Work out the speed of the snail in kilometres per hour and give your answer to an appropriate degree of accuracy.

5 (a) Shani is planning a holiday in Thailand. She has paid for her flights, hotel transfer and bed and breakfast each day in the hotel. She now needs to estimate how much money she should save for other expenses.

List some of the assumptions she should make.

(b) For two of the assumptions you have suggested in question 3 (a), describe what evidence might lead Shani to review and amend those assumptions.

Practice question

The cost of living in the fourteenth century
A labourer earns £2 per year
An infantry soldier earns 8d per day
A man-at-arms earns 1s per day
A master carpenter earns 3d per day
Kitchen servants earn 2s to 4s per year
A tunic for a serf costs between 1d and 6d
2 chickens cost 1d
2 dozen eggs cost 1d
1 pig costs 2s
1 cow costs between 6s and 9s
Pepper costs 4s per lb
Ale costs 1d per gallon
Wine costs 3d to 4d per gallon
Cheese costs 3s 4d for 80 lb
Rent for a cottage is 5s per year

12 pence (d) = 1 shilling (s)
20 shillings = £1
1 gallon = 8 pints
1 gallon = 4.5 litres
1lb (pound) = 16 oz (ounces)
1lb = 0.454 kg

Use the information about the cost of living in the fourteenth century to answer these questions. Remember that, in a feudal economy, many items may be supplied by lords or produced in the home, so calculations are necessarily estimates.

(a) Estimate how much a labourer earns per day, listing the assumptions you make.

(b) Estimate the cost of food for a labourer for a year. Comment on your answer.

(c) Estimate the total living costs of a labourer for a year. List the assumptions you make and justify a lower and an upper bound for your answer.

(d) Can the labourer live within his £2 wages for the year?

(e) Review your assumptions if necessary and amend your estimate of the labourer's total living costs for a year.

Now that you have completed this chapter, you should know:

- How to make a rough estimate of a quantity from available information.
- How to use estimates when checking calculations.
- How to make and justify upper and lower bounds for a calculation.
- How to interpret and present error bounds or tolerances.
- How to use the modelling cycle.

CHAPTER 5

Financial problem solving

Throughout your life you will have to make financial decisions. This chapter looks at some of the issues involved.

Which job should I apply for?

Should I buy a house or rent?

Can I afford to buy a new car?

Who should I bank with?

How much should I save for my pension?

Where should I go on holiday?

5A The financial problem solving cycle

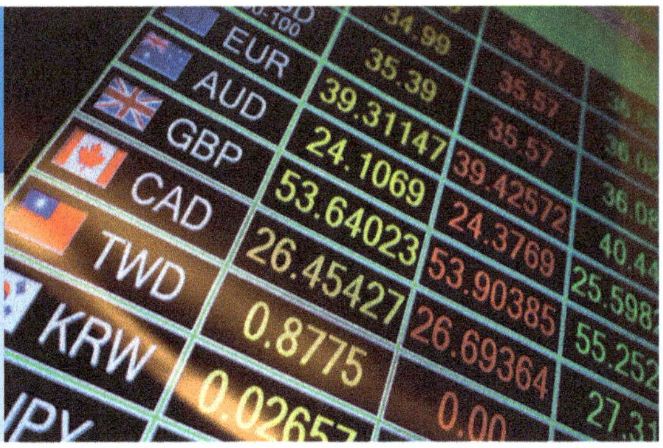

When you approach an important financial decision, you may well be faced with questions like those above. Answering them may not be straightforward. There are other considerations beyond simple financial gain that you will need to consider.
Your thoughts will go around in circles – in a cycle.

Activity 5A Inside the financial cycle

The boxes in this cycle are the steps you go through when addressing a financial problem.

Look at the following questions and decide which box in the cycle each one fits into.

(a) Is this really what I want or need?
- Will the less well-paid job be more interesting or lead to better opportunities?

(b) Will this have an adverse effect on the environment?
- Is it better to use green electricity even though it costs more?

(c) Are there other ethical considerations?
- Does the bank also invest in businesses that you don't approve of for moral reasons?

(d) Will it affect my way of life?
- Do you want to live closer to your work or closer to your friends and family?

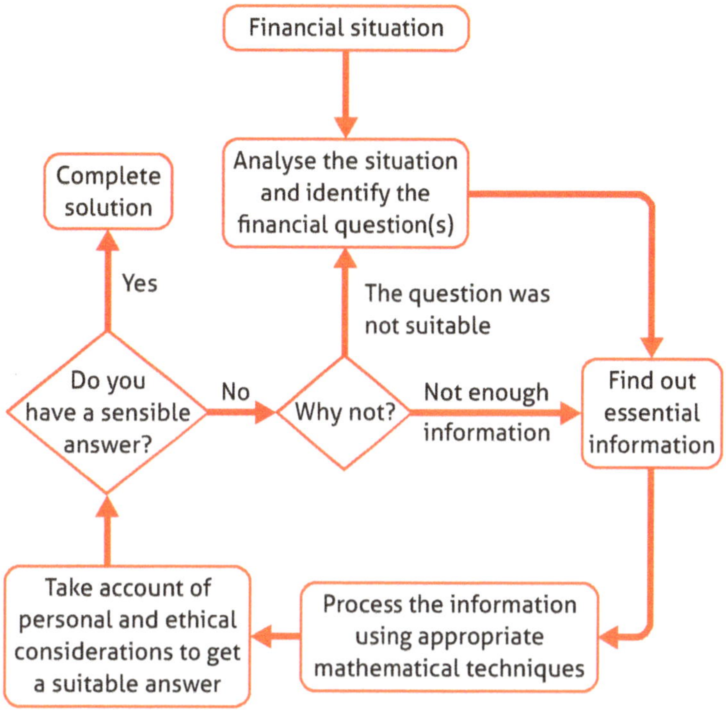

Worked example 5A

? Here are some further questions. Which box in the cycle would they fit into?

(a) What information do you need?
- Where can you find this information?
- Should you ask an expert?
- Can you use the internet?

(b) Can you afford it?

(c) What calculations do you need to carry out?
- Will a spreadsheet help?

(d) What problem are you trying to solve?
- What do you want to achieve?

(e) Can you cut costs further?
- Do you need to modify the problem?

(f) What other factors should you consider?
- Is the original proposal sound?
- Is the cheapest option the best?

Solution

(a) Find out essential information.

(b) Do you have a sensible answer?

(c) Process the information using appropriate mathematical techniques.

(d) Analyse the situation and identify the financial question(s).

(e) Do you have a sensible answer? Why not?

(f) Take account of personal and ethical considerations to obtain a suitable answer.

Exercise 5A

1 Katya and Jack want to go on holiday. They have a budget of £1500. They have researched some costs:

Flights to Paris: £75 per person return

Youth hostel: €17 per person per night

Food cost: Allow €15 per person per meal

Eurostar London to Paris: £50 per person one way

Hotel: €120 per person per night, includes breakfast

The travel agency has this sign for changing pounds to euros:

We sell: 1.11
We buy: 1.35

This means when you change pounds (£) to euros (€) then £1 = €1.11. But when you convert euros back to pounds then £1 = €1.35.

Katya and Jack decide they need 40 euros each in spending money per day.

(a) How much does their holiday cost in pounds if they:

  (i) fly and stay in a youth hostel for 5 days

  (ii) go by Eurostar and stay in a hotel for 3 days.

(b) Katya and Jack change all their money to euros after they have paid for travel.

Estimate how much they have after their holiday once they have changed their left-over euros back into pounds in (a)(i) and (a)(ii)?

What assumptions have you made?

(c) Suggest which holiday option Katya and Jack should take.

What further information would you need to know in order to advise them properly?

2 In 2021, students could apply for a means tested loan of up to £9203 per year to cover living expenses.

(a) Imagine you will be going to university this year. Make a list of all the living expenses you will have.

(b) Design a monthly budget for a university student, using your list of living expenses.

Decide whether the student will need to top-up the loan with part-time work.

5B Compound interest

Some bank or building society accounts pay interest on money invested with them. You also earn more interest on any interest already earned. This is called **compound interest**.

Activity 5B Receiving interest

When do you need to pay interest?

When is interest paid to you?

£300 is invested in a high interest account for 2 years. The account pays 10% interest per year.

How much more interest is earned if the account earns compound interest rather than simple interest?

Worked example 5B

? Jasmine's grandparents give her £5000. She invests it in a building society at 2% compound interest per annum.

Per annum means each year.

How much money does she have in her account after 4 years?

What assumptions have you made?

Solution

The value, A, of an investment after n years is given by the formula:

$$A = P\left(1 + \frac{r}{100}\right)^n$$

The letter A stands for Amount, which is another way of saying value.

where P is the initial investment (called the **principal**) and

r% is the interest rate.

Jasmine's investment grows by 2% (or 0.02) each year, giving an annual multiplier of $1 + 0.02 = 1.02$.
She invests the money for 4 years, so to find the final value she multiples the initial investment by 1.02^4.

$$\begin{aligned} \text{So, } A &= 5000(1 + 0.02)^4 \\ &= 5000 \times 1.02^4 \\ &= 5412.16. \end{aligned}$$

Jasmine has £5412.16 in her building society account after 4 years.
Assumptions: Jasmine hasn't withdrawn or invested any more money.
The interest rate doesn't change from 2%.

Exercise 5B

1. Mia invests £2200 in a building society account that pays compound interest at the rate of 2% per annum.
Calculate the value of her investment after 5 years.

2. George is about to start a 3-year university course. He plans to go travelling when he finishes it and estimates he will need £3000.
George invests some money in a savings account that pays compound interest at 3.5% per annum.
How much money should he invest so that he has £3000 after 3 years?

3. Jack invests £500 in some stock that pays compound interest.
After 10 years his investment is worth £640.
Calculate the rate of interest paid per annum.

5C Annual equivalent rate (AER)

Jane invests £100 in a company that pays 10% compound interest every 6 months.

After 6 months her account is credited with £10, so is now £110.

After another 6 months it is credited with £11, so its value is now £121.

So after one year Jane's £100 has received £21.

This equivalent interest rate is 21% per year.

This is called the annual equivalent rate, or AER.

AER is used to compare the interest rates on savings. As well as yearly, banks may pay interest at other intervals; for example, daily, monthly or quarterly (every 3 months). The AER tells you what percentage your investment would grow by if you just left it in the account for a year.

Activity 5C Savings accounts

(a) How would you choose which of these savings accounts to invest in?

ABC Bank	5 Star Savers	Halo Bank	Investa Bank
2.95% per annum with interest calculated quarterly.	2.9% per annum with interest calculated daily.	3.0% per annum with interest calculated monthly.	3.1% per annum with interest calculated annually.

(b) What factors other than interest rate should you consider when choosing a bank account?

Worked example 5C

(a) Work out the AER for ABC Bank and Investa Bank in Activity 5C.

(b) Amy's savings account pays interest of 1.8% every 6 months.
Work out the AER for her account.

Solution

(a) **Investa Bank:** The interest is calculated annually, so the AER is 3.1%.

Per annum means each year and annually means yearly.

ABC Bank: The interest is calculated quarterly (every 3 months).

So, the interest is $\frac{2.95\%}{4} = 0.7375\%$ per quarter.

As a decimal 0.7375% = 0.007375.

So, to find the interest for 1 quarter you multiply by 1.007375.

The interest is compounded, so the multiplier is $1.007375^4 = 1.02982$.

So, the AER is 2.98% (to 3 sf).

(b) 1.8% every 6 months gives a multiplier of 1.018.

So, the multiplier for 12 months is $1.018^2 = 1.036324$.

A multiplier of 1.0363 corresponds to a percentage of 3.63% (3 sf).

So, the AER = 3.63%.

Exercise 5C

1 Hannah invests £10,000 in a high interest savings account that pays interest monthly at an annual rate of 5%.

(a) What is the balance of the account at the end of 3 years?

(b) What is the AER for Hannah's investment?

2 A building society pays an annual rate of 4% on its cash ISAs (Individual Savings Accounts).
Calculate the AER if the interest is calculated:

(a) annually
(b) six monthly
(c) quarterly
(d) monthly
(e) weekly
(f) daily.

What is the advantage to savers in choosing an account that calculates interest more frequently?

5D Annual percentage rate (APR)

Compound interest is also used when you take out a loan, so you end up paying back more than you borrowed. Usually, the longer the term of the loan, the more you have to repay.

Most banks and building societies calculate interest more often than once a year.

The **annual percentage rate**, or **APR**, is the equivalent rate that you would have to pay if you were charged interest only once a year. It is the same idea as for AER but in this case you are paying out the interest rather than receiving it.

Activity 5D Taking out a loan

Danya wants to take out a loan to buy a laptop. He needs to borrow £1000.

(a) Which of these options would you advise?

  (i) A loan, with 24 monthly repayments of £50.

  (ii) Using a credit card at 1.5% interest per month.

(b) What assumption have you made to answer this question?

(c) What factors should you consider when purchasing an expensive item like a laptop?

As you can see, it can be hard to tell which interest rate is better. When you take out a loan, the bank or building society has to tell you the **APR**; it also takes into account the interest and any fees paid on the loan. APR makes it easier to compare loans, especially when the interest is calculated over different time periods.

Worked example 5D

(a) Work out the APR for both of the loan options in Activity 5D.

(b) A credit card company charges interest at 30% per year, calculated monthly. What APR is it charging?

Solution

(a) **Loan:** 24 monthly repayments of £50 = 24 × £50 = £1200.

Interest rate over the two years = $\frac{200}{1000} \times 100\% = 20\%$.

So, the annual percentage rate = 20% ÷ 2 = 10%.

> The APR doesn't take into account any repayments that Danya might make, so in reality he might pay back less interest on the debt.

Credit card: 1.5% per month gives a multiplier of 1.015 for each month.
So, the multiplier for 12 months is $1.015^{12} = 1.196$ to 3 dp.
A multiplier of 1.196 corresponds to a percentage of 19.6%.
So, the APR = 19.6%.

(b) Interest is $\frac{30\%}{12} = 2.5\%$ per month.

2.5% gives a multiplier of 1.025 each month.
So, the multiplier for 12 months is $1.025^{12} = 1.3448$.
A multiplier of 1.3448 corresponds to a percentage of 34.48%.
So, the APR is 34.5% (to 3 sf).

Exercise 5D

1 (a) Calculate the APR on loans (i), (ii) and (iii) below.

(b) For (i), (ii) and (iii) work out how much money would be repaid on a loan of £500, which is repaid in full after 1 year.

(i) Interest rate: 0.5% calculated weekly.

(ii) Interest rate: 2.5% calculated quarterly.

(iii) Interest of 25% per year calculated weekly.

2 Esme takes a payday loan of £200 at 0.8% per day for 2 weeks.

(a) Work out the APR.

(b) How much does Esme need to repay after 2 weeks?

A payday loan is a very short-term loan designed to tide you over until your next paycheck. Typically, a payday loan has a very high APR. They are often taken out by people who find it difficult to access a loan from a traditional bank.

3 Danya sets up a spreadsheet to work out how much he needs to repay if he uses the credit card at an interest rate of 1.5% calculated monthly for his loan of £1000. Danya repays £50 each month.

	A	B	C	D
1	Month	Interest	Repayment	Balance
2	0			£1,000.00
3	1	£15.00	£50.00	£965.00
4	2	£14.48	£50.00	£929.48
5	3	£13.94	£50.00	£893.42

(a) What formula does Danya enter in each of these cells?

(i) A3 (ii) B3 (iii) D3

(b) (i) Use a spreadsheet to work out how long it will take Danya to repay his credit card debt.

(ii) How much does Danya repay altogether?

(c) How long will it take Danya to repay the cost of the laptop if he pays £100 per month instead?

(d) Danya decides to make the minimum payment of £20 per month.

(i) How long will it take to clear his loan?

(ii) How much interest will he pay in total?

5E Costing

You often need to work out the costs involved in a project or a course of action.

Activity 5E Financing a new car

Natasha wants to buy a new car.

She doesn't have enough money to buy a car outright, so she explores two different options.

The car costs £15,500.

Option A: Take out a loan for £15,500 paid back at the end of 3 years at 6% APR.

Option B: Use hire purchase (HP):

- pay £2000 deposit
- make 36 monthly payments of £440
- After 3 years the car belongs to Natasha.

How much does Natasha pay altogether under each option available to her?

Worked example 5E

Natasha considers a third option, using personal contract purchase (PCP).

Using this she would pay £2000 deposit and make 36 monthly payments of £245.

As part of this package, she can opt to pay a further £8500. Provided she drives no more than 9000 miles a year, the car belongs to her at the end of the period. If she does not pay this, then she must return the car undamaged to the dealer.

This scheme is called GMFV, standing for guaranteed minimum future value.

If Natasha takes up the option, how much does she pay for the car?

Solution

Deposit	£2000
Monthly payments 36 × £245 =	£8820
Further payment	£8500
So, Natasha pays	£19,320

Exercise 5E

Archie wants to start a small business running a gourmet burger van.

A business proposal is often made to a bank to get a start-up loan for the business.

1 Archie uses this spreadsheet as part of his business proposal.

He decides to write-off the cost of the van over 8 years and the cooking equipment over 6 years.

	A	B	C	D	E	F	G	H
1	**Inputs**			**Annual debts**			**Annual credits**	
2								
3	**Van**							
4	Cost	£20,000		Van	£2,500		Sales of burgers	£26,000
5	Life (years)	8		Cooking equipment	£400			
6				Ingredients	£10,400			
7	**Cooking equipment**							
8	Cost	£2,400		**Total**	£13,300		**Total Profit**	£12,700
9	Life (years)	6						
10								
11	**Ingredients**							
12	Weekly cost	£200						
13								
14	**Burgers**							
15	Sale price of burger	£5						
16	Number sold per week	100						

To write-off a cost over 4 years means that Archie spreads the cost of the van over 4 years.

Archie inputs formulae into some of the cells in his spreadsheet.

What formula does Archie input into each of the following cells?

(a) E4 (b) E5 (c) E8 (d) H4 (e) H8

2 Archie decides to buy a more expensive van for £25,000, which he will write-off over 1 year.

(a) What happens to Archie's profit?

Archie decides that he can sell 150 burgers a week if he sells burgers at farmers' markets.

(b) Estimate Archie's profit if he attends farmers' markets.

(c) How would you advise Archie to deal with the cost of his van?

5F Demand curves

A demand curve shows the relationship between the price of an object and the number sold (the demand).

Activity 5F Costing bracelets

Isobel makes bracelets. She needs to decide how much to charge for them.

Each bracelet takes her 2 hours to make and costs her £7 in materials.

(a) What are the advantages/disadvantages of a very high or low selling price?

(b) How can Isobel decide on what the selling price should be?

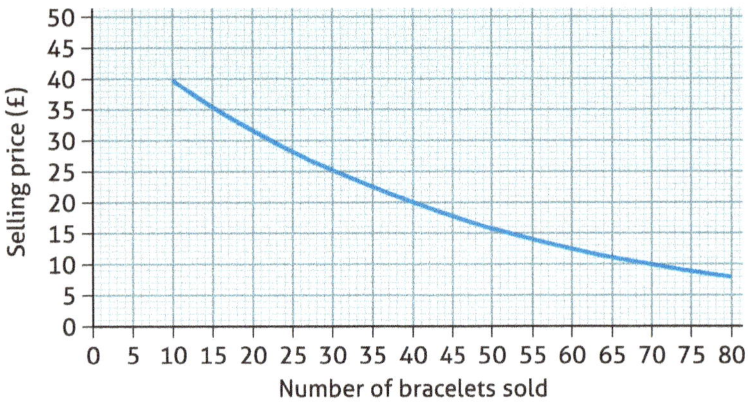

A demand curve doesn't have to be a curve – it could be a straight line. Usually, demand decreases as the selling price increases.

If Isobel makes her bracelets too expensive no-one will buy them. However, if she sets too low a selling price, she may make a loss even if she sells lots of bracelets.

She asks some potential customers how much they would be prepared to pay and uses her findings to draw the **demand curve** illustrated below.

A demand curve illustrates the relationship between the selling price (↑) of an object and the demand for it (→). It can be used to help decide on the optimum selling price.

Exercise 5F

Isobel uses her demand curve to work out the best price for selling her bracelets.

1 (a) Find how many bracelets Isobel expects to sell if the selling price is:

(i) £10 (ii) £30 (iii) £40.

(b) Find the sale price that Isobel should use if she wants to sell:

(i) 20 bracelets (ii) 40 bracelets (iii) 60 bracelets.

2 Isobel enters the expected sales for selling prices from £10 to £40, at £5 intervals, into a spreadsheet. She then uses the spreadsheet to work out the expected profit for each of these prices.

	A	B	C	D	E
1	**Selling price**	**Number of bracelets sold**	**Revenue**	**Cost of making bracelets**	**Profit**
2	£10	70	£700	£490	£210
3	£15	52	£780	£364	£416
4	£20				

(a) Isobel inputs formulae into some of the cells.
Write down the formula that she uses in each of the following cells:
(i) A3 (ii) C3 (iii) D2 (iv) E2

(b) (i) Recommend a selling price for her bracelets.
(ii) Estimate the maximum profit Isobel can make.
(iii) Isobel takes 2 hours to make each bracelet.
Is the selling price a viable business option? Explain your answer fully.

Test yourself 5G

1 Jess wants to buy a new mobile phone.
She finds these two options for the same phone.
Which mobile phone contract would you recommend?

Option 1	Option 2
£30 upfront cost £38 per month for 24 months	Sim only deal £729 for phone £8 per month

2 Chloe invests £4000 in a building society account, which pays compound interest at the rate of 2% per annum.
(a) Calculate the value of her investment after 3 years.
(b) (i) How much does Chloe have after 3 years if the interest is calculated quarterly rather than annually?
(ii) What is the AER when the interest is paid quarterly?

3 Adil goes clothes shopping.
He spends £400 on his credit card.
The credit card charges 2% interest every month.
(a) What is the APR?
Adil pays £100 per month to clear his credit card debt.
He makes a spreadsheet to keep track of his payments.
(b) What formula does Adil input into (i) cell B3 (ii) cell D3?

	A	B	C	D
1	Month	Interest	Repayment	Balance
2	0			£400.00
3	1	£8.00	£100.00	£308.00

(c) Set up your own spreadsheet. Copy and complete Adil's table.
(d) How much interest does Adil pay on his debt?

4 All employees in the UK pay income tax.
The table shows the tax rates for 2020.
The **personal allowance** is the amount that you can earn before you start paying any income tax.

Tax band	Taxable income	Tax rate
Personal allowance	£12,500	0%
Base rate	From £12,500 up to £50,000	20%
Higher rate	£50,000 up to £150,000	40%

(a) Calculate how much tax each of these people pay:
(i) Sam earns £300 per week.
(ii) Tom earns £2500 per month.
(iii) Sasha earns £70,000 per year.

All employees also pay National Insurance contributions:
- 12% of weekly earnings between £183 and £962
- 2% of weekly earnings above £962.

(b) How much does each person in part (a) pay in National Insurance contributions?

5 Kirsty earns £22,000 per year.
Her outgoings are shown on the right.

Pension contributions (tax exempt): £150 per month
Tax: _____
National Insurance: ______
Rent: £750 per month
Food: £200 per month
Bills: £60 per month

(a) Use the information on Income Tax and National Insurance in Question 4 above to complete the information for how much Kirsty pays in tax and National Insurance.
(b) How much money does Kirsty have left at the end of each month?

Practice question

Mia is starting a new business making cakes for special occasions.

Mia uses a demand curve to help her work out the price for one cake.

The demand curve shows the number of cakes Mia can expect to sell each week.

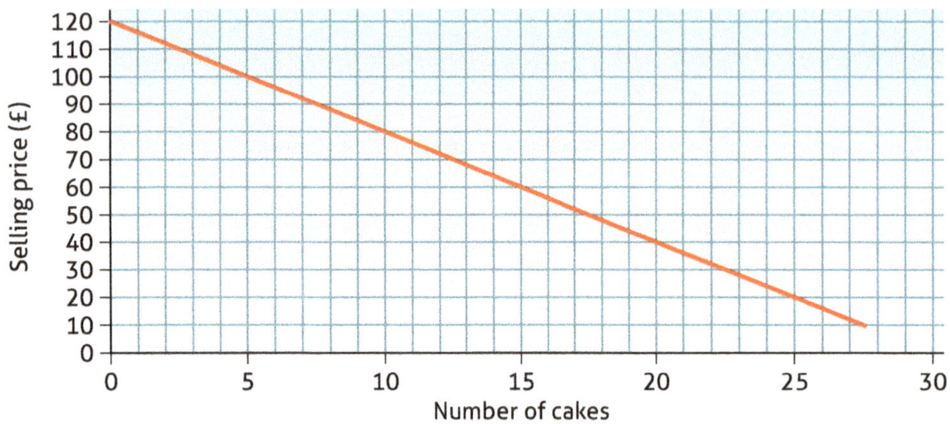

(a) How many cakes does Mia expect to sell in one week if she sells them for £60 each?

(b) How much should Mia sell her cakes for if she wants to sell 20 cakes per week?

Mia uses a spreadsheet to help her work out how much she should sell each cake for.

	A	B	C	D	E
1	Selling price (£)	Number of cakes sold	Revenue (£)	Cost of making cakes (£)	Profit (£)
2	10	27.5	275	275	0
3	20	25	500	250	250
4	30	22.5	675	225	450
5	40	20	800	200	600
6	50	17.5	875	175	700
7	60	15	900	150	750
8	70	12.5	875	125	750
9	80	10	800	100	700
10	90	7.5	675	75	600
11	100	5	500	50	450
12	110	2.5	275	25	250
13	120	0	0	0	0

(c) How much does each cake cost to make?

(d) Mia inputs a formula into some of the cells.

Write down the formula that she uses in each of the following cells:

(i) A3 (ii) C3 (iii) D2 (iv) E2.

(e) At what cake price will Mia just break even?

(f) Mia wishes to sell the cakes for a whole number of pounds (e.g. £5 not £5.40).

(i) Estimate the maximum profit Mia can make.

(ii) How many cakes does she sell a week in this scenario?

(iii) Mia takes 4 hours to bake and ice each cake.

With this information in mind, does the answer to part (i) suggest that this is a realistic business plan?

Now that you have completed this chapter, you should know:

- How to use the financial problem solving cycle in making financial decisions.
- How to solve problems involving percentages in context, including compound interest, AER and APR.
- How to use a demand curve.

CHAPTER 6
Working with data

Whether in everyday life, while studying or at work, you will forever find yourself coming across data. This chapter covers the basic skills you will need to understand and interpret the ideas involved and present them to others. More specialist work on statistics follows later in the book.

6A The statistical problem solving cycle

Statistics is a practical subject. Its purpose is to help you solve problems. The steps you go through are illustrated below, in the statistical problem solving cycle.

The first step is to state clearly, in simple English, the problem that you are trying to solve.

Activity 6A Thinking about the cycle

This diagram on the right shows the connections between the four basic activities in a statistics enquiry.

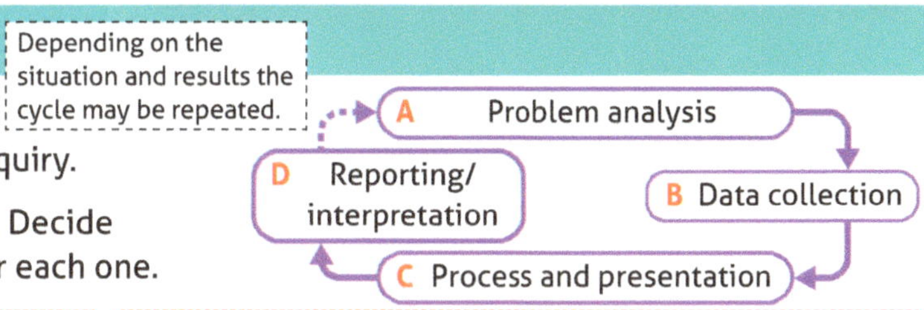

Here are a set of actions, decisions and questions. Decide which box, A, B, C or D, is the appropriate place for each one.

1	Record any collected data in a form easily used.	**13**	Decide what techniques to use.
2	Stick to the original plan.	**14**	Discuss the reliability of the results.
3	Are there any possible improvements?	**15**	Are there suitable data accessible and free from bias?
4	Analyse/interpret any diagrams and measures.	**16**	Are the conclusions clear to the intended audience?
5	Look for outliers.	**17**	Consider the resources available such as time, IT, software, etc.
6	Are the data complete?	**18**	What is the problem?
7	What needs to be measured?	**19**	Avoid misrepresentation of the data.
8	Are the data a sample?	**20**	Are any factors worth reconsidering?
9	Have the data been 'cleaned' and checked for omissions or typos?	**21**	Are diagrams clear, simple and relevant?
10	What is the origin of the data?	**22**	Consider any further work.
11	Collect sufficient data to draw reliable conclusions.	**23**	Select which factors may be important.
12	Draw together conclusions relating to the original problem.	**24**	Look for any patterns or trends.

A word often used in investigations or problems in statistics is **hypothesis**.

A hypothesis is an idea, theory or question that can be tested.

Hypothesis testing is covered in Chapter 19.

Exercise 6A

1 For each of the hypotheses below, make a bullet point list to show how you might investigate it. Group your bullet points into the four stages of the statistical problem solving cycle.

(a) It is quicker to put ten numbers in order than ten words in alphabetical order.

(b) London, England has more rain than London, Canada or London, USA. (There are almost 30 London's around the world.)

(c) People tend to overestimate short time intervals such as 1 minute.

(d) In Premier League football matches, the difference between a winning and losing score is less than that in lower divisions.

2 Richard is using the statistical problem solving cycle to investigate a hypothesis. For the 'Process and presentation' stage (C) he uses a spreadsheet to draw this scatter diagram and line of best fit. (The data are real.)

Suggest what Richard might have done in each of the other three stages.

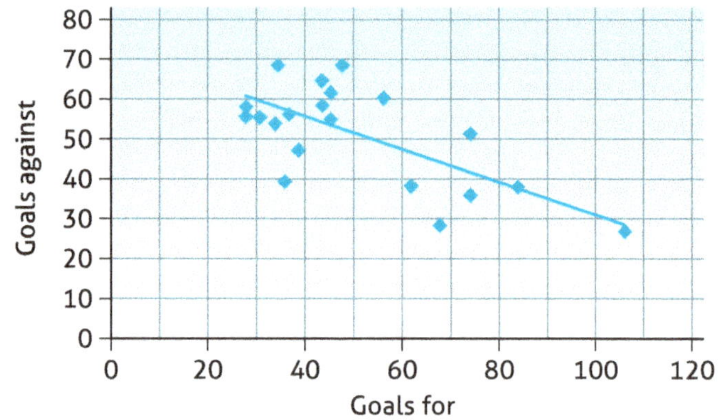

6B Looking quickly behind the numbers in data

Data are a collection of figures. When arranged in an organised form, they provide **information**.

- **Primary data** are collected for a particular investigation.
- **Secondary data** are already in existence.

This section introduces some of the first steps in converting data into information.

Activity 6B Old Faithful

Old Faithful is a geyser in the US. It erupts many times a day.

The data below are the times, to the nearest minute, between successive eruptions of Old Faithful over two days in May 1990.

They are in order of their occurrence.

What conclusions can you draw from the data?

51	69	66	74	47	84	78	54	79	85
77	78	73	54	83	88	47	79	76	84
55	52	62	79	85	62	52	83	51	74
85	78	55							

Geyser Times

Worked example 6B

? (a) Show the data given in Activity 6B on a dot plot.

(b) What does the dot plot suggest about the distribution of times between eruptions?

Solution

(a)

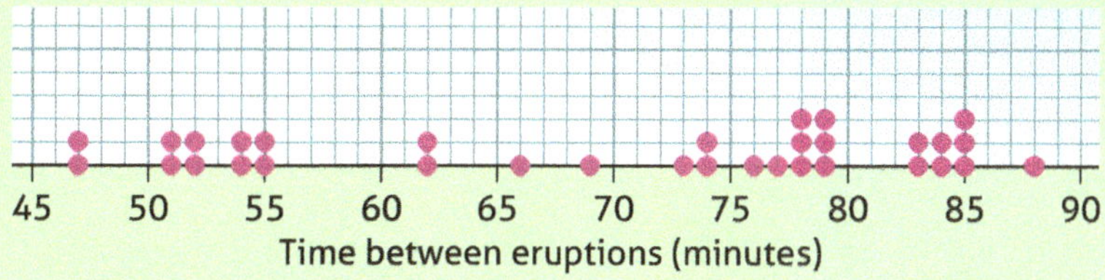

(b) The dot plot suggests that the times cluster around two figures, just over 50 minutes and 80 minutes.

Exercise 6B

The figures below are the times between eruptions of Old Faithful in May 2020.

94	96	89	97	106	100	102	78	91	98
94	86	87	102	93	90	90	94	94	96
88	97	89	96	86	93	86	98	97	93

Geyser Times

1 Draw a dot plot for the eruptions in May 2020.

2 What similarities and differences do the dot plots show in the distributions of times for May 1990 and May 2020?

3 Apart from the dot plots, what other summary information would be helpful in comparing the data for 1990 and 2020?

6C Summarising data

This section covers common measures that are used to summarise a data set.

Activity 6C Sunny Camborne

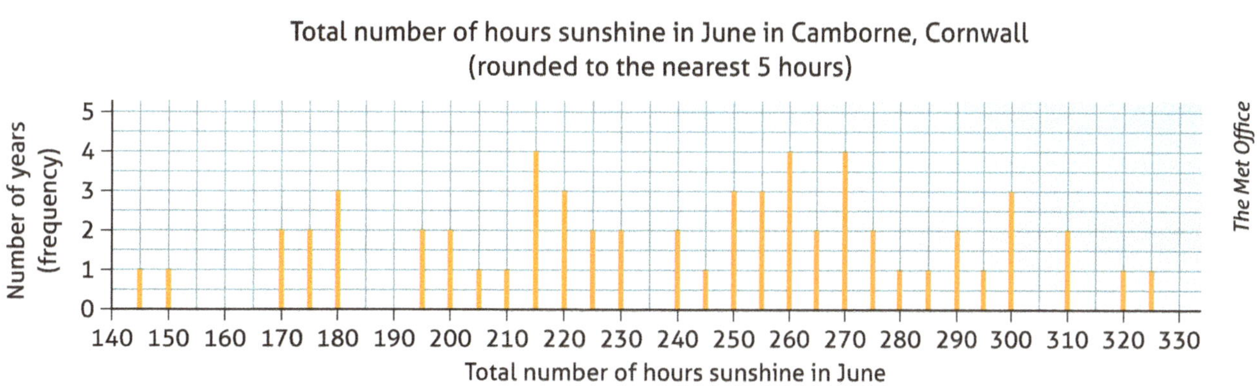

(a) How would you summarise the information in this vertical line chart in words?

(b) What numbers would you also want to use?

Worked example 6C (1)

? This bar chart shows the frequency of major earthquakes for each of the last 50 years.

Earthquake strengths are measured according to their magnitude. The term 'major earthquake' is used here to describe those of magnitude at least 7. Such earthquakes are very serious.

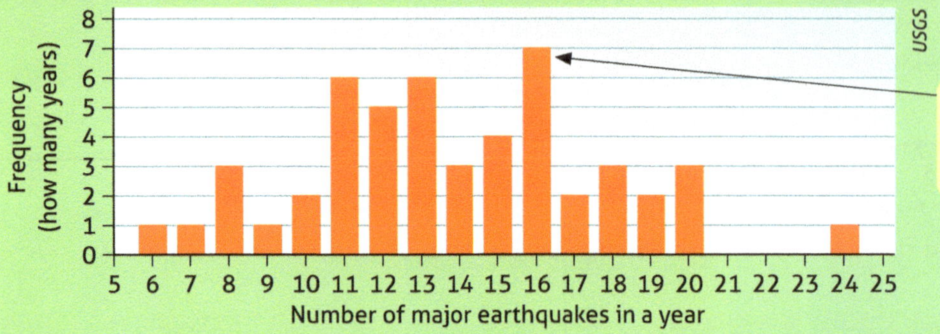

16 major annual earthquakes occurred in 7 years.

Find the (a) mean, (b) mode, (c) median and (d) range in the number of major earthquakes.

Solution

(a) Total of major earthquakes = 6 × 1 + 7 × 1 + 8 × 3 + 9 × 1 + 10 × 2 + ... 1 × 24 = 694.

There are 50 years, so the mean annual number of major earthquakes = 694 ÷ 50
= 13.9 major earthquakes a year.

(b) The mode is the most frequent annual number of major earthquakes. It is 16.

(c) The median is the middle point in an ordered data set.
There are 50 data points, so the middle point is midway between points 25 and 26.
The 25th point is 13 earthquakes, the 26th is 14 earthquakes.

The median is $\frac{13 + 14}{2} = 13.5$.

The numbers of years are found from the heights of the bars on the chart: 1 + 1 + 3 + 1 + ...

(d) The range of a set of data is the difference between the smallest and largest data points in the set. In this case it is 24 – 6 = 18 earthquakes.

It is not '6 to 24'.

Using a spreadsheet

These four data summaries (mean, median, mode and range) can be calculated using a spreadsheet. With the earthquake data in cells A1 to A50 of the spreadsheet below, the formulae in cells D1, D2, D3, D4 give these summary measures.

	A	B	C	D
1	6		Mean	=AVERAGE(A1:A50)
2	7		Median	=MEDIAN(A1:A50)
3	8		Mode	=MODE(A1:A50)
4	8		Range	=MAX(A1:A50)-MIN(A1:A50)
50	24			

Sometimes you needed to find the **weighted mean**. This is explained in the next Worked example.

Worked example 6C (2)

? The site Ali uses to shop online asks customers to rate their shopping experience.

They rate three features on a 1 to 5 scale, with 5 the highest rating.

Ali clicks on his ratings:

A	Did your purchase match the advert?	1	2	3	(4)	5
B	Would you buy from this Seller again?	1	2	3	(4)	5
C	Was your purchase attractively packaged?	(1)	2	3	4	5

However, not all the features are of equal importance.

Weightings of **3**, **2** and **1** are given for A, B and C respectively.

(a) What is the mean of Ali's rating without weighting?

(b) What is the weighted mean of Ali's rating?

Solution

(a) Mean of Ali's rating = $\frac{4 + 4 + 1}{3} = 3$.

(b) The weighted scores are for **3** × **4**, **2** × **4** and **1** × **1**,
A **B** **C**

giving a weighted mean of $\frac{3 \times 4 + 2 \times 4 + 1 \times 1}{3 + 2 + 1} = 3.5$.

The weighted mean is influenced by the high weighting given to ratings **A** and **B**, above.

Exercise 6C

1 A new mobile phone was given a star rating by 20 people.
These are the stars awarded by each person.

3	4	0	1	0	3	0	0	3	0
1	3	0	3	4	4	0	3	5	3

Don't forget that zero is a data point.

(a) Work out the mean, median and mode (modal) number of stars awarded to the mobile phone.

(b) Which would be used to show the phone in its best light?

(c) Which would a competitor choose to convince people that the phone was not worth buying?

2 Customer services asks shoppers to rate, on a points scale of 1 to 7, these four features: store cleanliness, staff attitude, value for money and choice of goods.

The four features are weighted and a weighted mean point score calculated using a spreadsheet. Alun's rating points are shown below, together with his mean weighted points score for the four features.

	A	B	C	D	E
1	Feature	Store cleanliness	Staff attitude	Value for money	Choice of goods
2	Points	4	6	7	2
3	Weighting	1	1	3	2
4	Weighted points	4	6	21	4
5					
6	Total weighted points =	35			
7	Mean weighted points =	5			

(a) By how much would Alun's weighted mean point score change if value for money was weighted as 5 rather than 3?

(b) Write down the formulae in cells:

(i) C4 **(ii)** B4 **(iii)** B6 **(iv)** B7.

6D Types of data

- If you cannot do calculations on the data they are **categorical**. For example, hair colour.
- With **discrete numerical data** the values are restricted to certain possibilities. For example, the number of passengers in a car.
- By contrast, **continuous numerical data** may take any realistic value. For example, a child's height.

Activity 6D Classifying data

Star rating is a popular method of rating people and services on the internet.

Amy says that star ratings are categorical data.

Beth says that they are numerical data.

(a) Discuss the two points of view.

(b) What about the data collected here?

How old are you? (Tick ✓ one box only)

Under 18	18–24	25–34	35–44	45–54	55–64	65 and over
☐	☐	☐	☐	☐	☐	☐

Worked example 5D

? Put the data sets below into one of the three following categories: categorical, continuous numerical or discrete numerical.

Question	Solution	Notes
Family name	Categorical	
Gender (M/F)	Categorical	
Height (cm)	Continuous numerical	Discrete if rounded
Weight (kg)	Continuous numerical	Discrete if rounded
Smoker (1 = No, 2 = Yes)	Categorical	
General health (Scale 1 [poor] to 5 [good])	Categorical	You cannot perform arithmetic with these numbers
Age last birthday (years)	Discrete numerical	Age last birthday is discrete Time lived is continuous

Exercise 6D

1 What type of data are being collected here?

2 In the 18th and 19th centuries, transportation to Australia was used as a punishment.
Here are some prisoner records from the time.

ID	Last name	First name	Sentence (years)	Date transported	Ship	Number of convicts on ship
btr807	Brindley	Joseph	7	19-06-1837	Blenheim	210
btr1098	Whitaker	Edward	21	29-11-1836	Sarah	254
btr100432	Ozeland	Elizabeth	21	01-01-1791	Albermile	106

British Transportation Registers

Categorise the data sets described by the headings in this table.

3 During the 18th and 19th centuries many offences carried the death penalty. However, not all convicts sentenced to death were executed.
These three pie charts show the final outcomes for convicts sentenced to death.

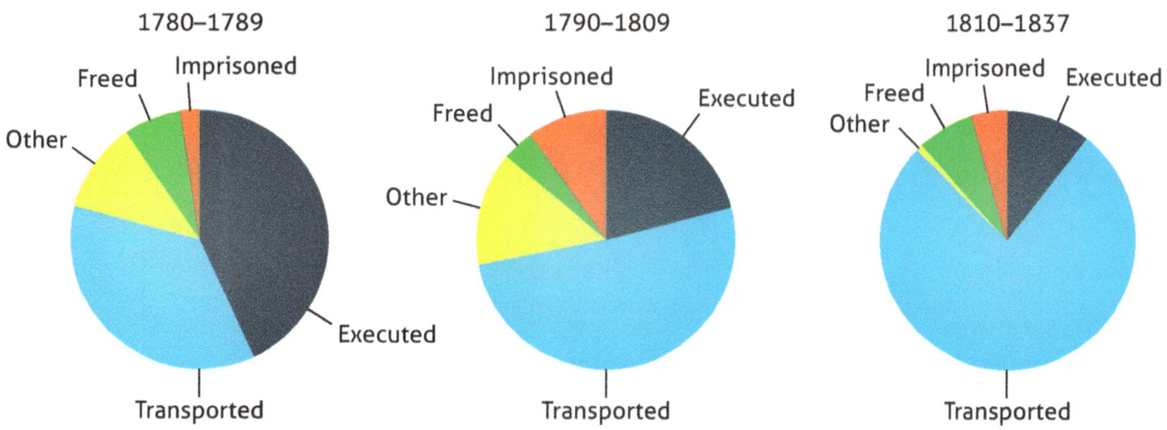

Digital Panopticon

'Other' includes military service and unknown (records lost).

Pie charts show relative proportions **not** actual numbers.

(a) What type of data are being shown?

(b) Write down three observations about how these final outcomes changed over this time.

6E Grouping discrete data

Activity 6E Grouping people by age

These are the ages (at last birthday, so discrete data) of 33 females sentenced to transportation at the Old Bailey in 1800. The great majority were sentenced for at least 7 years, for stealing relatively small items.

14	29	16	47	36	22	26	45	27	24	39
35	26	27	35	28	26	33	29	27	28	20
30	39	35	30	21	21	25	27	47	16	28

Old Bailey records for 1800

Construct a tally and a grouped frequency table.

Use equal group widths beginning with 10 to 14 years.

Worked example 6E

This grouped frequency chart shows the ages of 131 male convicts transported to overseas penal colonies.

Grouping items of data can make patterns or trends easier to see.

They were convicted at the Old Bailey in 1800.

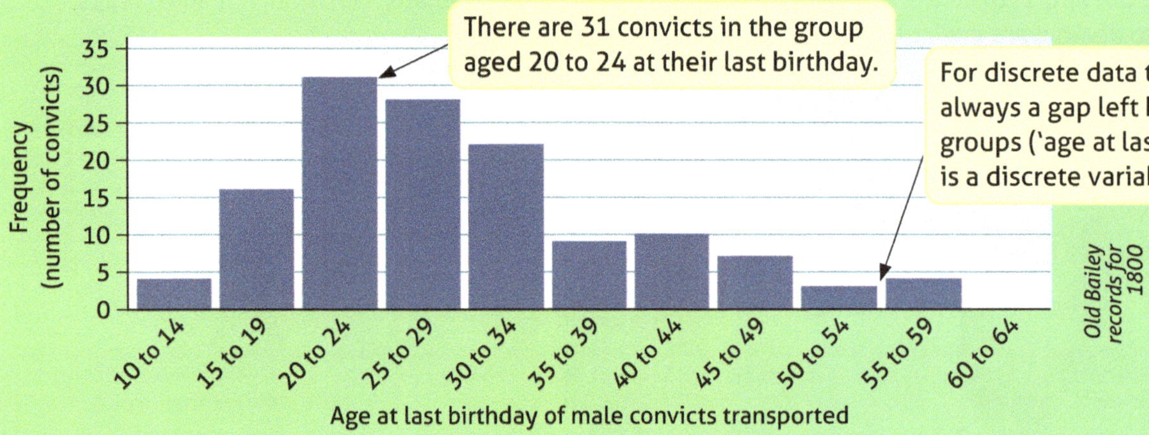

?

(a) How many of the transported convicts were under 20 years old?

(b) A student says that over half the convicts were 30 or more years old.

Is this true? Support your answer with suitable numbers.

(c) Find the modal age group of the male convicts.

(d) Another student said, 'The range in convicts' ages is 59 − 10 = 49 years.'

Criticise this statement.

Solution

(a) **Under 20:** There were 15 convicts aged 15 to 19 and 4 convicts aged 10 to 14.
The total was 15 + 4 = 19 convicts.

(b) **Aged 30 or more:** The number of convicts aged over 30 was 22 + 9 + 10 + 7 + 2 + 4 = 54.

In total there were 131 convicts and half of this is 65.5.

So, the statement that over half the convicts were 30 or more years old is not true.

(c) The most frequent group is that with the highest bar.

It is 20 to 24 years, so this is the modal group.

(d) You only know that there are four convicts aged 10 to 14 years old, not if there was a convict aged 10 years. Similarly, for the four convicts aged 55 to 59 years old, you cannot be sure there is one aged 59 years.

Exercise 6E

1 **(a)** Use the data in Activity 6E to answer these questions.

How many of the female convicts sentenced to transportation were:

(i) younger than 20; **(ii)** between 20 years and 24 years old; **(iii)** more than 49 years old?

(b) Show the information in Activity 6E in a grouped frequency chart.

2 Here are the numbers of convicts transported on ships leaving the UK from 1800 to 1809. Some shipping records have been lost, so the list does not cover all the ships transporting convicts.

1800	1801	1802	1803	1804	1805	1806	1807	1808	1809
301	299	252	1	15	122	308	202	99	59
50	32	31	295		156		114	280	63
295	200	402	338						73
		158							200

convictrecords.com.au

Write a few sentences, based on the above data and supported by suitable charts and tables, about the numbers of convicts carried in ships from 1800 to 1810. Mention anything interesting you may find.

6F Using grouped data with continuous data

Some people say that whenever a measuring instrument is used the resulting measurements are, in practice, discrete (the more so with digital displays). However, it is usual to treat all measurements as continuous.

Putting discrete data points into the correct intervals needs care.

For example, consider the two distance intervals 5 cm to 10 cm and 10 cm to 15 cm.

In which one does a distance of 10 cm fit? To resolve this question, inequalities are used.

So, for a distance d cm the two intervals are described as $5 \le d < 10$ and $10 \le d < 15$, and 10 cm fits in the second interval.

$\le$ means less than or equal to.

$<$ means less than.

Activity 6F Grouped continuous data

The results of 50 measurements of the force, P newtons, needed to separate two wooden rods glued together are shown on the grouped frequency chart in the image below. The data are continuous, so there are no gaps between the bars.

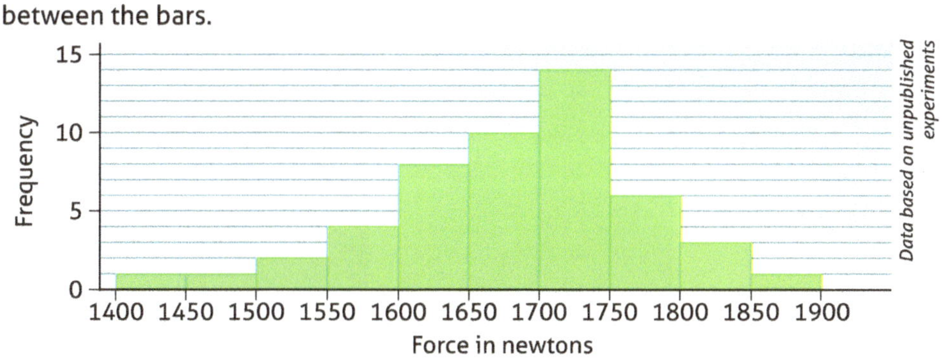

(a) Construct a grouped frequency table for the data, starting with the interval $1400 \le P < 1450$.

(b) The intervals used here were chosen when the original data were grouped.

Explain why they are more suitable than those beginning:

$1385 \le P < 1435$ or $1400 \le P < 1420$.

Worked example 6F

Use the data in Activity 6F to answer this question.

(a) How many of the measurements were less than 1550 newtons?

(b) What can you say about the largest measurement?

(c) Use your grouped frequency table from Activity 6F to estimate the mean force.

Solution

(a) The number of forces less than 1550 newtons (N) was 1 + 1 + 2 = 4.

(b) All that can be said about the greatest force is that it was at least 1850 N and less than 1900 N.

You would be wrong to say that the largest force had the middle value of 1875 N. You just don't know.

(c) Estimate the mean force.

The intervals in the chart are shown in Column 1.

This column gives the frequencies for the various intervals.

Force *P* N	Frequency	Mid-interval	Frequency × Mid-interval
$1400 \le P < 1450$	1	1425	1425
$1450 \le P < 1500$	1	1475	1475
$1500 \le P < 1550$	2	1525	3050
$1550 \le P < 1600$	4	1575	6300
$1600 \le P < 1650$	8	1625	13 000
$1650 \le P < 1700$	10	1675	16 750
$1700 \le P < 1750$	14	1725	24 150
$1750 \le P < 1800$	6	1775	10 650
$1800 \le P < 1850$	3	1825	5475
$1850 \le P < 1900$	1	1875	1875
Total	50		84 150

The estimated mean is $\frac{84\,150}{50} = 1683$ N.

Exercise 6F

This grouped frequency chart shows the best distances achieved by the 100 best female javelin throwers in 2020.

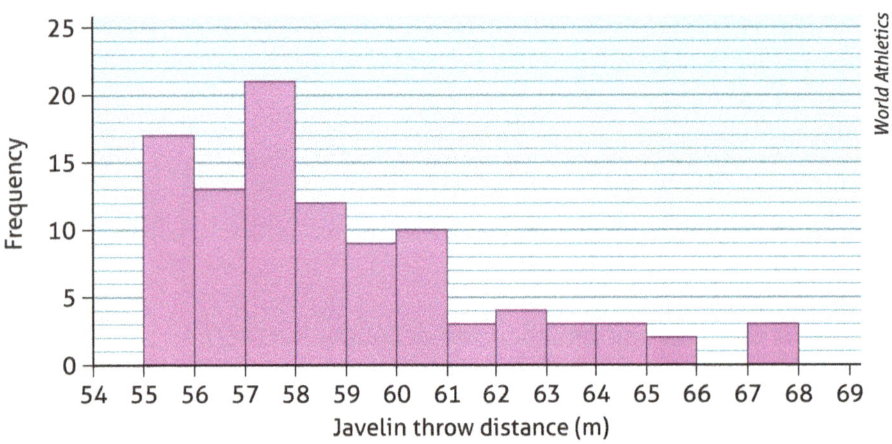

1 **(a)** How many of the distances were 62 m or longer?

(b) Write down the combined interval for the 5 longest distances.

2 **(a)** Calculate an estimate for the mean distance thrown by the top 5 female javelin throwers.

(b) The actual distances thrown by the top 5 javelin throwers were, in metres:

65.19	65.7	67.17	67.29	67.61

How close is your estimate in (a) to the actual mean?

(c) By how much would the 100th best javelin thrower need to improve to just get into the top ten?

6G Histograms

This section is about using different intervals when displaying grouped continuous data.

Activity 6G Unequal class intervals

A student, Ellie, wants to investigate how rainy it is in April, in Bradford.

She collects April rainfall figures for Bradford going back 100 years from the Met Office.

Here is a grouped table of her results together with a grouped frequency chart.

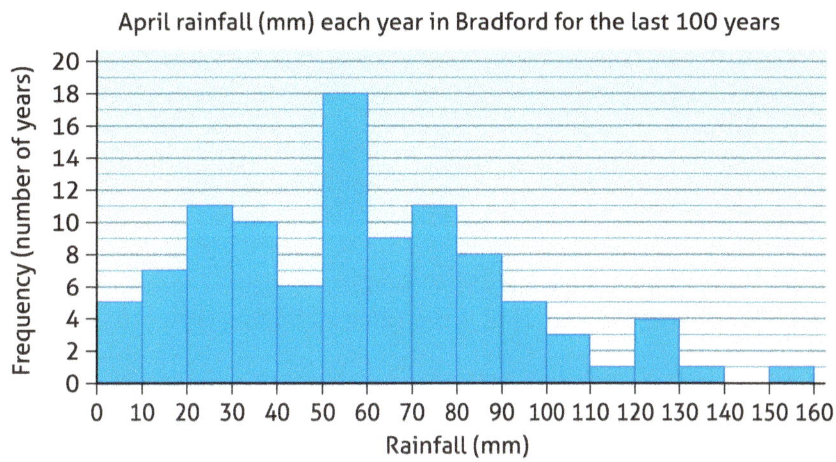

Met Office

Ellie thinks that the grouped frequency chart is a bit 'lumpy' and she redraws it by hand with some of the intervals grouped together.

This is how it looks.

Look carefully at Ellie's chart.

Why is it misleading?

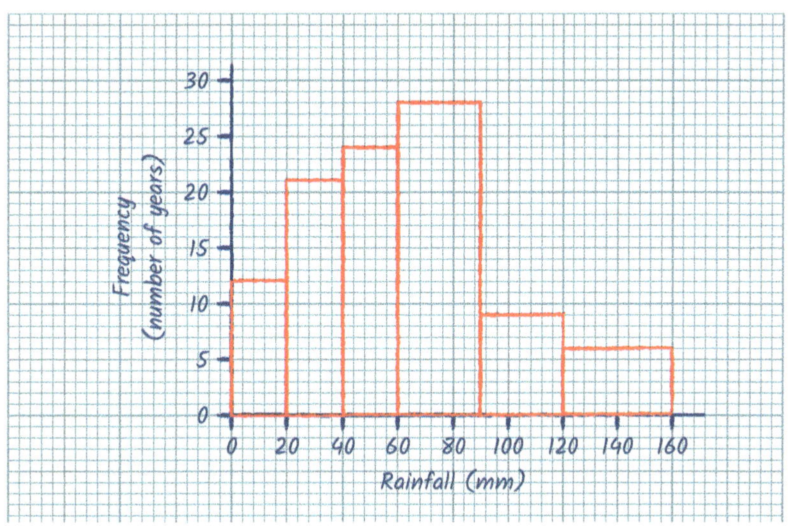

The chart below shows the number of hours of sunshine in Bradford in April for the last 100 years. You will notice that the intervals for the groups are not all the same width. This means that if you draw a frequency chart it will be misleading, like the one in Activity 6G, with the bars for wider intervals too high.

Instead, in this chart the frequency is represented by the areas of the bars rather than their heights. It is the sort of chart that Ellie should have drawn. In this case one square represents one year.

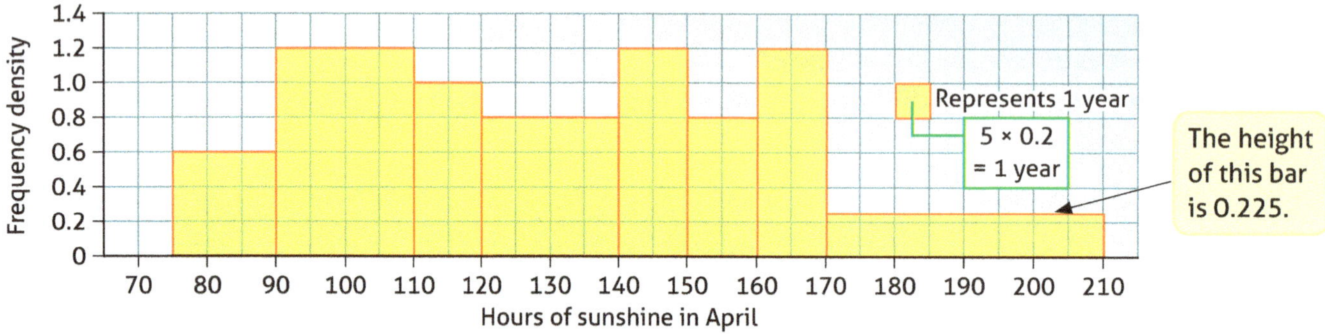

A chart like this is called a **histogram**.

- Histograms are used for grouped continuous data.
- They are particularly useful when the intervals do not all have the same width.
- The vertical scale is labelled **Frequency density** rather than Frequency.

Worked example 6G

Use the histogram above to work out the original grouped frequency table.

Solution

Each square represents one year, so the numbers of years represented by the various bars can be found by counting the squares in them. That is the same as finding their areas.

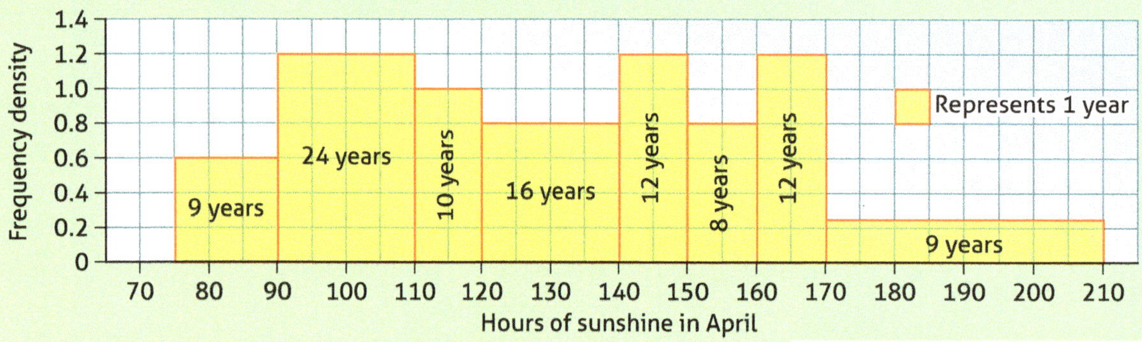

The information on the histogram can now be used to construct the original grouped frequency table.

The letter s denotes the number of hours of sunshine in April.

Sunshine, s, (hours)	Frequency (years)
$75 \le s < 90$	9
$90 \le s < 110$	24
$110 \le s < 120$	10
$120 \le s < 140$	16
$140 \le s < 150$	12
$150 \le s < 160$	8
$160 \le s < 170$	12
$170 \le s < 210$	9
Total	100

Exercise 6G

1 This histogram shows the April maximum average daily temperature for over 100 years in Bradford.

(a) Find for how many years were the maximum average daily temperatures:

(i) at least 13°C

(ii) less than 10°C

(iii) between 11°C and 12°C.

(b) Say whether this statement is true, explaining your reasoning.

'Over 50% of the maximum average daily temperatures were at least 11°C.'

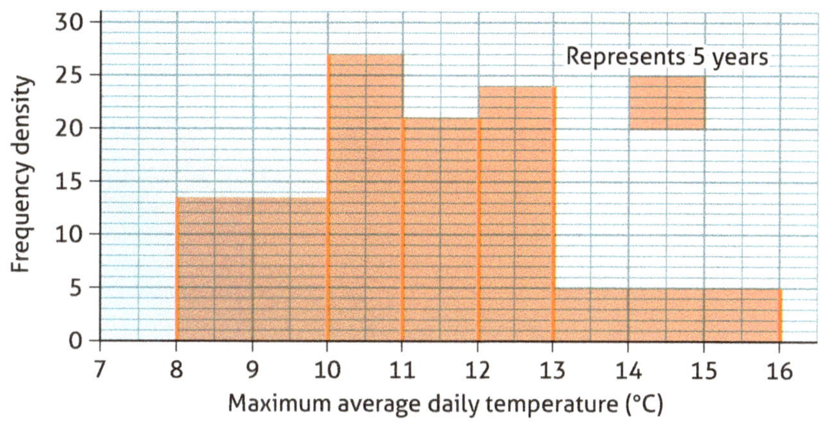

6H Cumulative frequency graphs

Activity 6H River flow

The Missouri is the longest river in the USA. The time series shown on this graph is the average water flow at a point near its source each year from 1970 to 2019.

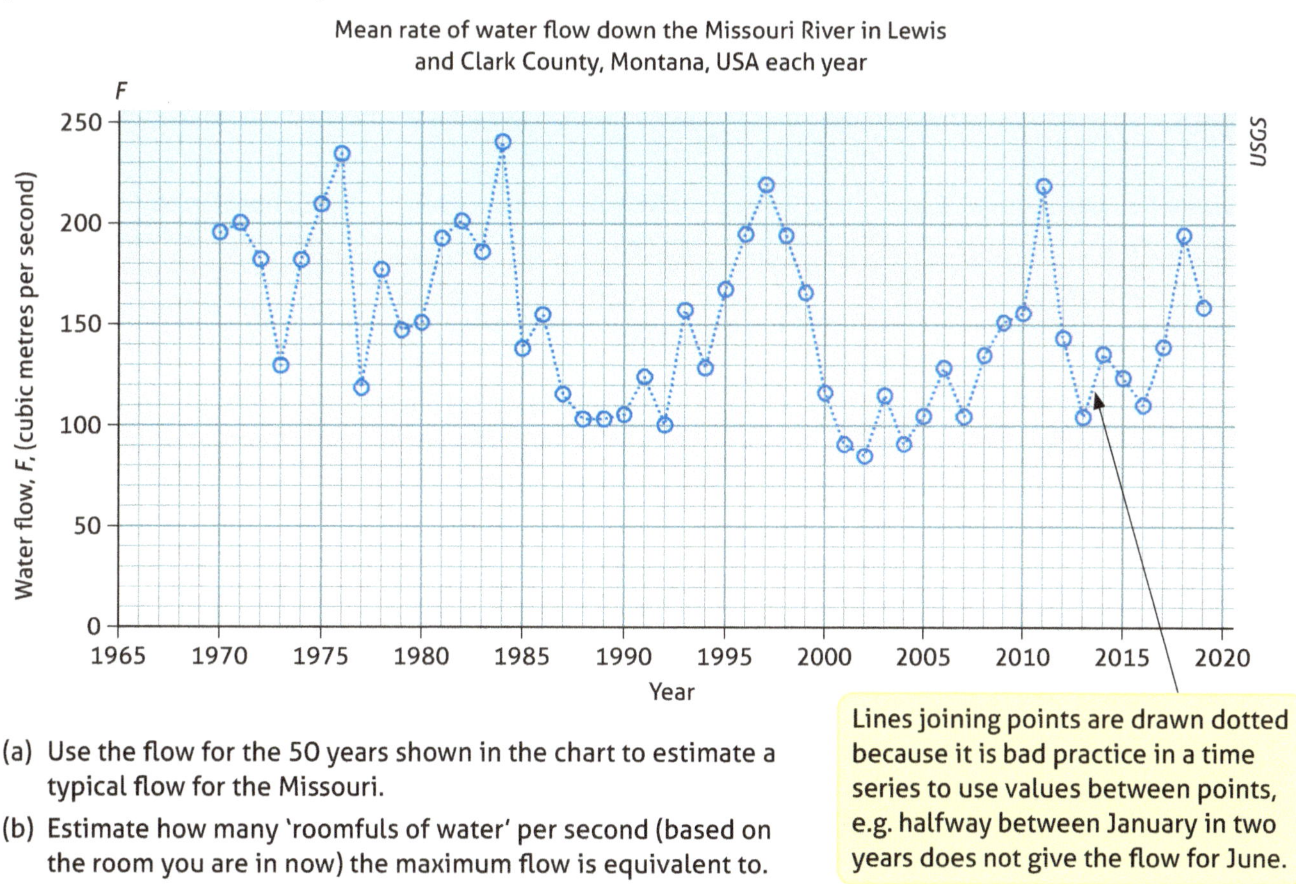

(a) Use the flow for the 50 years shown in the chart to estimate a typical flow for the Missouri.

(b) Estimate how many 'roomfuls of water' per second (based on the room you are in now) the maximum flow is equivalent to.

Worked example 6H

The table below left gives the data illustrated in Activity 6H as a frequency distribution. The table below right shows the corresponding **cumulative frequency** distribution, the number of years, with up to the given flow levels.

Flow, F (cubic metres per second)	Frequency (years)
$60 \le F < 80$	0
$80 \le F < 100$	3
$100 \le F < 120$	12
$120 \le F < 140$	9
$140 \le F < 160$	8
$160 \le F < 180$	3
$180 \le F < 200$	8
$200 \le F < 220$	5
$220 \le F < 240$	1
$240 \le F < 260$	1

Flow, F (cubic metres per second)	Cumulative frequency (years)
$F < 80$	0
$F < 100$	3
$F < 120$	15
$F < 140$	24
$F < 160$	32
$F < 180$	35
$F < 200$	
$F < 220$	
$F < 240$	
$F < 260$	

There were 24 years (0 + 3 + 12 + 9) when the flow was less than 140 cubic metres per second.

There were 12 years with flow from 100 up to 120 cubic metres per second.

(a) Complete the cumulative frequency column.

(b) Show the cumulative frequencies in a **cumulative frequency graph**.

(c) Use your cumulative frequency graph to **estimate** the median flow.

Solution

(a)

$180 \le F < 200$	8
$200 \le F < 220$	5
$220 \le F < 240$	1
$240 \le F < 260$	1

$F < 200$	43
$F < 220$	48
$F < 240$	49
$F < 260$	50

The final figure, in this case 50, in the cumulative frequency column should be equal to the total number of data points.

(b)

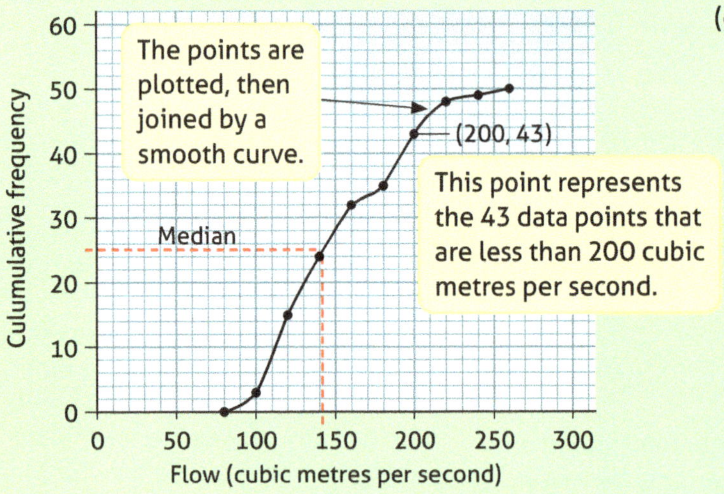

(c) There are 50 data points, so the median is the mean of the 25th and 26th points. This can be read from the curve as about 142 cubic metres per second.

Exercise 6H

1 This partially completed table gives the daily mean level of the river Vermillion in the USA, in 2019.

Level (*h* cm)			Frequency	Less than	Cumulative frequency
100	≤ *h* <	150	43	150 cm	
150	≤ *h* <	200	54	200 cm	
200	≤ *h* <	250	179	250 cm	
250	≤ *h* <	300	56	300 cm	
300	≤ *h* <	350	17	350 cm	
350	≤ *h* <	400	10	400 cm	
400	≤ *h* <	450	4	450 cm	
450	≤ *h* <	500	1	500 cm	
500	≤ *h* <	550	1	550 cm	365
			Total = 365		

USGS

(a) Complete the cumulative frequency column.
(b) For how many days was the level less than 200 cm?
(c) For how many days was the level at least 200 cm?
(d) Draw the cumulative frequency graph.
(e) Use your graph to estimate the median river level.

2 This partially completed cumulative frequency table shows the maximum water level, *h* cm, each year for the last 85 years at a point on the James River.

Maximum water level *h* (cm)			Frequency	Cumulative frequency
100	≤ *h* <	200	0	0
200	≤ *h* <	300	4	
300	≤ *h* <	400	12	
400	≤ *h* <	500		56
500	≤ *h* <	600	14	
600	≤ *h* <	700		
700	≤ *h* <	800	3	84
800	≤ *h* <	900		85
			Total = 85	

USGS

(a) Complete the cumulative frequency table.
(b) Draw the graph.
(c) Water engineers say that the flood defences will cope with all but 5% of the highest water levels. Use your cumulative frequency graph to estimate this level.

6I Box and whisker plots

This section is about using box and whisker plots to illustrate distributions.

Activity 6I Earthquakes

The area around the British Isles is subject to weak earthquakes.

Between January 1998 and September 2020 there were 200 earthquakes of magnitude greater than 1 but none above magnitude 5.

Draw up the cumulative frequency table that was used to draw the graph.

Then use it to construct the original frequency table.

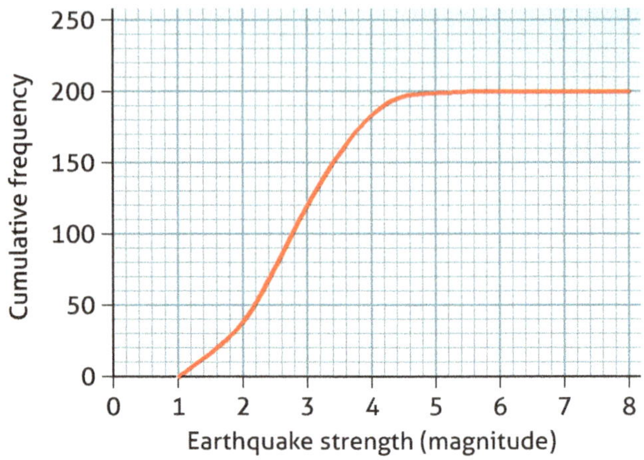

USGS

This cumulative frequency chart is based on putting the 200 results in order.

The median is such that 50% of these ordered results are below it, in this case 100 results.

You can see from the cumulative frequency graph that the median value of the magnitude is about 2.8.

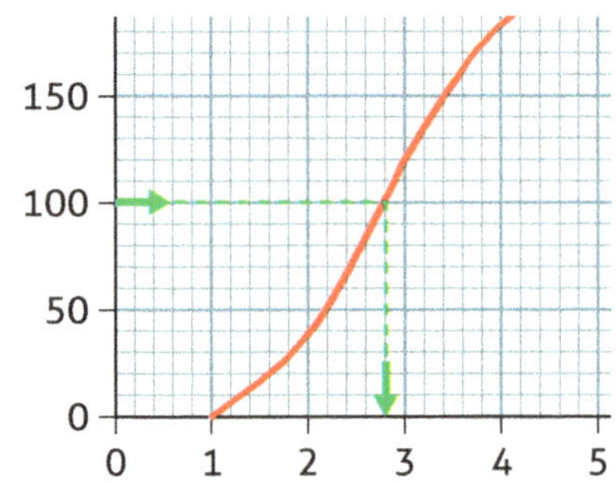

The **lower quartile** is the value at the top of the lowest quarter of the data. There are 200 data points, so the first 50 form the lowest quarter. The lower quartile is the value of the 50th data point. Using the graph this is about magnitude 2.2.

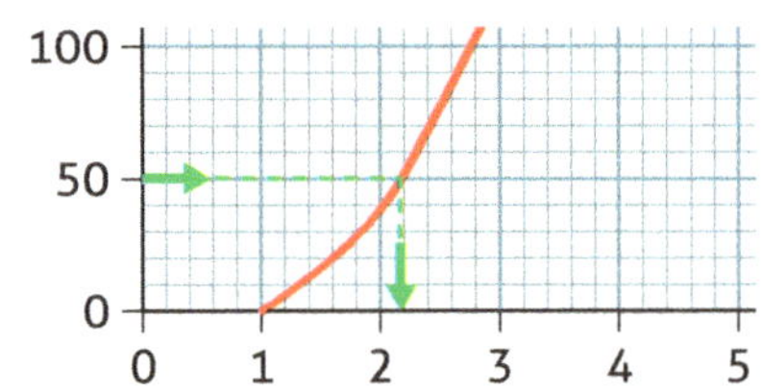

Similarly, the **upper quartile** is the earthquake strength, which 75% of the 200 earthquakes (150) are below. This is, from the graph, about magnitude 3.5.

The difference between the upper and lower quartiles is called the **inter-quartile range** or IQR.

For the earthquake data this is 3.5 – 2.2 = 1.3.

IQR gives a measure of the spread of the data.

The earthquake strength data has these measures.

Minimum	Lower quartile	Median	Upper quartile	Maximum
1.0	2.2	2.8	3.5	5.0

These can be shown on a **box and whisker plot**:

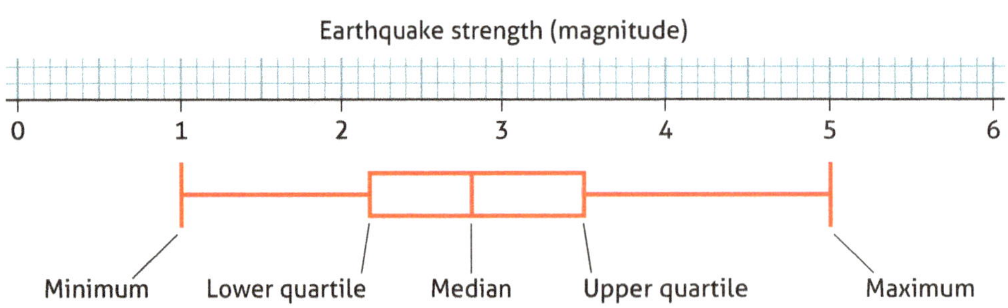

Worked example 6I

This cumulative frequency table shows the strengths (magnitudes) of 100 earthquakes recorded during a day in 2020.

(a) Find:

(i) the lower quartile

(ii) the median

(iii) the upper quartile

(iv) the IQR.

(b) Use the results to show the distribution on a box and whisker plot.

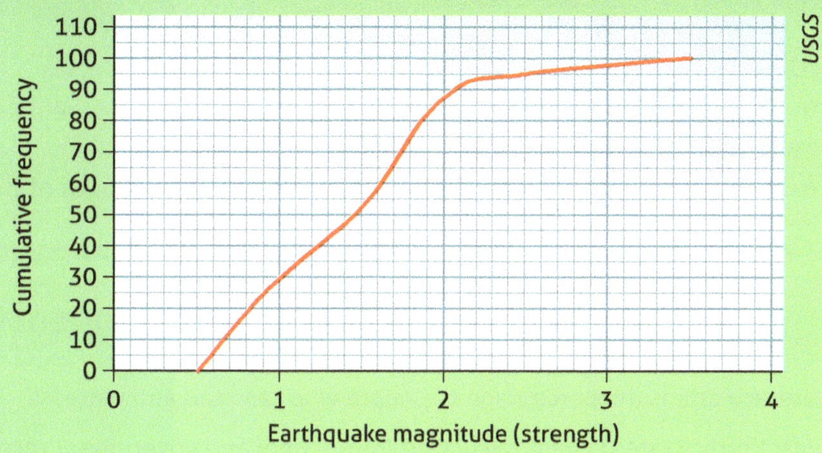

Solution

(a) There are 100 data points, so:

(i) the lower quartile is at 25 and has value about magnitude 0.9

(ii) the median is at 50 and has value about magnitude 1.5

(iii) the upper quartile is at 75 and has value magnitude 1.8

(iv) the IQR is 1.8 − 0.9 = 0.9.

The median of 100 data points really should be midway between numbers 50 and 51. However, with a large number of data points like this it is common to take it to be at half the number, so at 50 in this case.

(b) From the graph, the minimum value is magnitude 0.5 and the maximum value magnitude 3.5.

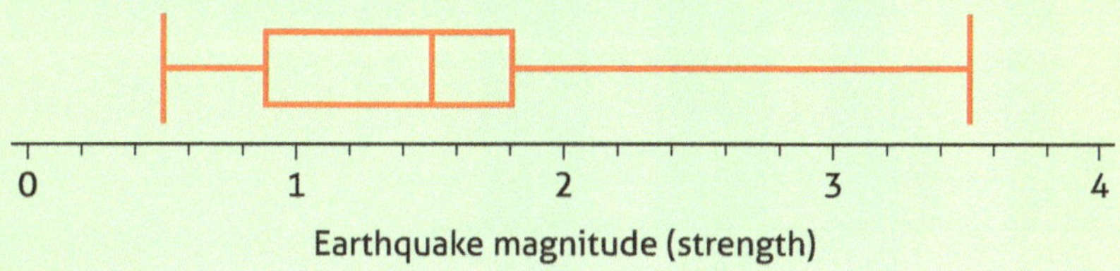

Exercise 6I

1 This box and whisker plot shows information about the strengths of the 2388 earthquakes recorded in the world during one week in 2020.

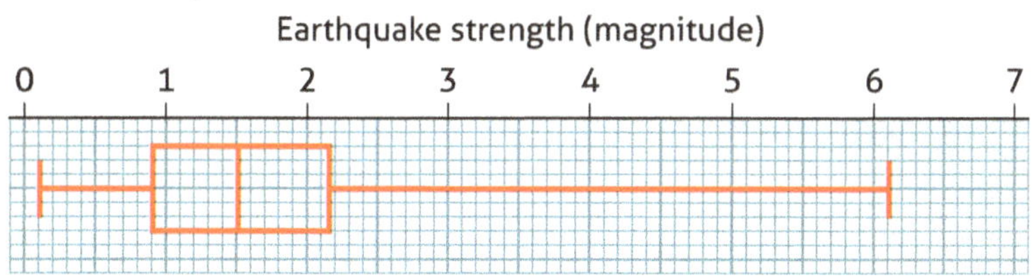

(a) What was the median earthquake strength?

(b) What was the maximum earthquake strength?

(c) What was the IQR for these earthquake strengths?

(d) What was the range of these earthquake strengths?

2 Use the box and whisker plot in Question 1 to answer these.

(a) What percentage of the earthquakes were stronger than 0.9?

(b) What percentage of the earthquakes were stronger than 1.5?

(c) How many earthquakes had a strength greater than 2.15?

6J Looking at charts closely – what's to be read in the data?

The point of a data display chart is to emphasise key points in the data. Always keep this in mind when deciding what form of data display to use.

You will usually use some form of data display in the later stages of the statistical problem solving cycle: 'Process and presentation' and 'Reporting/Interpretation'.

Activity 6J Literacy rates

Literacy rate is the percentage of people who can read and write.

Past literacy rates can be calculated from school and government records.

Records of births, deaths and marriages have been kept since 1538.

People who wrote their signatures on official documents are considered to have been literate, in contrast to those who made a mark (such as a cross).

Here are some literacy rates, England 1580 to 1880, calculated using the above method.

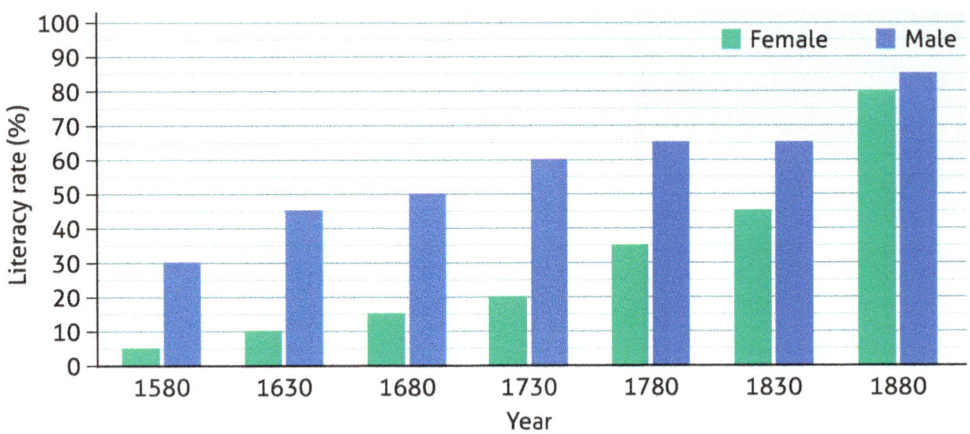

The Great Escape: The industrial revolution in theory and in history by G. Clark

(a) What was the difference in literacy rates between males and females in 1580?

(b) Comment on the overall trend in literacy rates for females and males 1580 to 1880.

(c) Would the source of the information count as a primary or secondary source? Justify your answers.

An 'overall trend' should not usually involve specific data points. It should just describe the story told by the chart.

Exercise 6J

1 For each of the following five charts:

(a) Name the type of chart it is, choosing from this list:

Bar chart	Percentage bar chart	Horizontal bar chart	Scatter diagram
Pie chart	Time series line graph	Vertical line chart	Frequency chart
Box and whisker plot	Histogram		

(b) Write down in bullet points what it is showing clearly and immediately.

(c) If it does not make a point straight away, suggest how it might be improved. Where appropriate, suggest a different chart for the data.

(i)

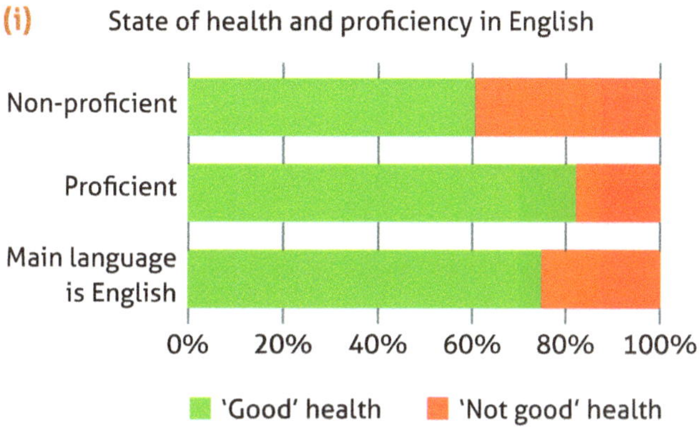

(ii)

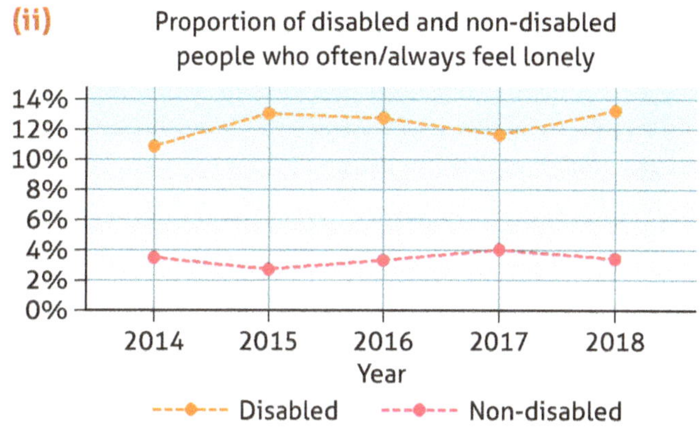

(iii)

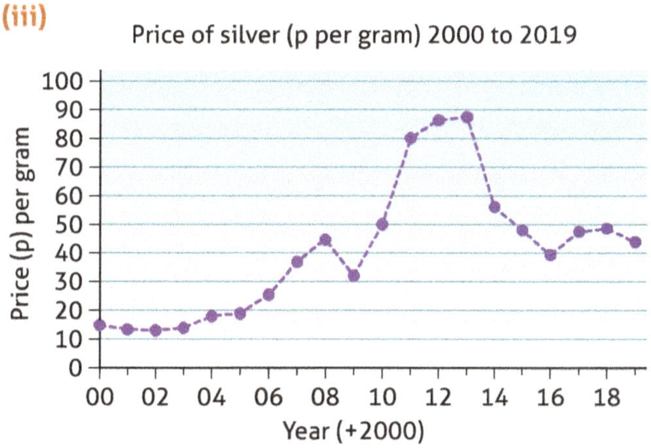

(iv)

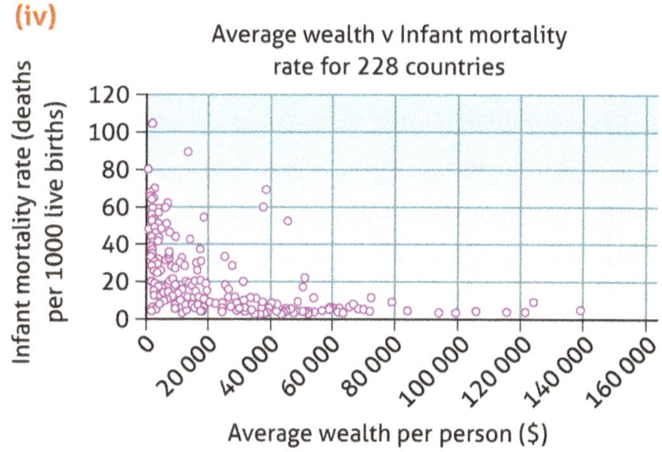

(v)

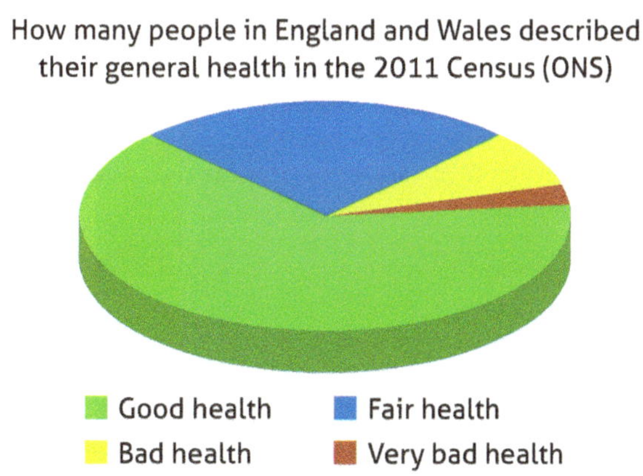

Test yourself 6K

1 This chart shows the points scored by 400 contestants in an online quiz. The maximum points score of 50 was only gained by one contestant.

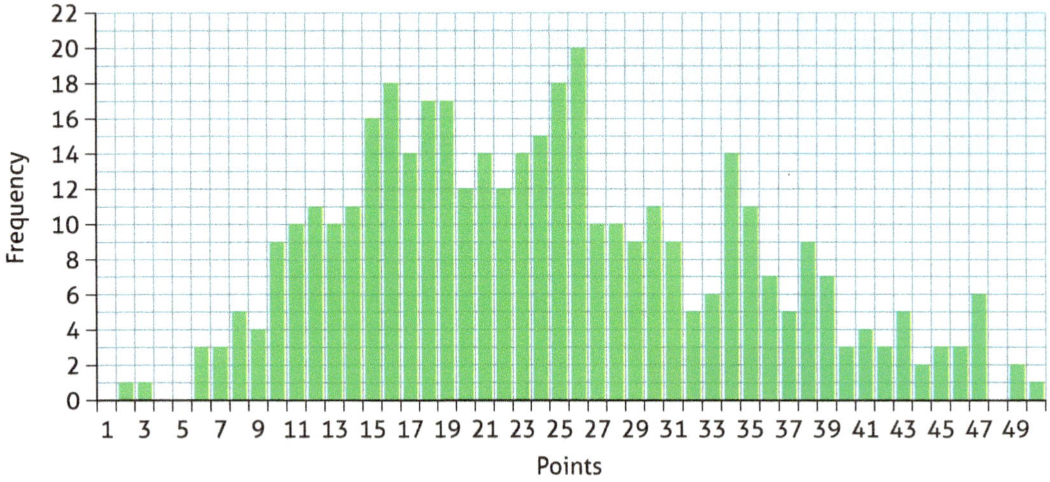

(a) How many contestants scored more than 45?

(b) How many contestants scored less than 10?

(c) Find the **(i)** modal score, **(ii)** median score, **(iii)** range.

(d) The top 8% of contestants will go through to the next round.
Suggest a possible cut-off points score. Justify your choice.

2 **(a)** Briefly summarise in three bullet points the key information shown in this chart.

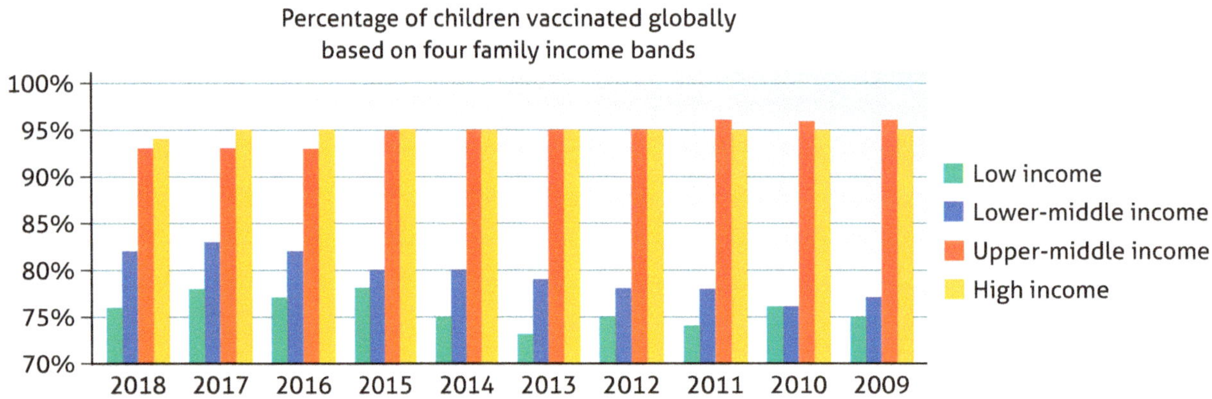

(b) State one way in which the chart is misleading.

(c) At first glance do you think the chart makes a fair point?

3 The proportion of those charged with crimes who are found guilty is called the conviction rate. The present conviction rate is about 80% for both genders in England and Wales.

These pie charts show the proportion of prisoners convicted or found not guilty of all crimes in the middle of the 18th century at the Old Bailey.

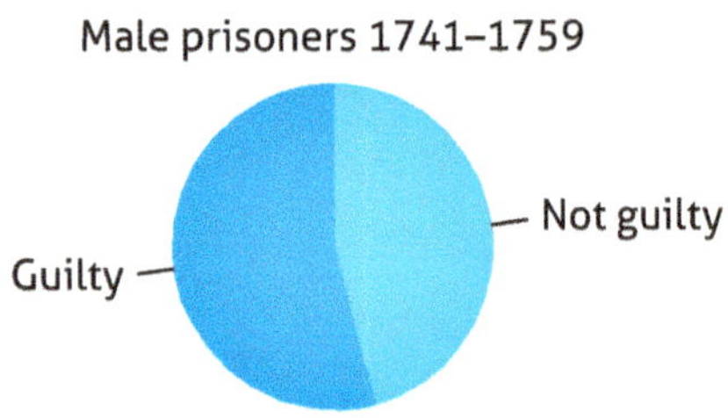

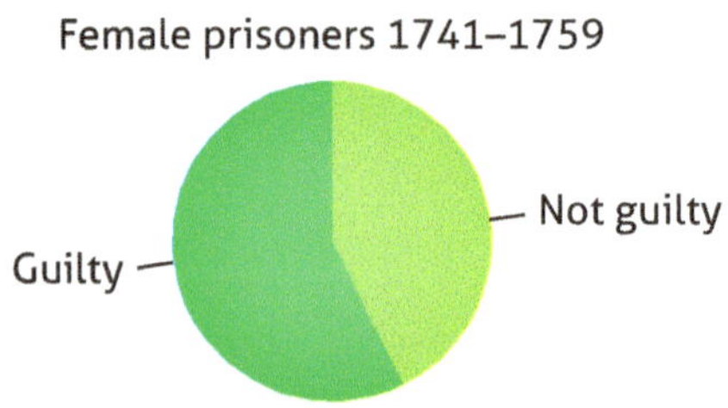

How do the mid-18th century conviction rates compare with the present ones?

4 The brightest star in the sky is Sirius. Tamsin goes online and finds this pie chart comparing the 10 brightest stars.

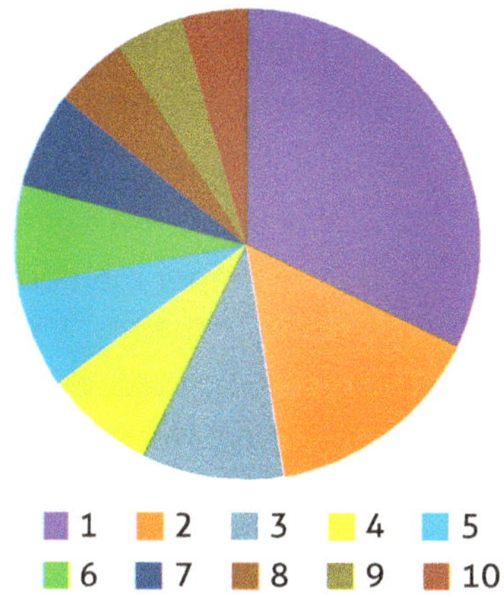

(a) Comment on whether this is a helpful way to display the information, giving your reasons.

Sirius it is also one of the closest stars.

Tamsin says, 'I have an idea that I want to investigate. It is that most of the brightest stars are also close to us.'

She then finds data for the brightness and distance of many stars. She enters the data for the 60 brightest stars onto a spreadsheet and uses it to draw this scatter diagram.

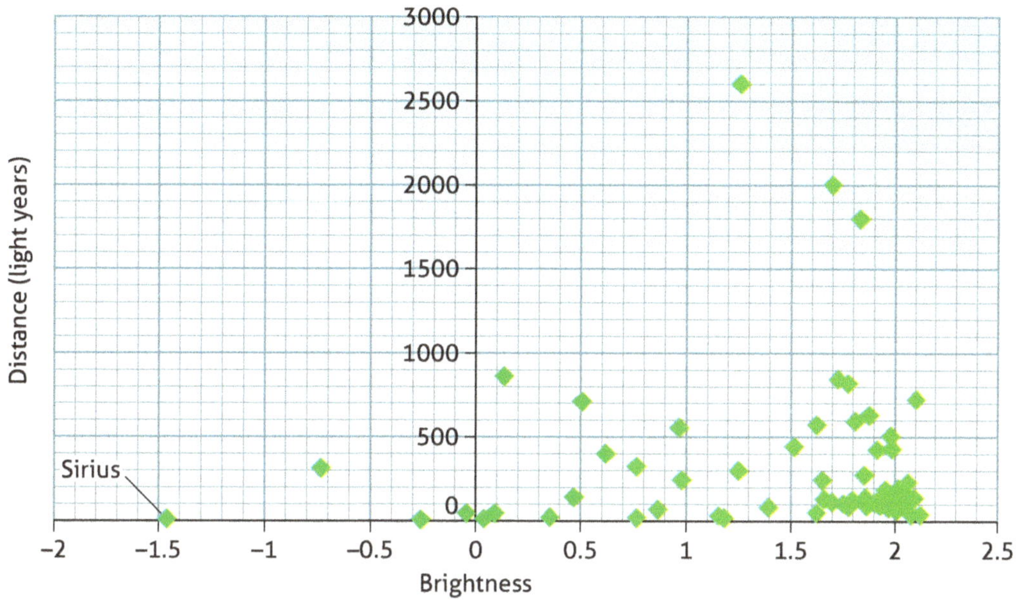

The brightest stars are on the left and the less bright on the right. This is the scale that astronomers use.

(b) Identify the work that Tamsin does in each of the first three stages of the statistical problem solving cycle.

(c) For the 'Reporting/Interpretation' stage, Tamsin says, 'It works for some stars but not for others. I want to investigate further.'

Suggest what Tamsin might do in each stage of a further round of the cycle.

5 Sixteenth-century records from Norwich have been used to compare literacy rates between people in different occupations.

Occupation	Sample size (signature + mark)	Signature	Literacy rate $\left(\frac{\text{Signature}}{\text{Sample size}}\right) \times 100$ to the nearest whole number
Gardener	11	3	27
Shepherd	10	1	10
Tailor	139	78	56
Weaver	225	131	58

D. Cressy, *The Historical Journal*, 1977

(a) What was the difference in literacy rate between weavers and shepherds?

(b) A history researcher wants to find the overall literacy rate for all four occupations.

Taking the mean of the four literacy rates shown in the table was suggested.

(i) Why is this not a good method?

(ii) Use another method to find the overall literacy rate for all four occupations. Does it give a different result to the first method?

(c) The researcher uses the information in the table to put the literacy rates for the four occupations in order:

Weaver > Tailor > Gardener > Shepherd.

Comment on the reliability of this ordering.

Practice question

This frequency table shows the distribution of wind speeds recorded on the top of a building at noon over a 6 month period.

Wind speed, w, m s^{-1}	Frequency
$0 \le w < 2.0$	12
$2.0 \le w < 3.0$	16
$3.0 \le w < 4.0$	20
$4.0 \le w < 4.5$	30
$4.5 \le w < 5.0$	60
$5.0 \le w < 6.0$	25
$6.0 \le w < 10.0$	10
Total	173

(a) Use these figures to estimate the mean wind speed.

State why your estimate may not be very accurate.

(b) Draw up a cumulative frequency table for the wind speeds.

(c) Draw a cumulative frequency graph.

(d) Use your graph to estimate the median, upper and lower quartiles, and the inter-quartile range of the wind speeds.

(e) Draw a box and whisker plot to illustrate the data.

(f) Compare your estimates of the mean and median.

Which do you consider to be the better 'typical' value of the wind speed?

Now that you have completed this chapter, you should know:

- How to use the statistical problem solving cycle.
- How to use and interpret dot plots, tallies, pie charts, bar charts, vertical line charts, time series graphs, line graphs and scatter diagrams.
- How to identify and summarise the main points made by a data display.
- How to calculate the mean, mode, median and range of a set of data and to use a spreadsheet to do so.
- How to calculate weighted means.
- How to distinguish between categorical and numerical data, and also between discrete and continuous numerical data.
- How to group both discrete and continuous data and draw grouped frequency charts.
- How to draw and interpret histograms.
- How to construct and use cumulative frequency tables and charts.
- How to find quartiles and use them in drawing box and whisker plots.

CHAPTER 7

Representing the real world graphically

You are probably familiar with the saying, 'A picture is worth a thousand words.' In this chapter you will see that this is not only true for words but also for numbers.

7A Using graphs

When you look at a graph you are seeing a representation of the information that would otherwise be locked up in a set of numbers.

Activity 7A Sunset times

Use this graph to answer the following questions.

(a) When does the sun set on 1st January?

(b) What is the latest sunset time?

(c) What is the reason for the vertical line in March?

(d) Estimate the sunset time on 21st August.

(e) What other questions can you ask about this graph?

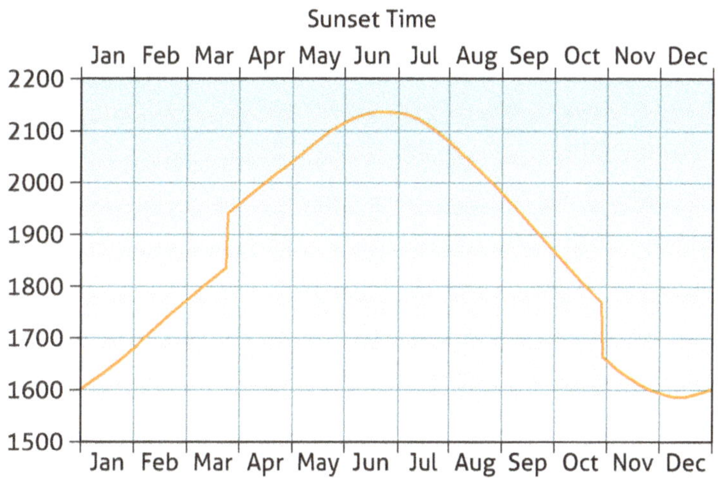

Worked example 7A

This graph shows the times of sunrise one year.

(a) Construct a table showing the sunrise time at the start of each month.

(b) What does the graph for the previous year look like?

(c) Describe the graph for 10 years in the future.

(d) Why are the months on the horizontal axis?

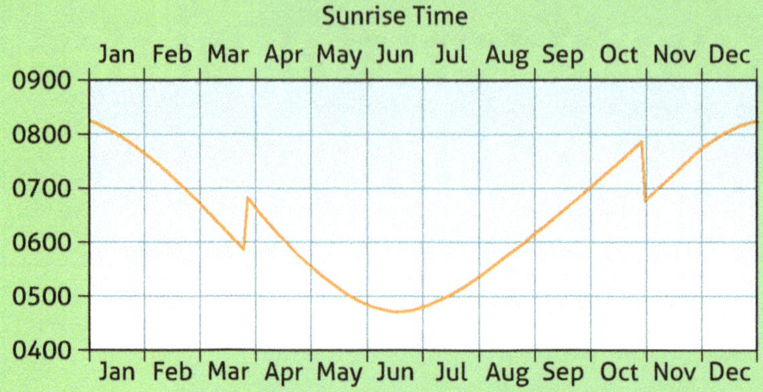

Solution

(a)

Month	Jan	Feb	Mar	Apr	May	Jun	Jul	Aug	Sep	Oct	Nov	Dec
Time	0805	0740	0650	0640	0535	0450	0445	0525	0610	0700	0650	0740

(Times have been rounded to the nearest 5 minutes.)

(b) Exactly the same.

The graph is periodic. It repeats every year. The effect of leap years is too small to notice on this graph.

(c) The same as this one, unless the clock changes no longer happen.

(d) The months are the **independent** (**explanatory**) variable, with regular steps on the scale, so are marked on the horizontal axis.

The sunrise time depends on the date, so is the **dependent** (**response**) variable and goes on the vertical axis. Time is usually on the horizontal axis.

Exercise 7A

1 (a) Using the graph below, construct a table to show the life expectancy for males and females in England between 2001 and 2014.

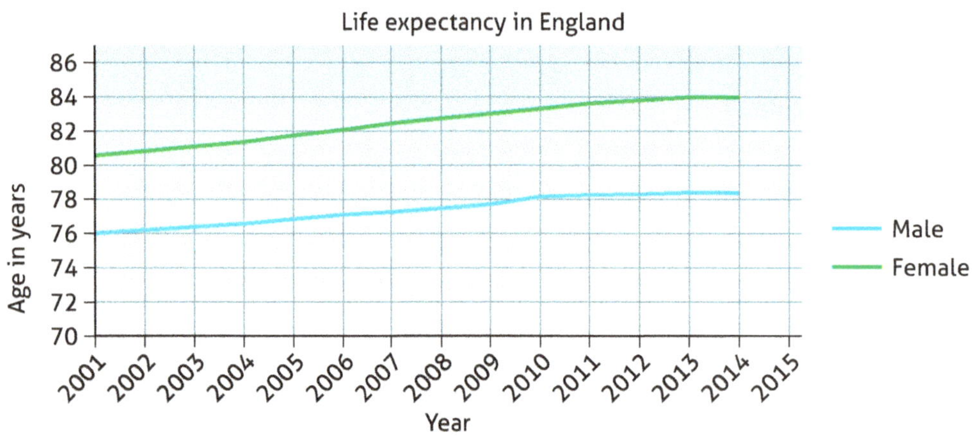

(b) Describe two trends that you notice from the graph.

2 (a) Use the graphs of sunset and sunrise times to construct a table of values for the length of the day at the start of each month.

(b) By roughly how much do the times of sunrise change over the months when the clocks change? What about the times of sunset?

7B Gradients

The steepness of a graph is called its **gradient** and this is often a very useful measure.

Activity 7B Fastest

Look at the four graphs on the right and assume they all have the same scale.

(a) Which graph shows the fastest rate of change?

(b) Arrange the graphs in ascending order of rate of change.

In fact, the graphs may not have the same scales.

(c) Explain why it is now not possible to arrange the graphs in order of their rate of change.

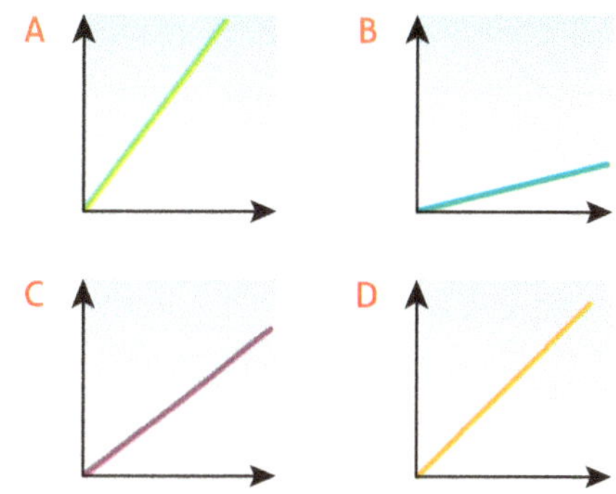

To measure the rate of change from a graph means working out the gradient of a straight line.

Worked example 7B

? The graph below shows the velocity (in metres per second) of an object against time (in seconds).

(a) Calculate the gradient of the graph.

(b) Explain what the gradient of the graph represents.

Solution

(a) Gradient = $\frac{4}{6} = \frac{2}{3}$.

(b) The gradient of the graph represents acceleration.

The units of the gradient are $\frac{m}{s} \div s = \frac{m}{s \times s} = \frac{m}{s^2}$.

So, metres per second squared are the units for acceleration.

The graph goes up 4 for 6 along.

That is $\frac{2}{3}$ for each 1 along.

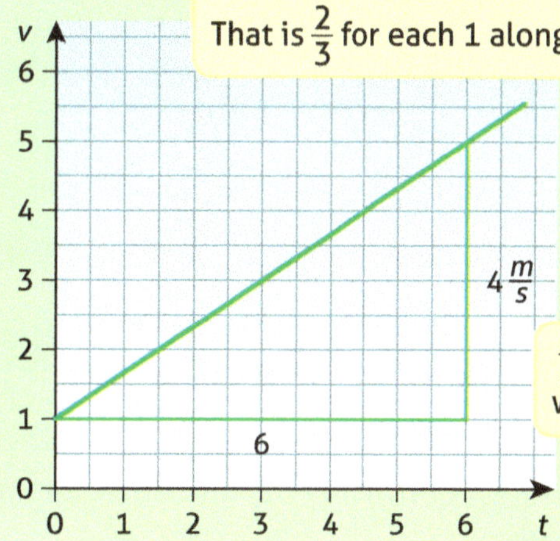

$\frac{m}{s^2}$ can also be written as ms^{-2}.

⚠ Lines that go down have negative gradients.

Exercise 7B

1 Which of the following is the gradient of the graph on the right?

(a) 2 (b) −2 (c) $\frac{1}{2}$ (d) $-\frac{1}{2}$

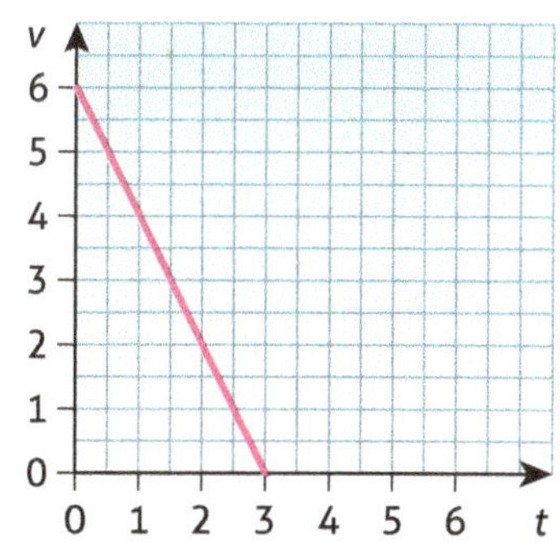

2 Calculate the gradients of the lines joining:

(a) (0, 1) and (6, 3) (b) (1, 1) and (4, 9) (c) (1, 7) and (5, 1).

3 State what the gradient represents for the graphs of:

(a) distance travelled in kilometres against time taken in hours

(b) volume of water in a leaking tank in litres against time in minutes.

7C Tangents

It is often helpful to draw the tangent at a particular point of a graph. It can give you an insight into what is going on.

Activity 7C Tell the story

1 Describe what is happening on the graph shown if the independent variable (→) is time and:

(a) the dependent variable (↑) is number of people in a theatre in the half hour before a performance

(b) the dependent variable (↑) is the volume of applause following a performance

(c) the dependent variable (↑) is the number of people in a population who are immune against a new infectious disease.

(d) Think of some other situations that could be described by this graph.

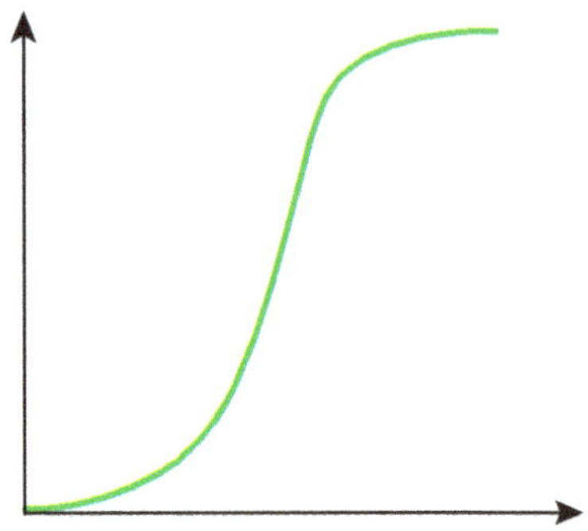

Independent and dependent variables are explained in **Chapter 6**.

Worked example 7C

The graph shows the height of a broad bean seedling from when the seed germinates.

(a) When is the plant growing fastest?

(b) How fast is it growing at 6 weeks?

(c) How fast is it growing at 12 weeks?

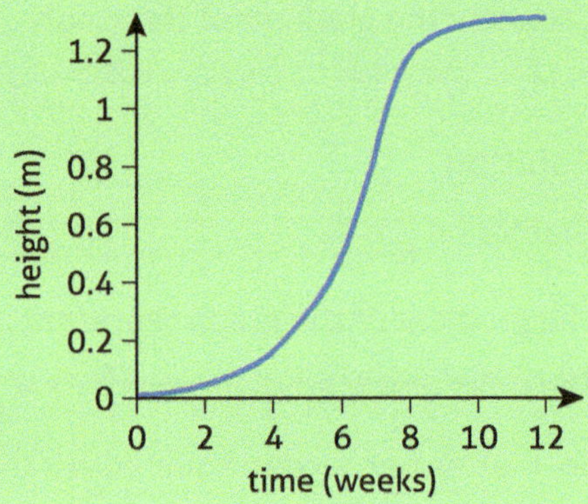

Solution

The graph is steepest between 6 and 8 weeks.

(a) Between 6 and 8 weeks after planting.

(b) The graph is curved at 6 weeks.

A straight line is drawn that has the same gradient as the curve at that point.

Gradient ≈ $\frac{1.2}{4.8}$ = 0.25 metres per week.

(c) At 12 weeks the plant is no longer growing.

The rate of change is zero.

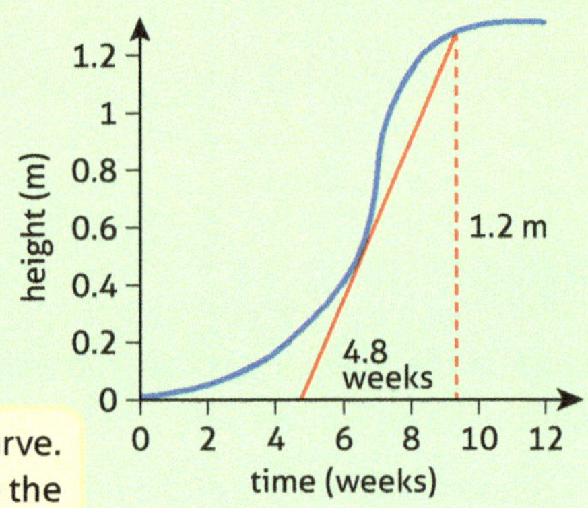

The line is a tangent to the curve. It is drawn by eye to estimate the rate of change (gradient).

Exercise 7C

1 The graph on the right shows the speed of an aeroplane as it takes off.

(a) Estimate the acceleration (in m s^{-2}) of the plane when the time is 10 seconds.

(b) When is the acceleration greatest?

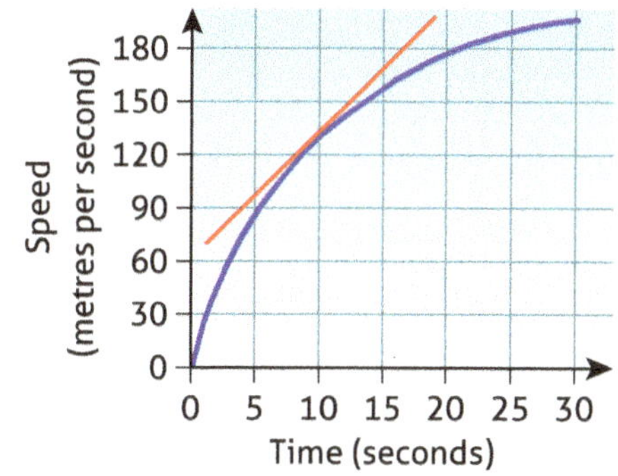

2 The graph on the right has a headline, 'Party membership soars', in the latest edition of a newsletter.

Comment on the accuracy of the headline for this graph.

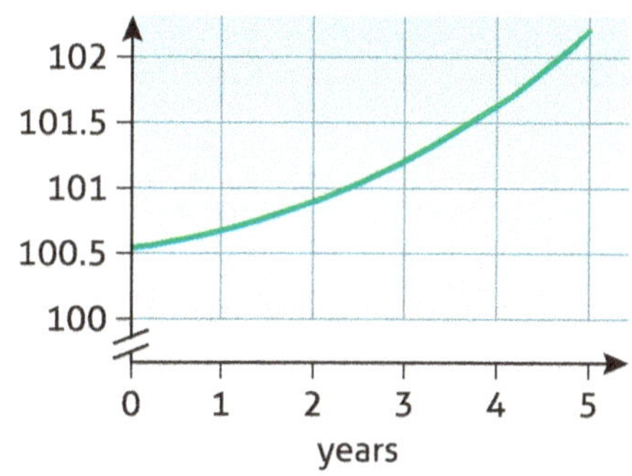

7D Plotting graphs

A graph consists of a set of points that are plotted on a sheet of graph paper, a calculator or a computer screen. You can use a spreadsheet to plot points. The points are usually joined up in some way, as a curve or a number of straight lines.

Activity 7D Does it fit?

A line has the equation, $y = 5 - 2x$.

(a) Which of these points lie on the line:

(i) (1, 3) (ii) (1, 7) (iii) (0, 5) (iv) (−10, 25) (v) (10, −25).

(b) (1, 2) does not lie on the line. Is it above or below it?

(c) Describe how to determine whether a point is on, above or below the line.

Worked example 7D

The formula,

$h = 1.6 \sin\left(\frac{36}{79}t + 71\right)° + 2.9,$

gives the height of the tide (in metres), minutes after midnight on 29/7/2020, in Falmouth.

(a) Plot the graph of tide height against time after midnight until 9:00 am.

(b) When is the tide at its lowest?

(c) What is the difference in height between the water levels when the tide is at its lowest and its highest?

Solution

Make sure your calculator is set to degrees.

You can save a lot of work by using the repeat entry facility. That way you only need to change the values of *t*.

(a) The independent variable is *t*.

Select a suitable interval for *t*, say every 45 minutes.

Use your calculator to construct a table of values.

t	0	45	90	135	180	225	270	315	360	405	450	495	540
h	4.4	4.5	4.4	4.1	3.6	3.1	2.5	2.0	1.6	1.5	1.3	1.5	1.8

Choose an origin and scales for the axes. The dependent variable, *h*, has a maximum value of 4.5 and the minimum is 1.3, so starting from 0 is reasonable.

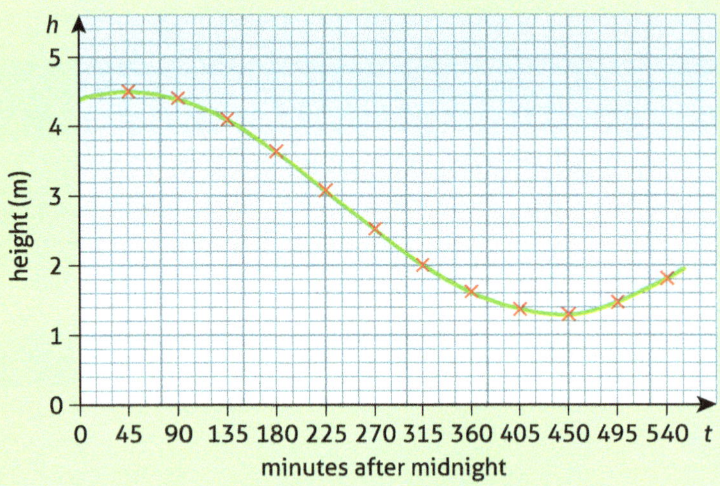

(b) The graph shows the the tide at its lowest at 450 minutes after midnight, 7.30 am.

(c) The difference between the water levels is 4.5 – 1.3 = 3.2 m.

Exercise 7D

1 The cost of a monthly electricity bill is given by the formula,

$C = 12.35 + 0.13426N$

where C is the cost in pounds (£) and N is the number of kilowatt hours of electricity used.

(a) Plot a graph of C against N from 0 to 600.

(b) Use the graph to work out how many units are used when the cost is £40.

2 Two beavers are introduced into a stretch of river. In the years that follow the beavers breed.

A model for their population is that if there are n beavers one year, the number the following year is given by $\frac{n(100-n)}{50}$.

(a) Set up a spreadsheet to give the number of beavers for each of the next 9 years. Since there must be a whole number of beavers, you should use Number with 0 decimal places for the column that gives the values of n.

(b) Use your spreadsheet to obtain this graph.

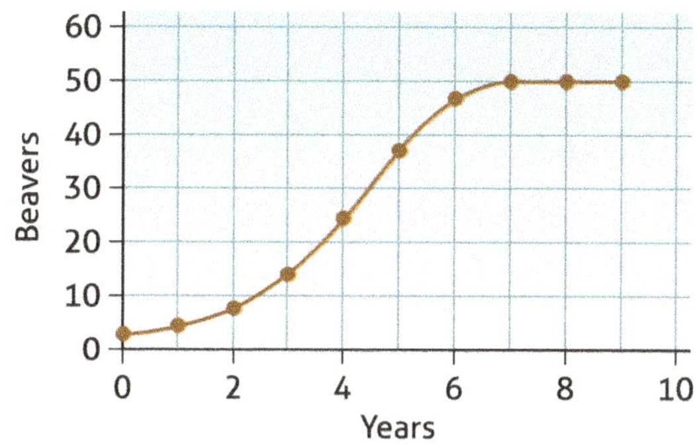

7E Graphs showing motion

It is common to use graphs to illustrate motion. Key quantities are: distance and displacement; speed and velocity; acceleration.

Activity 7E A v–t graph

The graph shows the velocity in metres per second of a car.

Time is measured in seconds.

Comment on its motion.

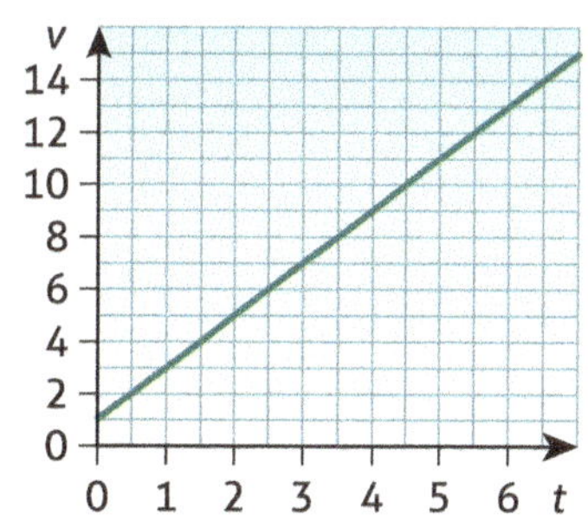

Worked example 7E

? The graph below shows the journey of a snail, in a straight line.

(a) What happens between 4 and 7 minutes?

(b) What is the snail's speed between 7 and 12 minutes? Give your answer in metres per second.

(c) Where is the snail at the start?

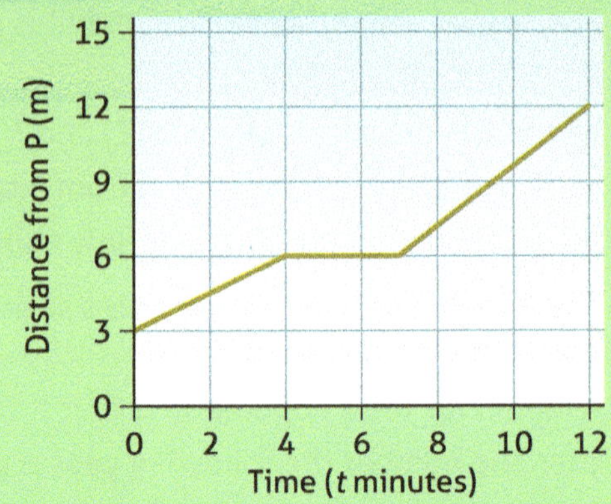

The distance, *d*, metres, of the snail from P between 0 and 4 minutes is given by the formula,

$d = 3 + \frac{3}{4}$

where *t* is the time in minutes.

(d) What is the significance of 3 in the formula for the graph and for the snail?

(e) What does $\frac{3}{4}$ in the formula represent on the graph?

Solution

(a) The snail is stationary.

(b) Speed = $\frac{6}{5}$ = 1.2 metres per minute.

$\frac{1.2}{60}$ = 0.02 metres per second. ← Divide by 60 to convert metres per minute to metres per second.

(c) 3 metres from P.

(d) 3 is where the line crosses the vertical axis (dependent variable). It is the distance of the snail from P at the start.

(e) $\frac{3}{4}$ is the gradient of the line and is the speed of the snail in metres per minute.

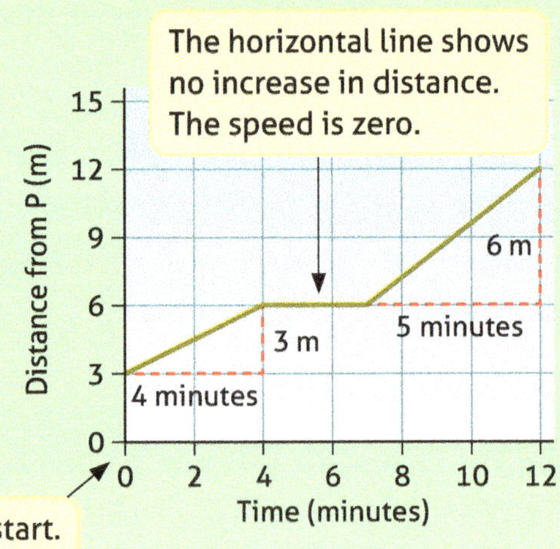

Exercise 7E

1 (a) Which of the following is the equation of the line on the graph?

(i) $d = \frac{3}{4}t$ (ii) $d = 40 + \frac{2}{3}t$ (iii) $d = 40 - \frac{2}{3}t$ (iv) $d = 40 - \frac{4}{3}t$

The graph shows the distance, *d* m, of a mouse from its nest at time *t* s. It is travelling in a straight line.

(b) How far from its nest is the mouse initially?

(c) What is the speed of the mouse in m s^{-1}?

(d) What is the gradient of the line?

(e) What happens after 30 seconds?

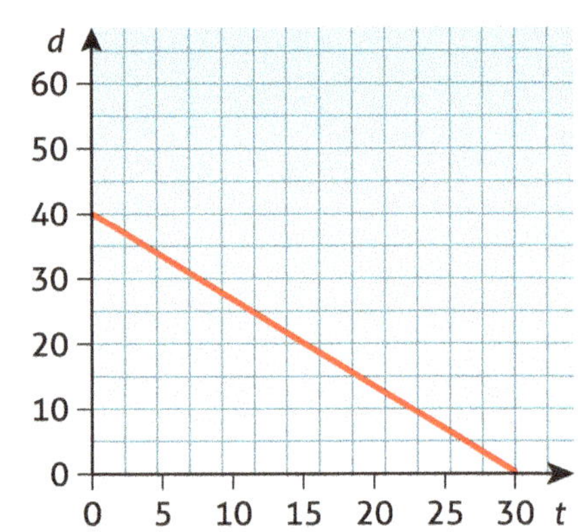

2 The graph shows a model of the speed of a jaguar in pursuit of its prey.

(a) What is the final speed of the jaguar?

(b) What is happening between 4 and 7 seconds?

(c) What does the gradient of the graph between 0 and 4 seconds represent?

(d) What is the acceleration of the jaguar between 7 and 12 seconds?

(e) What distance does the jaguar travel between 4 and 7 seconds?

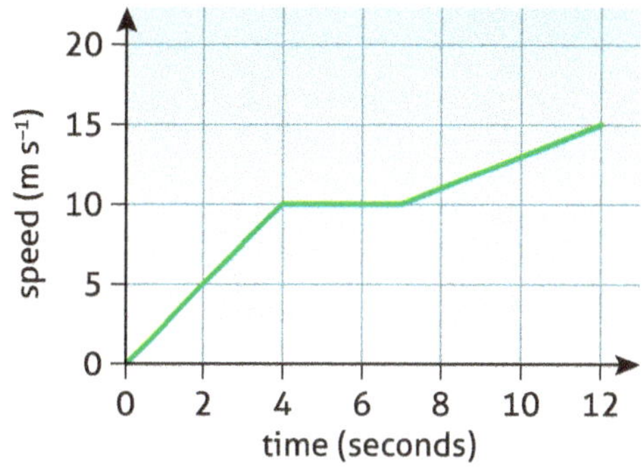

7F Proportion

You will often come across situations where one quantity is proportional to another. This can be seen from their graphs.

Activity 7F Are there different types of proportionality?

Which is the odd one out?

(i) Value in Euros compared with value in £s.

(ii) Number of people sharing a pizza and the size of each portion.

(iii) Number of pizzas bought and the cost.

Justify your answer.

Worked example 7F

? Explain the shape of the two graphs and state what type of proportion they are showing.

(a) total cost, c

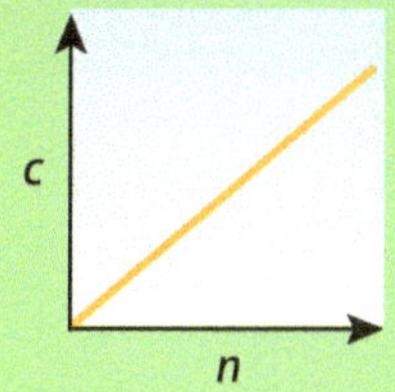

number of pencils bought, n

(b) time to complete job, t

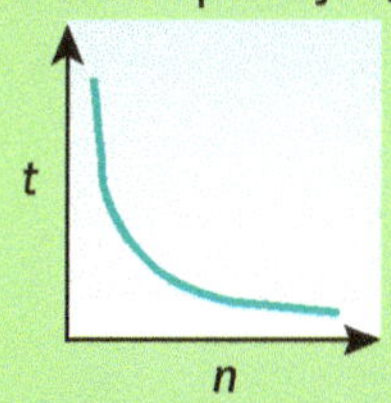

number of people working on a job, n

Solution

(a) The number of pencils bought is multiplied by the same scale factor each time to get the total cost. The scale factor is the cost of each pencil.

> This is called **direct proportion**. The graph is a straight line through the origin. The scale factor is the gradient of the line.

(b) The time it takes to complete a job is usually measured in people-hours. If there are double the number of people it should halve the time taken, if you have five times as many people then the time it takes should be divided by 5.

> This is called **inverse proportion**. The graph is a curved line that gets closer and closer to each axis.

Exercise 7F

1 For each of these graphs, state whether it shows direct proportion, inverse proportion or neither.

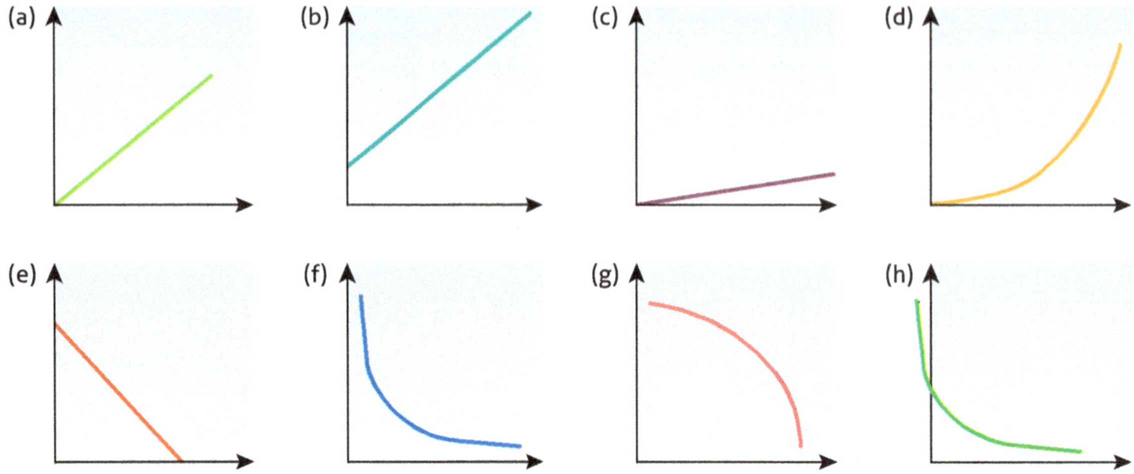

7G Linearising data

You will often come across data, for example from an experiment, for which you want to find an algebraic relationship between the variables. This is much easier if the points lie on a straight line graph.

Activity 7G Spot the difference

(a) Using suitable axes, plot a graph using the data in the table.

x	0	1	2	3	4
y	0	0.5	2	4.5	8

(b) Now plot the graph of y against x^2.

x^2	0	1	4	9	16
y	0	0.5	2	4.5	8

(c) What is the same in each graph?

(d) What are the differences between the two graphs?

(e) How is each graph linked with $y = \frac{1}{2}x^2$?

Worked example 7G

? An oil spill is increasing in size. The table shows the width of the spill in metres and its area in square metres. It is suspected that the relationship between the area and the width is of the form $A = kw^2$.

Width, w	1	2	3	4	5	6
Area, A	1	3.5	8	14.5	22.5	32.5

(a) Plot a graph of A against w^2 using suitable axes.

(b) Draw a line of best fit.

(c) How does the graph confirm that the relationship is of the form $A = kw^2$?

(d) Estimate the value of k.

(e) Write down the formula for the relationship between A and w.

Solution

(a) Work out values of w^2.

Width, w	1	2	3	4	5	6
w^2	1	4	9	16	25	36
Area, A	1	3.5	8	14.5	22.5	32.5

Plot A against w^2.

(b) The line of best fit is shown on the graph.

(c) The graph is a straight line.

(d) The gradient, $k = \frac{18}{20} = 0.9$.

(e) $A = 0.9w^2$.

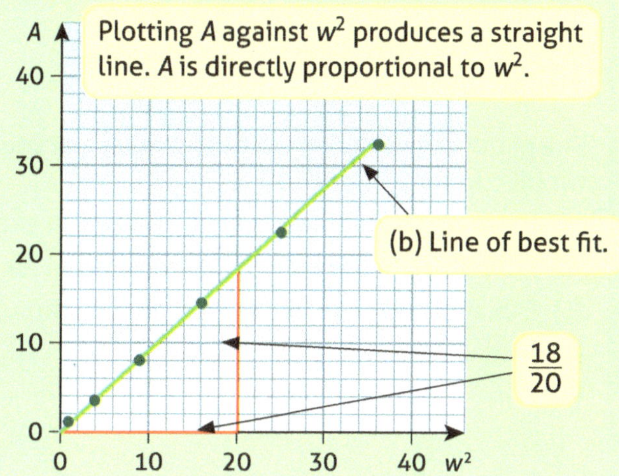

In this example A was plotted against w^2. This turned a curved graph into a straight line, making it easy to write the formula connecting the variables. This is an important technique; it is called **linearising the data**.

Exercise 7G

1 Petroc plans to estimate the depth of old Cornish tin mines by dropping stones down them and timing how long they take to reach the bottom. First, he records the times for stones in a number of mines of known depth. His results are recorded in the table on the right.

t (seconds)	0.8	1.3	2.2	2.9	3.5	4.7
d (metres)	2.3	6.5	18.6	30.0	47.1	83.0

It is believed that the relationship between time and distance is of the form $d = kt^2$.

(a) Draw a suitable graph and estimate the value of k.

(b) Estimate the depth for a drop time of 3.6 seconds.

(c) Estimate the time it would take the stone to reach the bottom of a 65 m deep mine.

2 A tank is filled with water but has a small hole. The volume, V litres per hour, of water being lost to the leak is known to be related to the depth, x metres, of the water in the tank, by a formula of the form $V = c\sqrt{x}$.

The table shows some data on the leakage from the tank.

(a) Draw a suitable graph and estimate the value of c.

(b) Estimate the depth of water when the leakage rate is half a litre per hour.

x (metres)	0	1	2	3	4	5
V (litres)	0	0.26	0.42	0.47	0.58	0.62

Test yourself 7H

1 For each of the following graphs (a)–(f), state whether it shows direct proportion, inverse proportion or neither.

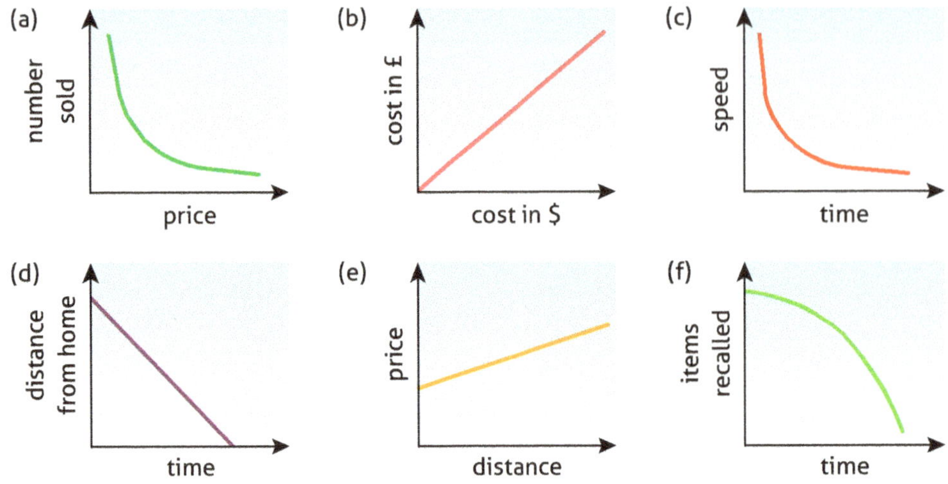

2 The graph shows the amount of Albanian lek you receive in exchange for UK currency £0 to £100.

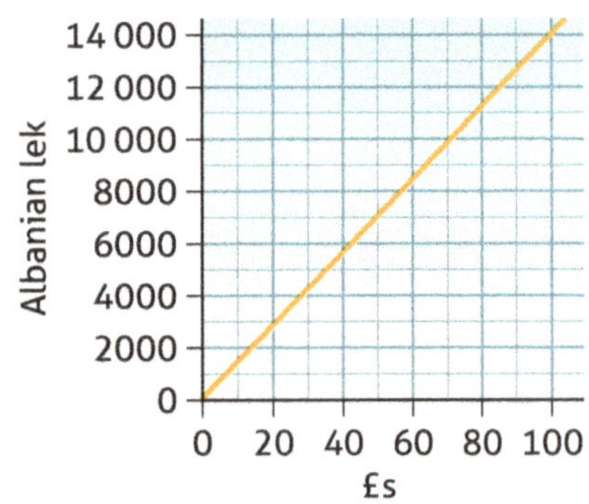

(a) How many Albanian lek do you receive in exchange for £50?

(b) How many £s do you receive in exchange for 10,000 Albanian lek?

(c) Explain why the graph is a straight line through (0, 0).

(d) Find the gradient of the graph.

(e) What does the gradient of the graph represent?

3 Helen is a psychology student. She is researching recall in school students. She has gathered data on the percentage of items, P, forgotten each day since being learned.

Day, *d*	1	2	3	4	5	6
***P* (%)**	38	55	66	76	84	93

Helen thinks that the formula connecting P and d is of the form $P = k\sqrt{d}$.

(a) Helen enters her data in a spreadsheet in order to calculate $\sqrt{d}$.

What formula is in cell B3?

	A	B
1	Day, *d*	
2	1	1
3	2	1.414214
4	3	1.732051
5	4	2
6	5	2.236068
7	6	2.44949

(b) Plot an appropriate graph for Helen to estimate k.

(c) Use the graph to estimate k.

(d) Write down the formula relating P and d.

(e) Use your formula to calculate P when $d = 7$.

(f) Comment on your answer to (e).

4 The table below shows the amount of caffeine (mg) in the bloodstream for up to 18 hours after drinking coffee.

Time in hours	0	2	4	6	8	10	12	14	16	18
mg of caffeine	125	99	76	60	48	38	29	24	18	14

(a) Plot a graph to show the data.

(b) Estimate the rate of change of caffeine in the bloodstream 8 hours after drinking the coffee.

5 Sonali drives to visit a friend for a week's holiday. She leaves home at 8 am. The graph illustrates her journey.

(a) Find what her speed was during the first two hours.

(b) How far does she travel altogether?

(c) The speed limit on the motorway is 70 miles per hour. When is she travelling on the motorway?

(d) Find her average speed for the whole journey.

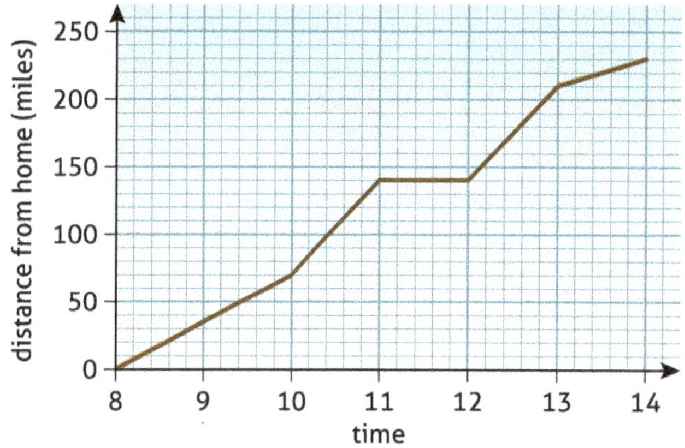

Practice question

Jane is carrying out a survey into weeds in lawns. She makes a random selection of 2-metre squares from ten different lawns and counts the numbers of daisies, x, and dandelions, y. She presents her results on a scatter diagram and draws a line of best fit by eye.

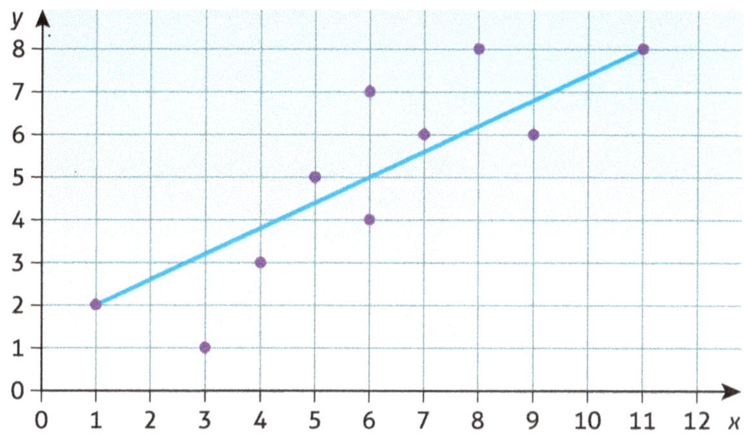

(a) Write down the coordinates of 2 points that Jane's line of best fit passes through.

(b) Find the gradient of the line.

(c) Show that the equation of the line is $y = 0.6x + 1.4$.

(d) Find the coordinates of the mean point of the data and show that the line of best fit passes through it.

(e) Find the coordinates of the point where the line would cross the x-axis if it were drawn further to the left. Explain why there is no sense to do this.

(f) The data suggest that there is a strong correlation between the numbers of daisies and dandelions in patches of lawns. Suggest one reason why this might happen.

Now that you have completed this chapter, you should know:

- How to plot graphs and to work with them.
- How to measure and interpret the gradient of a graph at a point.
- How to linearise data.

CHAPTER 8
It's Normal

In this chapter, you'll be looking at symmetrical charts for continuous data that are not grouped but still do have frequency represented by area. As frequency is represented by area, so the vertical axis is frequency density.

8A The bell-shaped curve

As you will see in this section, the Normal distribution has a distinctive bell-shaped curve.

Activity 8A Students' heights

Sketch a histogram to show the heights of:

(i) 100 male students

(ii) 100 female students.

(a) How are your histograms similar?

(b) How are they different?

(c) Give another example of a set of data that will also have the same shape.

Once you have collected a set of data you can draw a **histogram** to show the shape of the distribution. The histogram might look something like these:

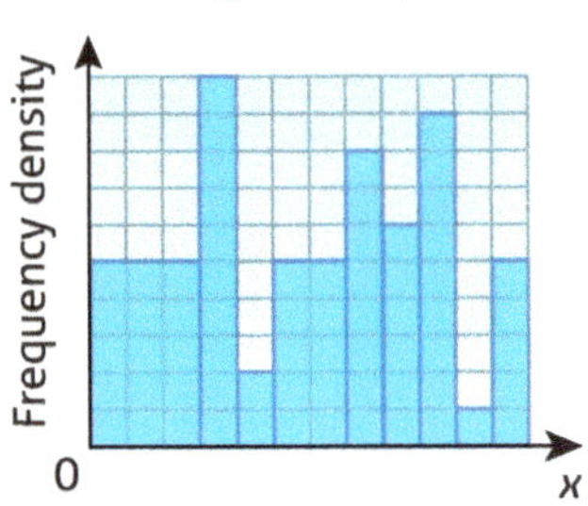

No particular shape to the distribution

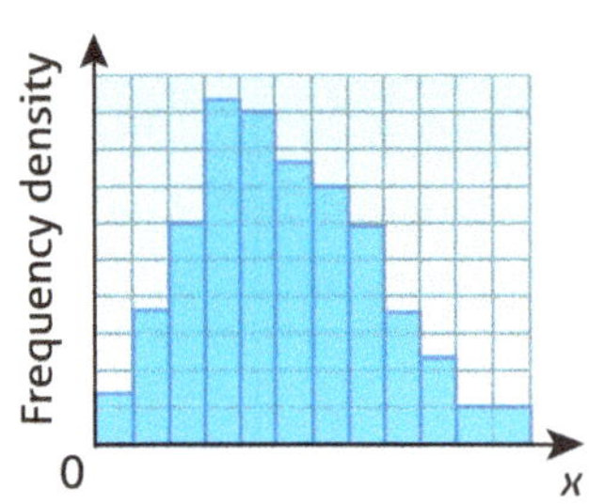

Positive skew; longer 'tail' on the right

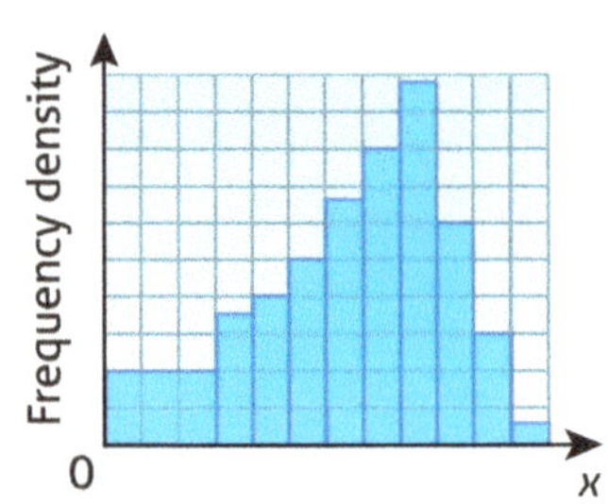

Negative skew; longer 'tail' on the left

The histogram might also be symmetrical, like this one:

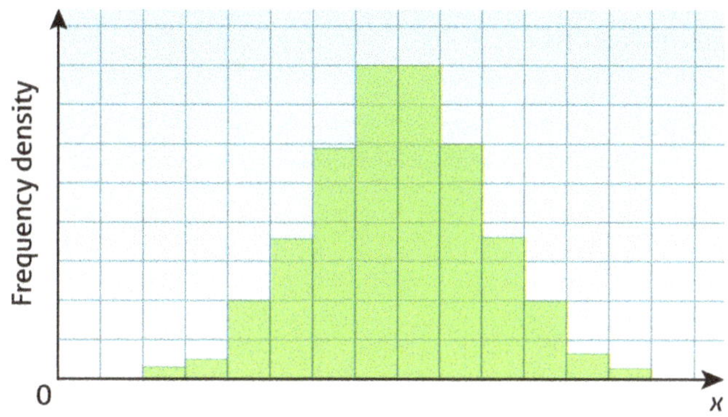

- Histograms are for continuous data.
- They are for grouped data and the groups may have different widths.
- The frequency is represented by the area under the curve rather than its height.
- The variable on the vertical axis is frequency density.

See **Chapter 6** for a reminder.

You can draw a bell-shaped curve over this histogram like this one showing the heights of female students:

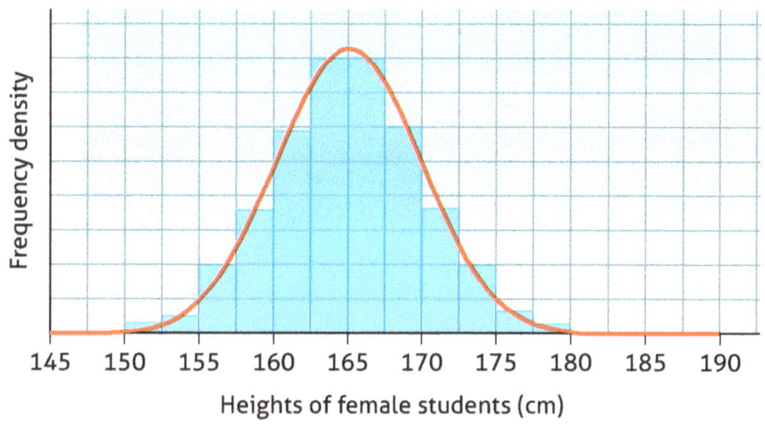

You would expect that most female students are around average height and that there are few very short or very tall students.

As the curve is symmetrical the mean, median and mode are all the same.

Mean, median and mode = 165 cm

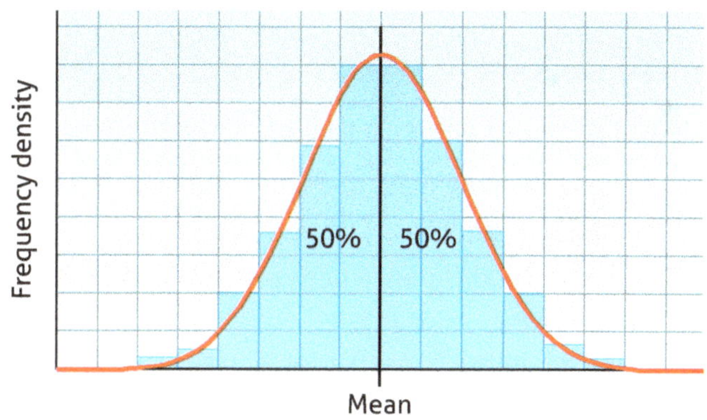

The bell-shaped curve shows the Normal distribution. The Normal distribution is important because it describes the population of many naturally occurring variables.
It has the following properties:

- The curve is symmetrical about the **mean**.
- 50% of the values are below the mean.
- 50% of the values are above the mean.

The Greek letter mu, μ, is used to represent the mean.

So, for the distribution of heights of female students, the mean μ = 165 cm.

The word **distribution** means the way in which the values of a variable occur. Here are examples of some other types of distribution.

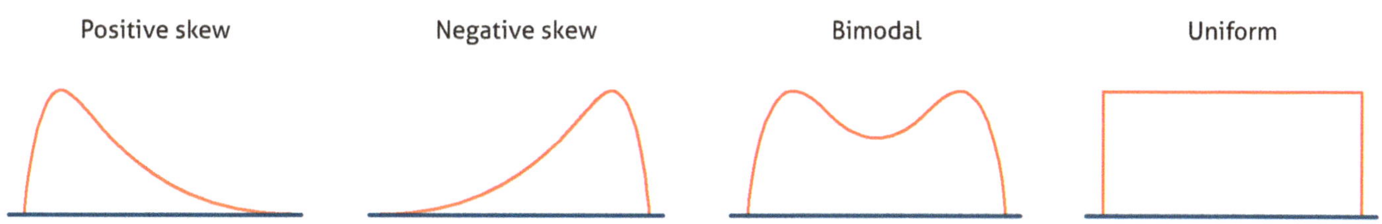

Exercise 8A

1 Each of the following diagrams illustrates a distribution that is Normal.

Write down the mean for each one.

(a)

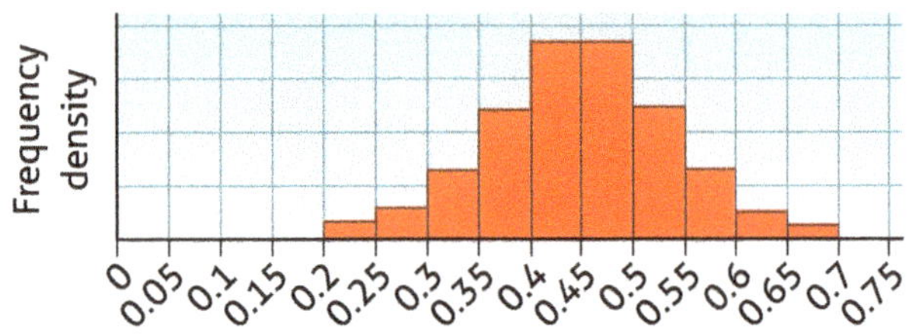

(b)

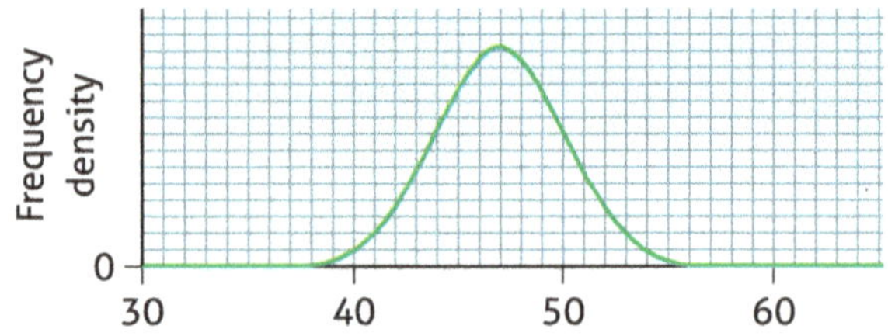

2 One year, the weights of the apples from an orchard are Normally distributed with mean 90 grams. A sack contains 150 of these apples.

(a) How many of the apples in the sack are heavier than 90 grams?

(b) What are (i) the median weight and (ii) the modal weight of these apples?

8B Standard deviation

When you have a set of data, they are a sample of an underlying population.

You often use your data to find:

- a typical value, called an average, that is representative of the population
- how spread-out the population is.

Activity 8B The ducklings

A wildlife conservationist weighs all the ducklings at her local pond.

She finds the mean mass of a week-old duckling is 100 grams and the range is 40 grams.

(a) What is the difference between an average value and a measure of spread?

(b) Why might the conservationist want to know both?

Standard deviation is a measure of how spread-out a set of data is. The larger the standard deviation, the more spread-out the data are. The Greek letter sigma, σ, is used for standard deviation and it has the same units as the set of data.

If your data are masses of ducklings in grams, then σ has units of grams.

The conservationist finds that the masses of the ducklings have

$\mu = 100$ grams and $\sigma = 5$ grams.

For a Normal distribution:

- 68% of values are within 1 standard deviation of the mean.

68% of ducklings are between 95 grams and 105 grams.

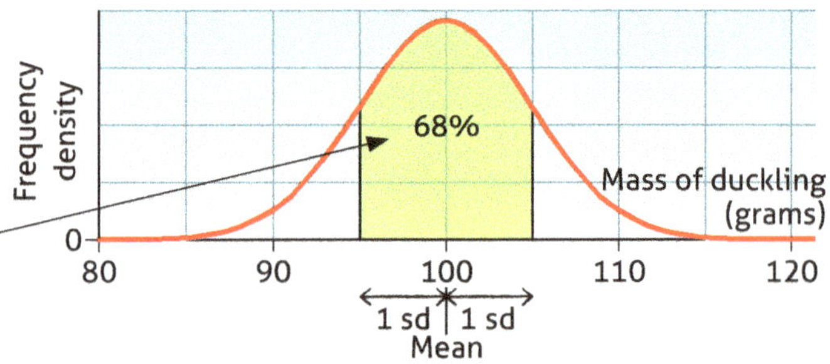

- 95% of values are within 2 standard deviations of the mean.

95% of ducklings are between 90 grams and 110 grams.

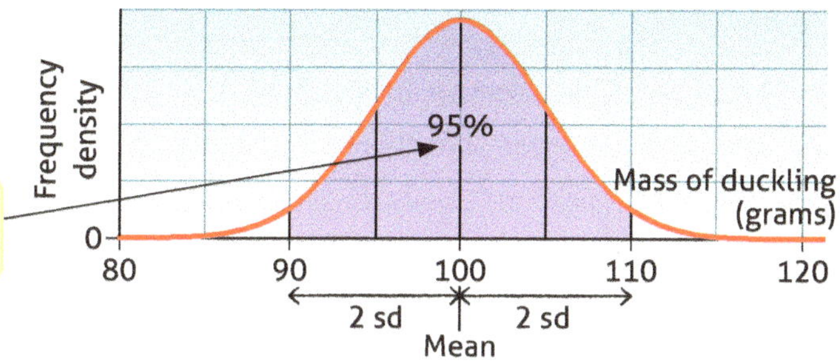

- 99.7% of values are within 3 standard deviations of the mean.

99.7% of ducklings are between 85 grams and 115 grams.

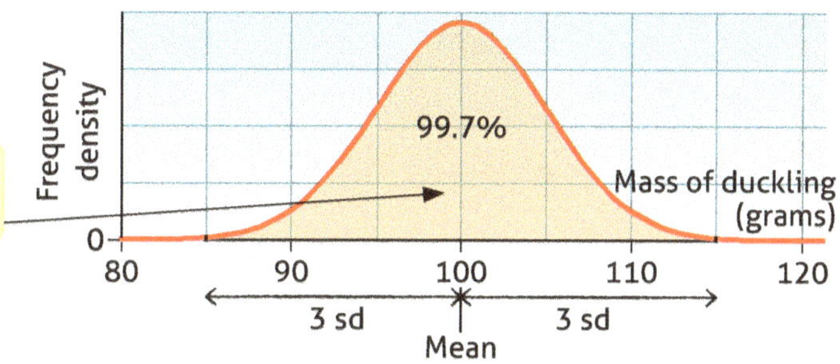

Worked example 8B

? The weights of some ducklings are Normally distributed with μ = 100 grams and σ = 5 grams.

(a) The conservationist selects a duckling at random.
Find the probability that its weight is:
(i) more than 100 g
(ii) between 95 g and 105 g
(iii) between 100 g and 105 g
(iv) more than 115 g.

(b) A duckling is underweight if it is 2.5 standard deviations or more below the mean.
A duckling is weighed and found to be just on the border of being underweight.
What is the weight of the duckling?

Solution

(a) (i) Half the ducklings have a weight above the mean.
So, the probability that the weight is more than 100 g is 50%.

(ii) 68% of the ducklings are within 1 standard deviation (1 sd) of the mean:
mean + 1 sd = 100 g + 5 g = 105 g
mean – 1 sd = 100 g – 5 g = 95 g
So, the probability that the weight is between 95 g and 105 g is 68%.

(iii) 68% ÷ 2 = 34%
So, the probability that the weight is between 100 g and 105 g is 34%.

(iv) Find the difference between 115 g and the mean: 115 g – 100 g = 15 g.
So, 15 g = $\frac{15}{5}$ = 3 standard deviations.
99.7% of ducklings have a weight within 3 standard deviations of the mean.
So, 100% – 99.7% = 0.3% of the ducklings are more than 3 standard deviations from the mean.
0.3% ÷ 2 = 0.15%
So, 0.15% of the ducklings have a weight of more than 115 g.

The question says that 5 g = 1 sd.

So, 0.15% are 3 sd below the mean and 0.15% are 3 sd above the mean.

(b) 2.5 sd = 2.5 × 5 g = 12.5 g
12.5 g below the mean is 100 g – 12.5 g = 87.5 g.
So, the duckling has a weight of 87.5 g.

How do you use your calculator to find the standard deviation of a set of numbers?

Exercise 8B

1 A farmer takes a sample of his crops of carrots and cucumbers and measures their lengths.

The Normal curves show his results for each crop.

Crop 1:

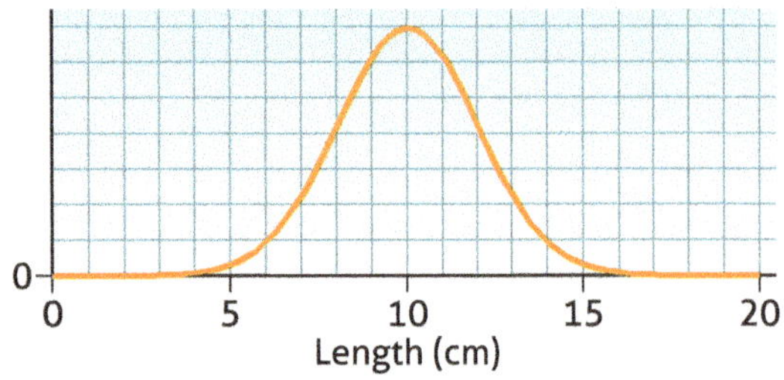

Crop 2:

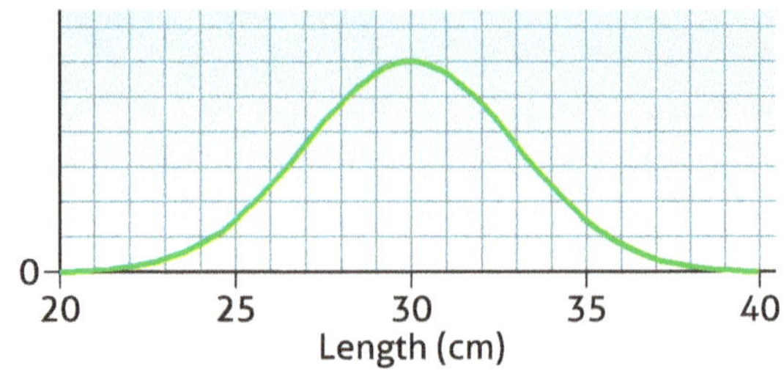

(a) Estimate the mean and standard deviation of each crop.

(b) Which curve represents the crop of cucumbers? Give a reason for your answer.

2 A child psychologist finds that the times taken for children to solve a puzzle have a Normal distribution.

The psychologist finds that the mean time to solve the puzzle is 6 minutes and the standard deviation in the times is 30 seconds.

(a) What percentage of children take longer than 6 minutes to solve the puzzle?

The psychologist finds that a child whose time is more than 2 standard deviations below the mean time is academically gifted.

(b) What times are 2 standard deviations below the mean?

Jake takes 6.5 minutes to solve the puzzle.

(c) **(i)** How many standard deviations above the mean is Jake?

(ii) What percentage of children take longer than Jake to solve the puzzle?

8C Using z-scores

All Normal distributions have exactly the same bell-shaped curve if their variables are scaled so that they all have the same mean and standard deviation.

Activity 8C IQ tests

An IQ (intelligence quotient) test is used to measure someone's intelligence.

The mean IQ is 100 and the standard deviation is 15.

Archie takes an IQ test.
The test shows he has an IQ of 148.
Is Archie unusually gifted?

The number of standard deviations from the mean is called the ***z*-score**.

A *z*-score of 1 means that the data value is 1 standard deviation **above** the mean.

A *z*-score of −2 means that the data value is 2 standard deviations **below** the mean.

To calculate a *z*-score:

- subtract the mean
- then divide by the standard deviation.

The *z*-score for a data value x is $\frac{x-\mu}{\sigma}$.

z-scores make it easier to compare values in a data set – they are useful in quality control, as you can see straightaway if a result is unusually high or low.

A *z*-score is also sometimes called a ***z*-value** or a **standardised score**.

Jess and Nish take an IQ test. Jess has a *z*-score of 0 and Nish has a *z*-score of 2.4. What are their IQs?

Worked example 8C

A running coach records the distances in kilometres that each of his trainees runs in a week.

The mean distance for the trainees was 65 km and the standard deviation was 10 km.

The sports coach decides that any trainee who is more than:

- 1 standard deviation below the mean should attend extra training
- 2 standard deviations above the mean should represent the team at the next running event.

(a) Calculate the *z*-score for each of these trainees:

(i) Tamsin: 58 km
(ii) Amir: 88 km
(iii) Peter: 47 km
(iv) Gemma: 84 km.

(b) Which trainees need to attend extra training?
Do any of these trainees get to represent the team at the next event?

(c) Kai has a *z*-score of 1.4.
What distance did he run that week?

Solution

(a)

Student	Calculation	*z*-score
Tamsin: 58 km	$\frac{58-65}{10}$	−0.7
Amir: 88 km	$\frac{88-65}{10}$	2.3
Peter: 47 km	$\frac{47-65}{10}$	−1.8
Gemma: 84 km	$\frac{84-65}{10}$	1.9

(b) Peter needs extra training as his *z*-score is less than −1.
Amir will represent the team as his *z*-score is more than 2.

(c) Kai has a *z*-score of 1.4, so: $\frac{x-65}{10} = 1.4$

$$x - 65 = 14$$
$$x = 79$$

Kai ran 79 km that week.

Exercise 8C

1 **(a)** Calculate z-scores for the following data values.

  (i) $x = 5, \mu = 7$ and $\sigma = 1$ **(ii)** $x = 30, \mu = 40$ and $\sigma = 5$ **(iii)** $x = 110, \mu = 106$ and $\sigma = 2.5$

(b) Calculate x for the following z-scores.

  (i) $z = 1, \mu = 10$ and $\sigma = 1$ **(ii)** $z = -2, \mu = 25$ and $\sigma = 3$ **(iii)** $z = -2.4, \mu = 112$ and $\sigma = 30$

2 A machine in a factory is calibrated to weigh out bags of crisps.

The mean mass of crisps in a packet is 40 g.
The standard deviation is 0.5 g.

An inspector checks the machine and finds the last 8 bags have a mass of:

39.4 g	40.2 g	41 g	39.8 g
38.0 g	40.3 g	38.4 g	40.6 g.

(a) Calculate z-scores for each bag of crisps.

(b) Does the machine need re-calibrating? Give a reason for your answer.

8D The Normal probability plot

In reality, it can be hard to tell if a set of data has a Normal distribution. The histogram may not be perfectly symmetrical and the data may not precisely follow the 68%–95%–99.7% rule.

A Normal probability plot is used to check whether a set of data fits a Normal distribution.

The data are plotted against a theoretical Normal distribution. If the data fit this theoretical distribution then the resulting graph is a straight line.

Activity 8D Is a distribution Normal?

Lily is an ornithologist. She measures the wingspan of a sample of golden eagles.

Here are her figures (in metres):

1.97	2.05	1.98	2.00	2.05	1.83
2.05	1.99	2.01	1.90	1.85	2.08
1.99	2.10	1.94	1.84	1.92	1.94.

Lily uses a spreadsheet to process her data, including drawing the Normal probability plot.

What does the graph tell Lily?

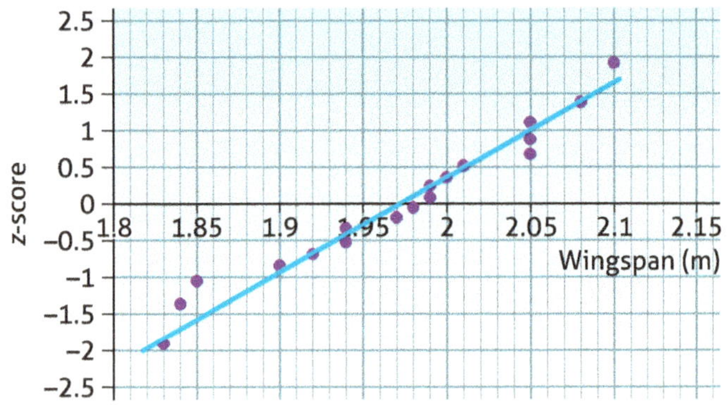

Worked example 8D

Lily's colleagues give her some more data. Lily adds the new data to the figures that are already in the spreadsheet and uses them to redraw the Normal probability plot, as shown in (a).

(a) Use the Normal probability plot to estimate (i) the mean and (ii) the standard deviation of the wingspan of a golden eagle.

(b) Suggest a reason why the Normal probability plot does not fit a straight line as well for extreme values.

Solution

(a)

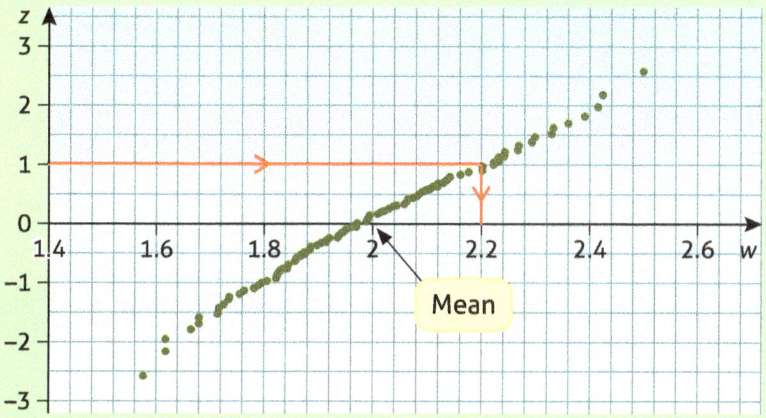

(i) The mean wingspan has a z-value of 0, so the mean is about 2 m.

(ii) Reading off from $z = 1$ gives 2.2 m.
So, the standard deviation is about 2.2 m − 2 m = 0.2 m.

(b) The points representing large wingspans are not far off the straight line but two of those for small wingspans are to the right of it. The smallest eagles in the sample were not quite as small as might have been expected.

Drawing a Normal probability plot is not very straightforward and you would expect to use suitable software to do it. Enter the data and the software gives you the graph. At this stage, what matters most is that you can interpret the graph once you have it.

A word of caution. Not all software programs number the axes in exactly the same way. In the graph above, the horizontal axis is marked in actual values and the vertical axis in z-values. Sometimes you may see z-values on the horizontal axis. The really important point is that a straight line indicates that your data are drawn from a Normal distribution.

However, it is also helpful to know what the graph is telling you when the points don't lie on a straight line. Here are some examples.

Distribution | **Typical Normal probability plot**

Positively skewed

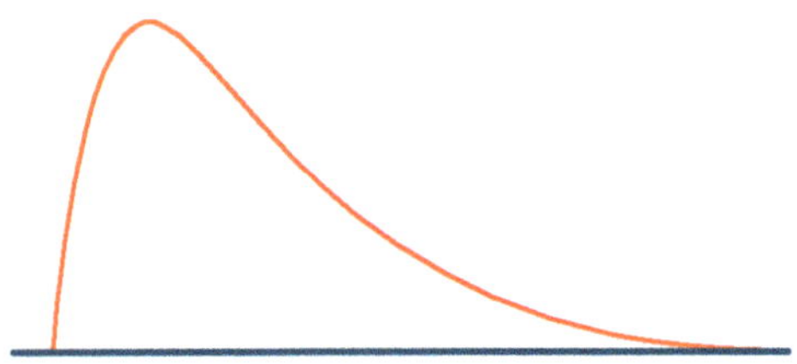

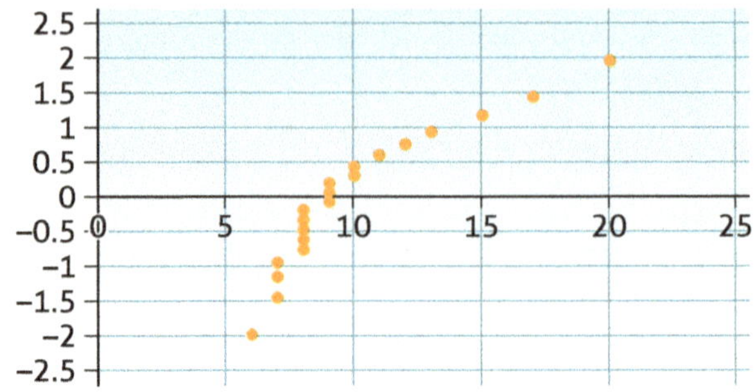

Negatively skewed

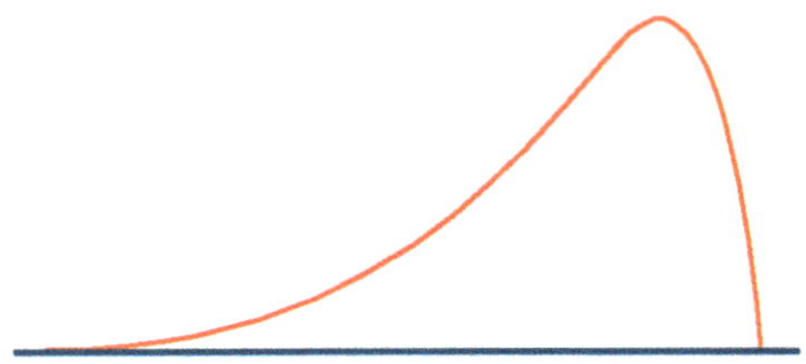

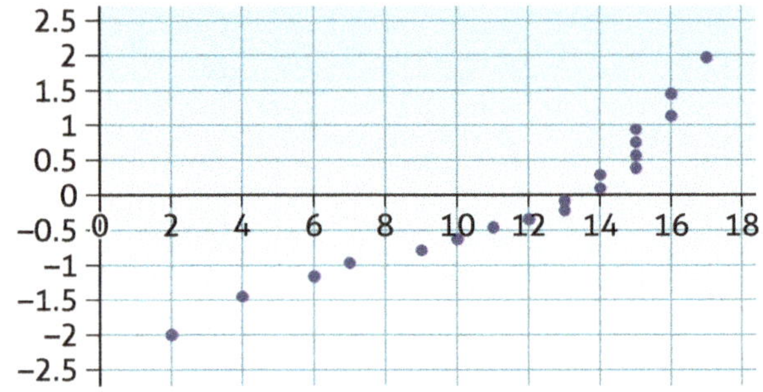

Bimodal

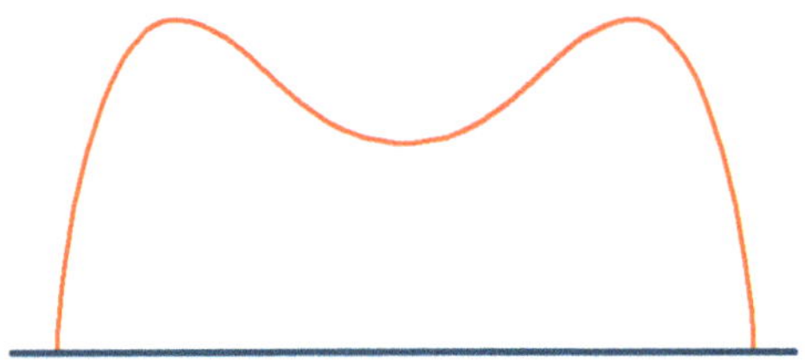

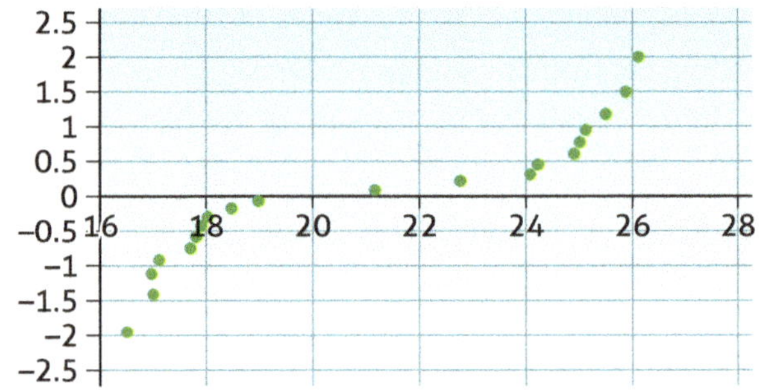

Exercise 8D

1 A personal trainer records the resting heart-rate of her clients. She draws a Normal probability plot of her data.

(a) What feature of the graph indicates that the distribution of h is Normal?

(b) Estimate the mean resting heart rate.

(c) Estimate the standard deviation.

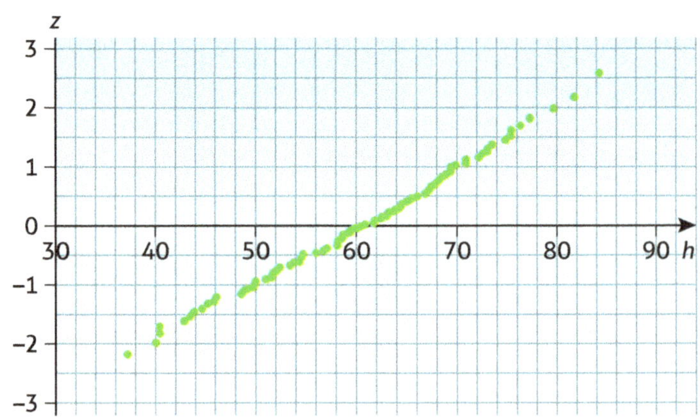

Test yourself 8E

1 (a) Which of the charts below could indicate a Normal distribution?

(b) What features of the other distributions tell you that they are not Normal?

Chart A

Chart B

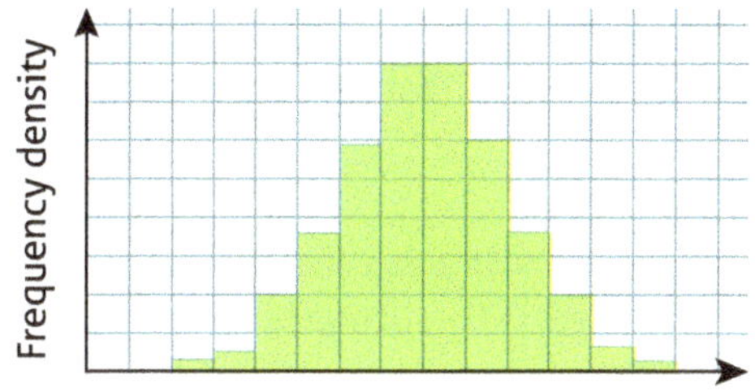

Chart C

Chart D

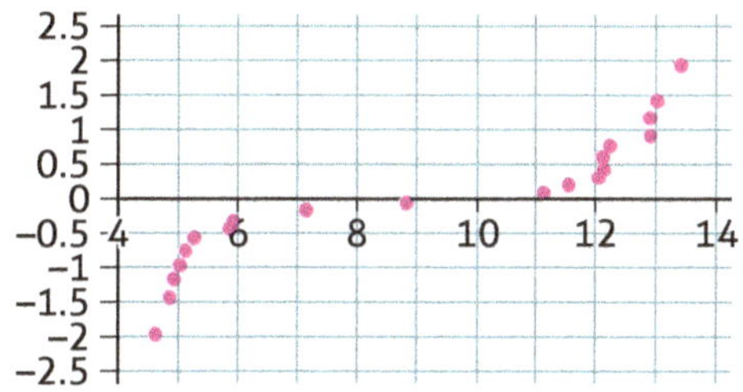

Chart E

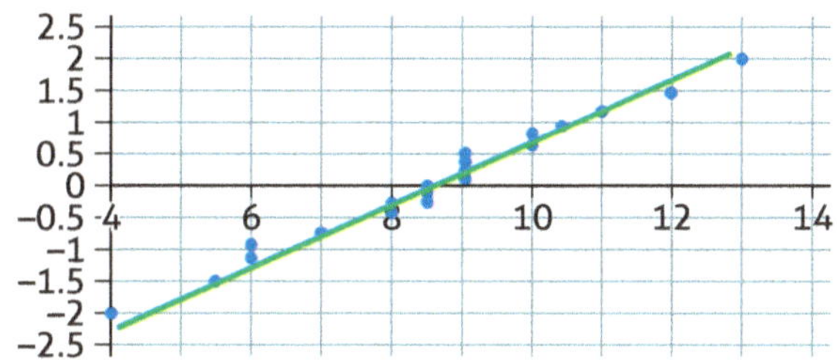

Chart F

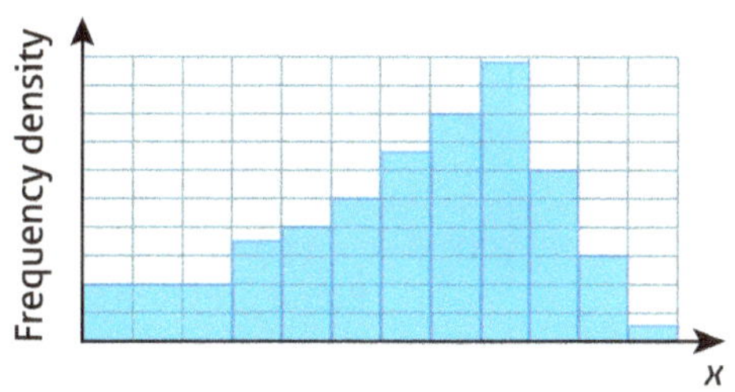

2 A survey of some basketball players suggests their heights have a Normal distribution.

A sports coach draws a frequency chart to show the heights of 60 players.

(a) State why this chart is not a histogram.

(b) Identify two features of the graph that are consistent with a Normal distribution.

(c) Write down the mean of the distribution.

(d) Show that the heights of the bars in the frequency chart are consistent with a standard deviation of 5 cm.

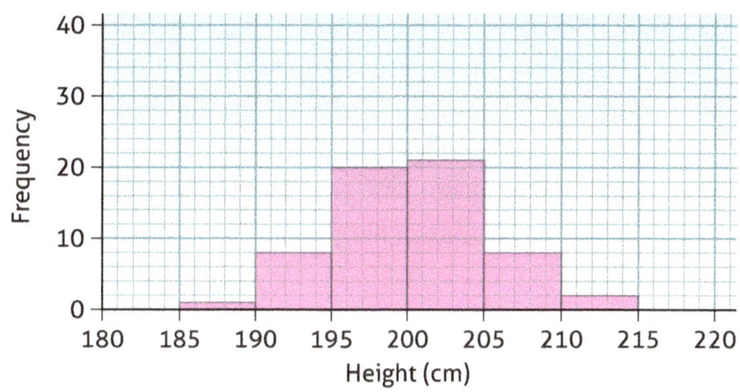

3 A tyre-fitting company times how long its workers take to fit tyres on a car.

It finds the mean time to change a tyre is 7 minutes and the standard deviation is 1.5 minutes.

(a) Calculate the z-scores for workers who take:

(i) 4 minutes (ii) 9 minutes.

(b) Sam can change a tyre in 3 minutes.

Do you think that Sam should get a pay rise? Explain your answer.

4 A doctor is assessing the memory of some of her patients.

She shows each patient a series of 40 cards with pictures of common objects on them.

After 10 minutes she checks to see how many objects each patient can remember.

The number of objects, n, recalled by each patient is illustrated in this Normal probability plot.

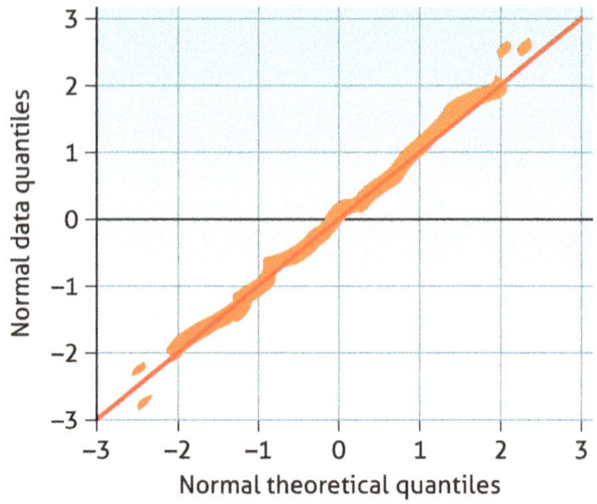

In this graph both axes are marked with z-values. It was drawn using different software from the other plots in this chapter.

(a) What feature of the probability plot suggests that the Normal distribution is an appropriate model?

The mean number of items her patients recalled was 20 with standard deviation 4.

(b) Mr Doyle remembers 11 objects correctly.

Should the doctor be concerned about his memory?

Strictly speaking, the Normal distribution is for continuous data rather than discrete. However, where the values are large, as here, it is common to use it as a model for discrete data.

5 A doctor collects data on her patients' blood pressure.

She finds that the systolic pressure (when your heart pushes blood out) of her patients has a Normal distribution.

The Normal curve shows the distribution of blood pressure in her patients.

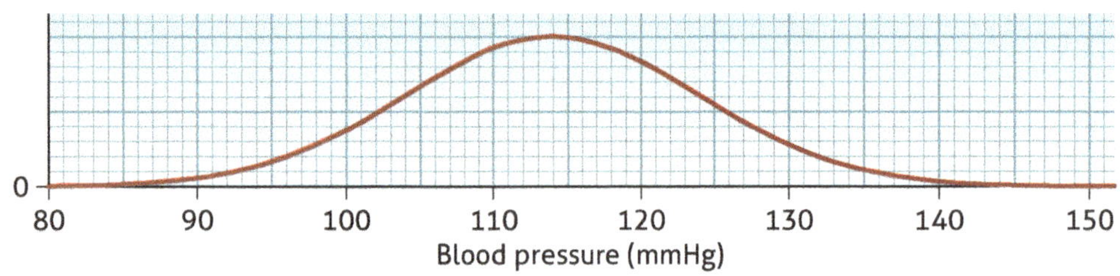

(a) Estimate the mean blood pressure.

(b) Estimate the standard deviation.

A patient whose blood pressure is more than 2.5 standard deviations above mean has high blood pressure.

(c) Mrs Smith has high blood pressure.

What is the lowest blood pressure she could have?

(d) The doctor decides to monitor any patient whose blood pressure is more than 2 standard deviations above the mean.

What percentage of her patients does the doctor need to monitor?

Practice question

Tom is a poultry farmer. He observes that the weights of the eggs his chickens are laying have a Normal distribution with mean 55 grams and standard deviation 6 grams.

He uses *z*-scores to classify the eggs.

z-scores	Size class of egg
$-2 \leq z < -1$	Small
$-1 \leq z \leq 1$	Medium
$1 < z \leq 2$	Large

(a) Tom selects an egg at random.

Write down the probability that the egg is medium.

Tom rejects any eggs that have a *z*-score of more than 2 or less than −2.

(b) What percentage of eggs does he reject?

(c) Calculate the *z*-score for each of the following eggs and classify them.

Weight of egg	z-score	Size class
55 g		
62 g		
68 g		
46.6 g		

(d) Find the weights of eggs with z-scores of (i) 1.8 (ii) −0.4.

Now that you have completed this chapter, you should know:

- That a histogram may be an approximation to a Normal distribution.
- The shape of the Normal distribution.
- The meaning of standard deviation.
- How to calculate *z*-scores.
- That for a Normal distribution, about 68% of the data lie within 1 standard deviation of the mean, 95% within 2 standard deviations and almost all within 3.
- How to interpret a Normal probability plot.

CHAPTER 9
Working with exponentials

This chapter is about working with very large and very small numbers.

9A Standard form

Large and small numbers are often written in standard form. Standard form is $A \times 10^n$, where A is a number between 1 and 10 and n is a whole number that may be positive or negative.

Standard form is also called scientific notation.

Activity 9A Large and small

(a) The closest star to Earth, Proxima Centauri, is 40 trillion kilometres away. 40 trillion is a very large number!

What other quantities can you think of that are very large?

(b) The diameter of a hydrogen atom is 0.000 000 000 12 metres. 0.000 000 000 12 is a very small number!

What other quantities can you think of that are very small?

Worked example 9A

(a) The distance from the Earth to the Sun is 93 million miles. Write 93 million in standard form.

(b) The volume of a sperm cell is 0.000 000 003 mm^3. Write 0.000 000 003 in standard form.

Solution

(a) 93 million = 93 000 000 = 9.3×10^7. ← 9.3 is a number between 1 and 10. 9.3 is multiplied by 10 seven times.

(b) 0.000 000 003 = 3×10^{-9}. ← 3 is the number between 1 and 10. 3 is divided by 10 nine times.

On a **calculator** display you may see standard form as 9.3E7 or 9.3EXP7; similar representations of negative powers of 10 are 3.1E-9 and 3.1EXP-9.

On a **spreadsheet** these would appear as 9.3E+07 and 3E-09.

	A	B
1	9.3E+07	3.1E-09

Exercise 9A

1 Write these numbers in standard form:

(a) 3500 (b) 1 million (c) 0.000 450.

2 Write out the numbers shown in cells A1 and B1 both in standard form and in full.

	A	B
1	5.67E+08	1.23E-06

9B Calculating with standard form

When you are doing calculations with numbers in standard form, always remember that they have two parts that need to be treated differently.

Activity 9B How would you do it?

Savid does not have a calculator.

He has to calculate $6.4 \times 10^{11} \times 5 \times 10^{-5}$.

(a) How would you advise Savid to work it out?

Dom lends him his calculator.

(b) How would you advise Savid to work it?

Worked example 9B

? Saturn's mean distance from the Sun is 1.434×10^9 km. Assume that Saturn's orbit is approximately circular, and its orbital speed is 3.49×10^4 km per hour. Work out the length of a year on Saturn. Give your answer as a number of Earth days.

Solution

A planet's year is the time it takes to complete one orbit of the Sun. In this case the planet is Saturn.

Length of orbit = $2 \times \pi \times 1.434 \times 10^9$ km

Time taken = $\dfrac{2 \times \pi \times 1.434 \times 10^9}{3.49 \times 10^4}$ hours

= $2.581... \times 10^5$ hours

$\dfrac{2 \times \pi \times 1.434}{3.49} = 2.581...$

$\dfrac{10^9}{10^4} = 10^5$

= $\dfrac{2.581... \times 10^5}{24}$ days

= 10 757.0... days, so 10 760 days, to the nearest 10 days.

⚠ Make sure that you know how to use your calculator for numbers in standard form. You may well come across more complicated calculations where you need to use it.

Exercise 9B

1 The speed of light is 3×10^8 m s^{-1}. A light year is how far light travels in a year.
How far is a light year?
Give your answer in km.

2 On a particular beach the mass of a grain of sand is approximately 2×10^{-6} kg.
There are 8×10^8 grains of this sand in a cubic metre.
What is the mass of a cubic metre of sand?
Give your answer in kg.

9C Exponential growth

Sometimes amounts grow so that after a particular time they have doubled in size.

They double again after the same time.

This is an example of exponential growth.

The amount increases more and more quickly with time.

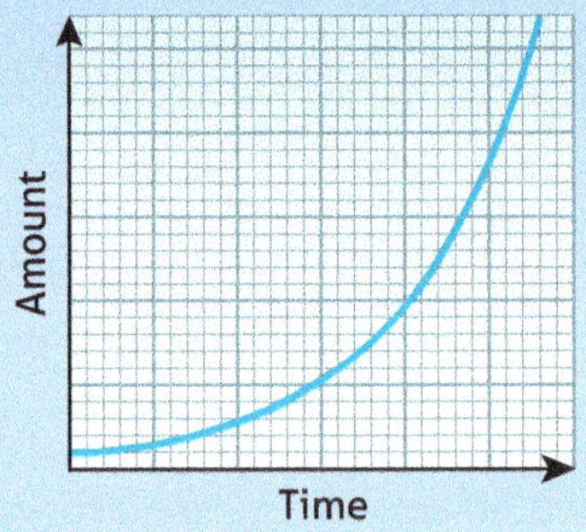

Activity 9C Granny's present

Granny has offered Tom and Sophie a choice of two gifts.

Which would you recommend to Tom and Sophie?

Explain your choice.

Granny's offer 1:

Month 1	£1
Month 2	£2
Month 3	£4

and doubling each month for the rest of the year.

Granny's offer 2:

£100 per month for a year.

Worked example 9C

The population of a small town is growing by 10% every 5 years. Initially, the population is 1000 people.

(a) Plot the graph of $y = 1.1^x$ for $0 \le x \le 8$.

1.1 is the multiplier for an increase of 10%.

Solution

(a) The table of values is obtained by substituting values of x into $y = 1.1^x$ to work out y.

x	0	1	2	3	4	5	6	7	8
y	1	1.1	1.21	1.33	1.46	1.61	1.77	1.94	2.14

$1.1^4 = 1.4641$. This is 1.46 when rounded to 2 decimal places.

Plot the points and join with a smooth curve.

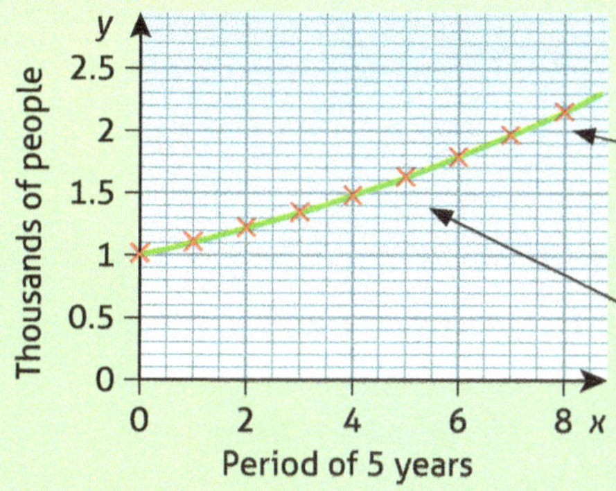

The vertical axis represents 'Thousands of people' and the horizontal axis represents 'Periods of 5 years'.

The graph crosses $y = 2$ here. $y = 2$ represents the amount doubling.

This curve shows exponential growth. The greater the value of x, the steeper it becomes.

(b) Use the graph to find how many years it takes for the population to double. Give your answer to the nearest year.

(b) The graph crosses $y = 2$ when $x = 7.4$. Each value of x represents 5 years. The population doubles after 37 years.

Exercise 9C

1. An investment of £500 grows at an annual rate of 2.5%.
   (a) Plot a graph of $y = 500 \times 1.025^x$. Use values of x from 0 to 30.
   (b) How many years does it take for the investment to double in value?
2. The incidence of a particular disease increases at a weekly rate of 25%. Initially the number of cases is 5.
   (a) Plot a graph of $y = 5 \times 1.25^x$ to show how the number of cases increases over the first 6 weeks.
   (b) How many weeks does it take for the number of cases to triple?

9D Using a spreadsheet for exponential growth

In this section you will see how using a spreadsheet can help you to compare different rates of exponential growth.

Activity 9D Comparing growth rates

The graphs show $y = 1.05^x$, $y = 1.1^x$, $y = 1.2^x$.

(a) Which graph is of $y = 1.05^x$?

(b) Which graph is of $y = 1.1^x$?

(c) Compare how quickly each graph doubles.

(d) Comment on how straightforward it was to answer part (c).

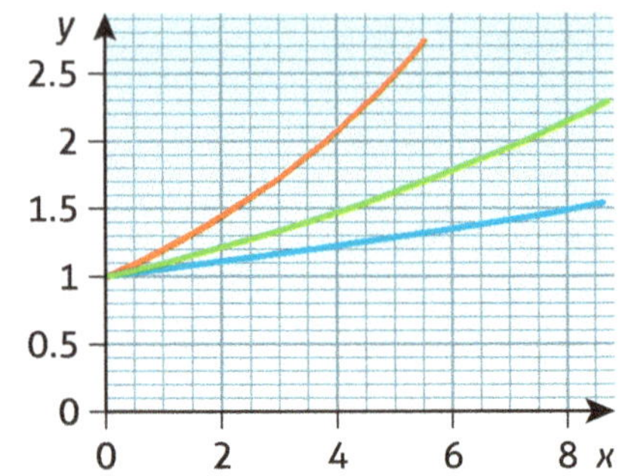

Worked example 9D

Yohsina is comparing the growth of bacteria in different environments.

She estimates that one colony grows at a rate of 1%, another at 2% and a third at 5% each day. Construct a spreadsheet to compare how quickly the number of bacteria in each colony increases by 10%.

Solution

	A	B	C	D
1	Day	1%	2%	5%
2	0	1	1	1
3	1	1.01	1.02	1.05
4	2	1.0201	1.0404	1.1025
5	3	1.0303	1.0612	1.1576
6	4	1.0406	1.0824	1.2155
7	5	1.0510	1.1041	1.2763
8	6	1.0615	1.1262	1.3401
9	7	1.0721	1.1487	1.4071
10	8	1.0829	1.1717	1.4775
11	9	1.0937	1.1951	1.5513
12	10	1.1046	1.2190	1.6289

The spreadsheet shows that the first bacteria in the first colony take 10 days to increase by 10%. In the second colony they take 5 days. In the third colony they take 2 days to increase by 10%.

An increase of 10% is indicated by the number in the cell passing 1.10.

Exercise 9D

1 This spreadsheet shows how an investment, at a rate of 3.5% compound interest per annum, grows over the first 10 years.

(a) After how many years has the investment increased by 10%?

(b) The investment grows by a quarter. During which year does this happen?

(c) Write down the formula in cell B4.

	A	B
1	Year	
2	0	1
3	1	1.035
4	2	1.071
5	3	1.108
6	4	1.146
7	5	1.186
8	6	1.227
9	7	1.269
10	8	1.313
11	9	1.358
12	10	1.405

9E Exponential decay

Sometimes amounts reduce, so that after a particular time they have halved in size.

They halve again after the same time.

This is called exponential decay.

The amount decreases ever more slowly, getting closer and closer to zero but never quite getting there.

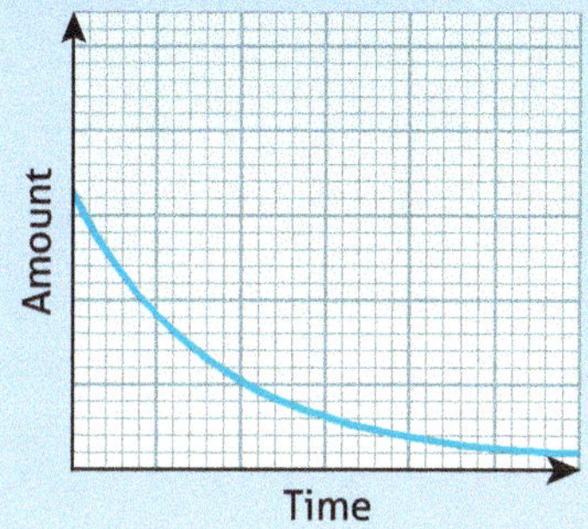

Activity 9E Comparing decay rates

This graph shows $y = 0.5^x$, $y = 1.5^{-x}$, $y = 0.9^x$.

Which graph is which?

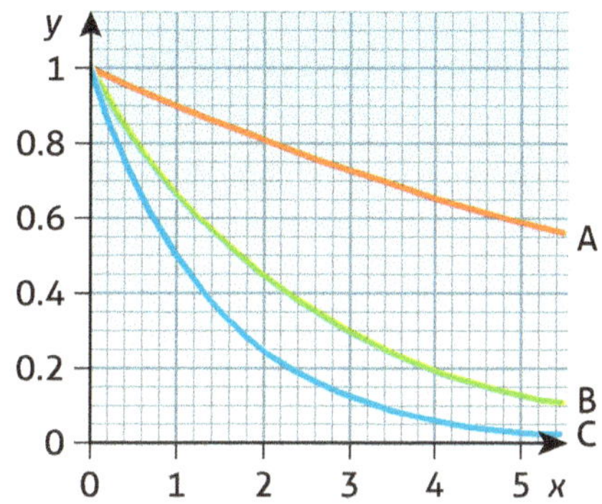

Discussion

(a) Use your calculator to find $\left(\frac{1}{7}\right)^3$ and 7^{-3}. Explain what you find.

(b) How would you use your calculator to find the 7th root of 3?

Worked example 9E

? The graph below shows the percentage of the original amount of radioactive carbon-14 remaining in an object over time.

(a) The half-life of carbon-14 is when 50% of the radioactive material has decayed. Estimate the half-life of carbon-14.

(b) A skull is found with other objects that are known to be 8000 years old. What percentage of radioactive carbon-14 would you expect to find in the skull?

(c) There is 10% of the carbon-14 remaning in an old femur. What does this suggest about the age of the femur?

Solution

(a) 5800 years (actually 5730 years)

(b) 38%

(c) 19 000 years

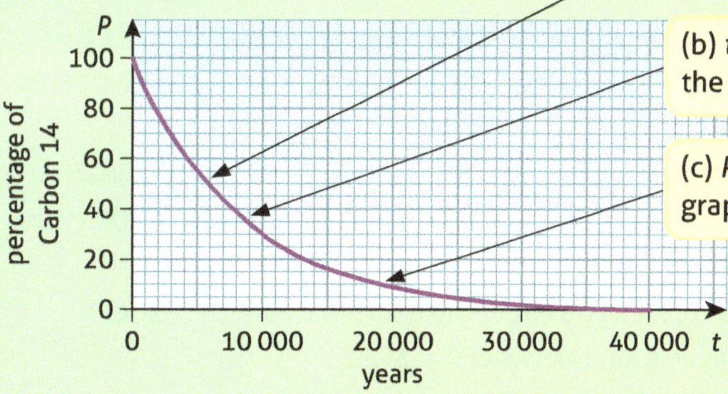

(a) $P = 50$ crosses the graph when $t \approx 5800$.

(b) $t = 8000$ crosses the graph when $P \approx 38$.

(c) $P = 10$ crosses the graph when $t \approx 19\,000$.

Exercise 9E

1 (a) State which of the graphs, A–D, corresponds to each of the equations, (p)–(w).

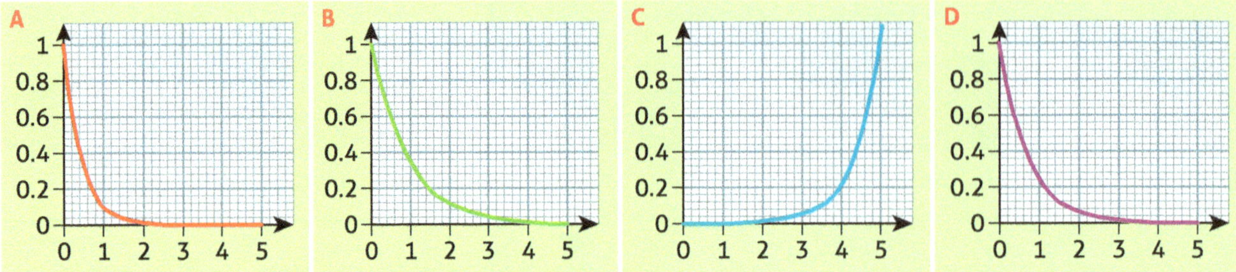

(p) $y = 3^{-x}$	(q) $y = 0.2^{-x}$	(r) $y = 4^{-x}$	(s) $y = 5^{x}$
(t) $y = (\frac{1}{3})^{x}$	(u) $y = 10^{-x}$	(v) $y = 0.25^{x}$	(w) $y = 0.1^{x}$

(b) Which equation has the same graph as $y = 100^{-x}$?

2 The population of Japan is decreasing. The graph on the right shows its predicted size for the next 100 years, starting in 2020. It is based on the assumption that the present annual growth rate of −0.30% continues.

(a) What was Japan's population in 2020?

(b) In what year is the population predicted to fall below 100 million?

(c) (i) Use the graph to predict the percentage reduction in the population from 2020 to 2120.

(ii) Use your calculator to find $(1 - 0.997^{100}) \times 100$.

(iii) Explain why your answers to parts (i) and (ii) are almost the same.

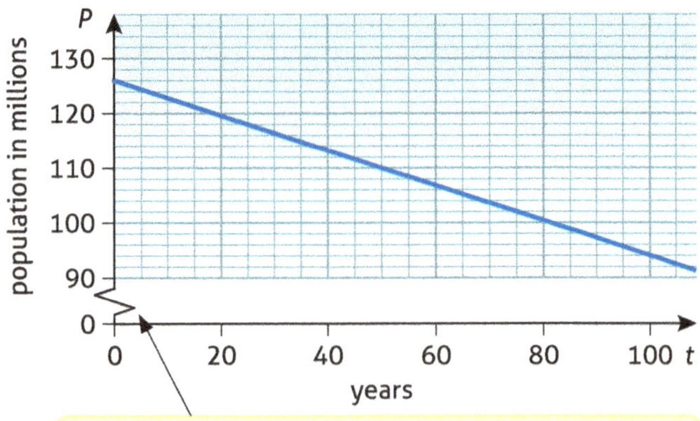

It is usually good practice to have a continuous vertical scale starting at zero. However, sometimes, as here, you only need to see the part of the graph with high values. In such cases use a broken axis symbol like that shown above.

9F Solving exponential equations

So far this chapter has looked at the outcomes from given rates of exponential growth. In this section you will see how to determine the growth rate that produces a given effect.

Activity 9F Half-life

The formula $P = 100 \times 0.945^t$, where t is in years, gives the percentage, P, of a particular substance that remains radioactive at time t.

(a) Find the value of P after: (i) 1 year; (ii) 10 years; and (iii) 100 years.

(b) Estimate the percentage that will still be radioactive after 1000 years.

(c) (i) Find P when $t = 12.25$.

(ii) What does this tell you about the half-life of the substance?

(iii) Explain why P is not zero when $t = 24.5$.

The half-life of a radioactive substance is the time it takes for half of the remaining radioactive material to decay.

Worked example 9F

Afzal invests £1000 in a savings account that offers 3% compound interest per year.
He wants to know how long it will take for his money to double.

He writes down the equation he needs to solve:

$$1000 \times 1.03^x = 2000.$$

He simplifies it to $1.03^x = 2$.

Solve the equation, giving your answer to the nearest year.

Solution

This uses the method of trial and improvement.
$x = 1$ is clearly too small.

x	1.03^x	
10	1.3439	Too small
20	1.8061	Too small
30	2.4273	Too large

For this equation, the target is 2.

x must be between 20 and 30.

Recording the results in a table makes the process more efficient.

25	2.0938	Slightly too large
24	2.0328	Slightly too large
23	1.9736	Slightly too small

x must be between 23 and 24.

23.5	2.003	

23.5 is too large, so the answer is 23 years to the nearest year.

There are other ways to solve this equation. However, they use mathematics that is not part of this course.

Exercise 9F

1. Use trial and improvement to solve the equation $x^4 = 42$, giving your answer correct to 2 decimal places.
2. Kenya has a population of 54 million that is increasing at an annual rate of 2.2% (2020 figures). Assuming this rate continues, the population, P (in millions), t years after 2020, will be given by $P = 54 \times 1.022^t$.
   (a) Use trial and improvement to find when the population will reach 100 million, correct to the nearest year.
   (b) Do you think that the rate is likely to stay at 2.2% for so many years?
3. A radioactive substance decays so that the percentage, P, of radioactive atoms remaining after t years is given by $P = 100 \times 0.86^t$. Use trial and improvement to find its half-life, correct to the nearest 0.01 of a year.

9G Logarithmic scales

When a quantity grows or decays exponentially, it may be difficult to use an ordinary graph; part of it may be too squashed. This section shows how using a different scale for the graph allows you to overcome the problem in cases like this.

Activity 9G What scale?

The table below shows the world population from AD 1 to 2020. Hans wants to draw a graph to show the data.

Year AD	1	1000	1500	1650	1750	1850	1900	1950	2000	2020
Billions	0.2	0.275	0.45	0.5	0.7	1.2	1.6	2.55	6	7.8

(a) What scale would you advise him to use for the population axis?

(b) Hans now wishes to find data for the world population between 50 000 years ago and AD 1. He has to use estimates as the data do not exist!

Is the scale you suggested in (a) still appropriate?

There is no Year 0. The year after 1 BCE is AD 1 with no 0 BCE or AD 0 in between.

Worked example 9G

The graph shows the Richter scale for measuring earthquakes. An earthquake of magnitude 5 is ten times as powerful as one of magnitude 4.

(a) How many times as strong is an earthquake of magnitude 6 as an earthquake of magnitude 4?

(b) Xavier says that the answer to 'What does it mean for an earthquake to have magnitude 2.5?' is $\frac{100 + 1000}{2} = 550$.

Why is he wrong?

(c) How much stronger is an earthquake of magnitude 2.1 than one of magnitude 2.0?

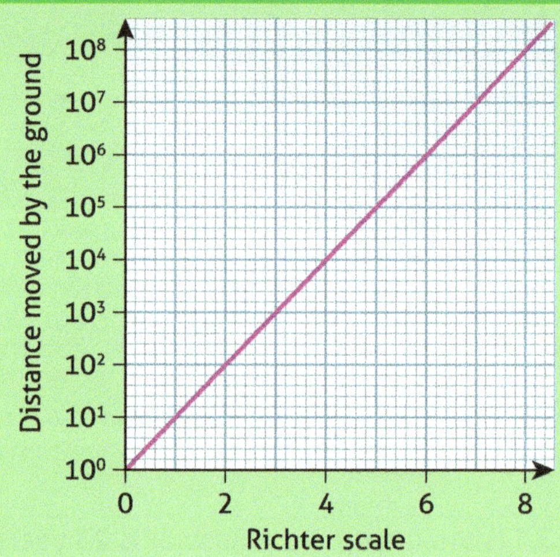

Solution

(a) $\frac{\text{magnitude of 6}}{\text{magnitude of 4}} = \frac{10^6}{10^4} = 10^2$

The earthquake of magnitude 6 is 100 times as strong as an earthquake of magnitude 4.

(b) A magnitude of 2.5 is the average of magnitudes 2 and 3.

The logarithmic scale requires using the geometric mean to calculate this average.

Geometric mean $= \sqrt{10^2 \times 10^3} = \sqrt{10^5} = 10^{2.5} = 316$

Xavier's mistake was to calculate the arithmetic mean.
He got $\frac{1}{2}(100 + 1000) = 550$ and this was wrong.

On a logarithmic scale the halfway point between values a and b is the **geometric mean**
$= \sqrt{a \times b}$
$= \sqrt{ab}$.

(c) Comparing a magnitude of 2.1 with a magnitude of 2:

$\frac{\text{magnitude of 2.1}}{\text{magnitude of 2}} = \frac{10^{2.1}}{10^2} = 10^{0.1} = \sqrt[10]{10}$

An earthquake of magnitude 2.1 is $\sqrt[10]{10} = 1.26$ times as strong as an earthquake of magnitude 2.

Sound is another quantity that is measured on a logarithmic scale. The loudness of a sound is measured in decibels (dB). A noise of 60 dB is 10 times louder than one of 50 dB. Exposure to a sound louder than 85 dB risks damaging your hearing over time.

Exercise 9G

1 A rock concert is planned and nearby residents raise objections as they know that a rock concert is likely to exceed the safe limit, with noise around 120 dB.

The organiser claims that 120 dB is less than half as much again as the normal safe limit of 85 dB.

Explain why the organiser is wrong.

2 The growth of a colony of bacteria is shown on the graph on the right:

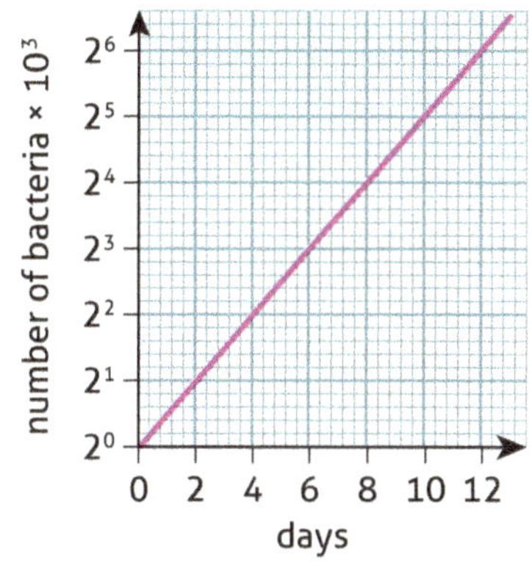

(a) How many bacteria are there at the start?

(b) How long does it take for the number of bacteria to double?

(c) How many bacteria are there after 3 days?

(d) How many bacteria are there after 9 days?

3 Marie uses a spreadsheet to calculate the arithmetic mean, and the geometric mean, of two numbers.

The two numbers are in cells A1 and A2.

	A	B	C
1	6		
2	8	7	6.928203

(a) What formula is in cell B2?

(b) What formula is in cell C2?

The means are similar, with the arithmetic mean larger than the geometric mean.

(c) Is this still the case for the numbers 6 and 9? Justify your answer.

(d) The value of A1 remains 6. Investigate whether you can find a new value for A2 such that C2 > B2.

Test yourself 9H

1 A particular virus is spherical with diameter 80 nanometres.

(a) Write the radius of the virus in metres, as a decimal.

(b) Work out the volume of the virus, giving your answer in cubic metres.

1 nanometre = 10^{-9} metres.

The volume of a sphere of radius r is given by $V = \frac{4}{3}\pi r^3$.

2 This graph illustrates the radioactive decay of uranium 238.

(a) Use the graph to estimate the half-life of uranium 238.

(b) What percentage of the original amount of radioactive material remains after 9×10^9 years?

Half-life is the time taken for half the radioactive material to decay.

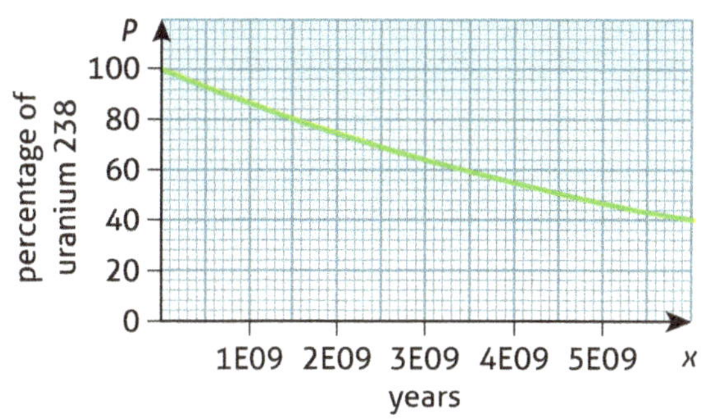

3 A disease spreads among a population of wild rabbits.
The number of cases doubles every 5 days.

(a) Sketch a graph to show the number of cases for up to 20 days, given there were 10 cases initially.

The equation of the curve is $N = 10 \times 2^{0.2d}$, where N is the number of cases and d is the number of days.

(b) Explain what the number 10 in the equation represents.

(c) Explain what the number 0.2 in the equation represents.

(d) Write down the equation that you would solve to work out when the number of cases would pass 1 million.

(e) Solve the equation.

4 Jeff starts a business. It grows by 30% each year. The graph shows the turnover for the first 20 years.

(a) What type of scale is used on the vertical axis?

(b) Why is it used in this case?

(c) Use the graph to estimate how long it took the business to reach a turnover of £10 million.

(d) Estimate the turnover of the business after 13 years.

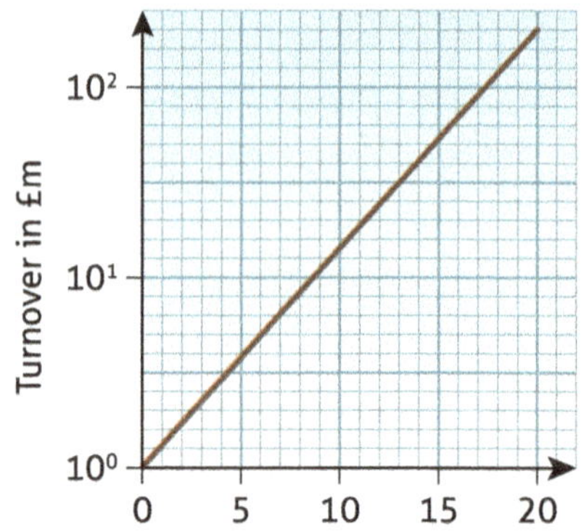

5 A colony of bats has declined over recent years. A naturalist models their population on this spreadsheet, with a reduction of 10% each year.

(a) According to this model, what percentage of bats remain after 10 years?

(b) In which year are there only half the original number of bats?

(c) What formula is in cell B4?

(d) The numbers in column B are given to 1 decimal place. Comment on whether this is appropriate.

(e) The actual figure for Year 8 turns out to be 70%. What can you conclude from this?

	A	B
1	**Year**	**Percentage of bats**
2	0	100
3	1	90
4	2	81
5	3	72.9
6	4	65.6
7	5	59
8	6	53.1
9	7	47.8
10	8	43
11	9	38.7
12	10	34.9
13	11	31.4
14	12	28.2
15	13	25.4
16	14	22.9
17	15	20.6

Practice question

A typical adult human contains 39 trillion bacteria.

The average length of one of these bacteria is 1 micrometre.

(a) Write 39 trillion in standard form.

(b) Write 1 micrometre in metres, in standard form.

The distance from the Earth to the Moon is 400 thousand kilometres.

(c) If all the bacteria in an adult human were laid end to end, would the line reach the Moon? Give suitable calculations to justify your answer.

A trillion is a million million. There are 1,000 micrometres in 1 millimetre.

Now that you have completed this chapter, you should know:

- How to interpret large and small numbers in standard form.
- How to do calculations with numbers in standard form.
- How to interpret output from a spreadsheet including numbers in standard from.
- How to represent and interpret exponential growth and decay in a graph.
- How to solve equations of the form $x^5 = 35$ and $1.05^x = 8.2$.
- How to use and interpret a logarithmic scale on a graph.

CHAPTER 10

Measures and scaling

This chapter is about the shapes you are likely to meet in everyday life and how to interpret typical measurements associated with them.

How big is a cubic metre?

Think of the outline of a cube that measures 1 metre on each side.

- How many centimetres are there along each edge?
- How many square centimetres are there on each face?
- How many cubic centimetres are there in the cube?
- Could you fit inside a cubic metre?

10A Similar shapes: Area and volume

This section is about the relationship between the lengths, areas and volumes of shapes.

Activity 10A Are chunky chips better for you?

The amount of fat absorbed by chips depends on their surface area.

Model chunky chips as cuboids measuring 1.5 cm by 1.5 cm by 6 cm, and skinny fries as cuboids measuring 0.5 cm by 0.5 cm by 8 cm.

(a) Here is the outline of a calculation for the surface area of 100 grams of chunky chips. Fill in the missing numbers.

150 cm^3 of cooked potato weighs 100 g.
Volume of one chunky chip = $1.5 \times 1.5 \times 6 = 13.5$ cm^3.
Number of chunky chips in 100 g = 150÷ ... = 11(.1).
Surface area of one chunky chip = $4 \times 1.5 \times 6 + 2 \times 1.5^2 = \ldots$ cm^2.
Surface area of 100g of chunky chips = $40.5 \times 11(.1) = \ldots$ cm^2.

Volume of a cuboid is length × width × height.

Surface area of a cuboid is the sum of the areas of its 6 faces.

(b) What is the total surface area of 100 grams of skinny fries?

(c) What do your answers tell you about which is better for you?

When a shape is enlarged, the shape and its image are similar: corresponding lengths are in the same ratio; corresponding angles are equal.

- The **scale factor of the enlargement** is the amount by which the lengths in the image are larger than the corresponding lengths in the original object.
- The **scale factor for the area** is the square of the scale factor for the enlargement.
- The **scale factor for the volume** is the cube of the scale factor for the enlargement.

Worked example 10A

? Cathy is designing a range of toy dinosaurs to sell. She plans to offer them in 3 sizes and needs to determine the cost of materials for each in order to establish a price point for the toys.

The heights of the dinosaurs will be 10 cm, 15 cm and 20 cm, and they will be all the same design. For the 10 cm dinosaur, the paint costs 60p and the resin costs 80p.

Determine the cost of materials for the larger sizes.

Solution

The scale factors for the heights are:

$\frac{15}{10} = 1.5$ for the 15 cm dinosaur and $\frac{20}{10} = 2$ is for the 20 cm dinosaur.

Cost of paint: The amount of paint depends on the area of the dinosaur.
15 cm dinosaur $1.5^2 \times 60\text{p} = £1.35$
20 cm dinosaur $2^2 \times 60\text{p} = £2.40$

The scale factor for the area is the square of the scale factor for the height.

Cost of resin: The amount of resin depends on the volume of the dinosaur.
15 cm dinosaur $1.5^3 \times 80\text{p} = £2.70$
20 cm dinosaur $2^3 \times 80\text{p} = £6.40$

The scale factor for the volume is the cube of the scale factor for the height.

Total cost of materials:
10 cm dinosaur £0.60 + £0.80 = £1.40
15 cm dinosaur £1.35 + £2.70 = £4.05
20 cm dinosaur £2.40 + £6.40 = £8.80

Exercise 10A

1 Dan is a scientific photographer. He has a rectangular photograph of an ant, taken using a microscope, where the image measures 2.5 cm wide. He enlarges it to a width of 17.5 cm. The printed area of the original photograph is 10 cm². What is the printed area of the enlargement?

2 Ayodele is commissioned by Monty to make a 1-metre tall sculpture of his pet dog. He makes a scale model 20 cm high, and it weighs 1.8 kg. What does the finished sculpture weigh?

3 Apinya plans to create a statue of Buddha. The statue is to be made from 50 kg of clay and covered with gold. She calculates the cost of the gold and discovers it is far too expensive. She says, 'If I only use 25 kg of clay it will halve the cost of the gold.'

Explain why Apinya is wrong.

10B Using suitable units

The importance of getting your units right was demonstrated to the world in 1999 when NASA lost a $125 million Mars Orbiter because one team working on it used imperial units of measurement while another used the metric system.

Activity 10B Happiness

This chart appeared in a national social survey, together with the caption:

'Wealth and happiness are connected but only up to a point.'

(a) Is the caption statement supported by the chart?

(b) What exactly is the chart showing?

(c) Is any information missing?

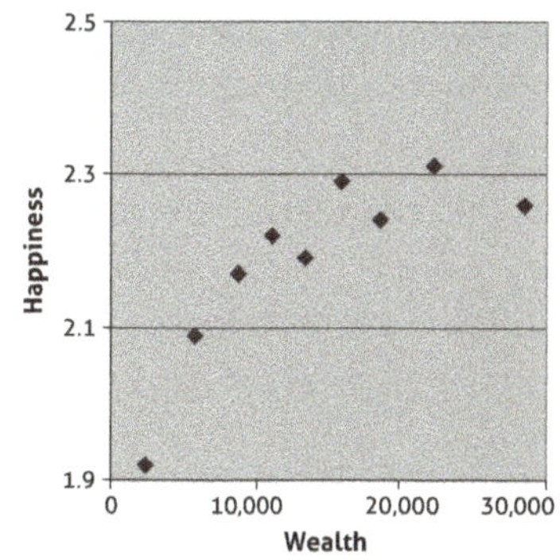

Adapted from Income and happiness in the US, 1972–4. Reprinted from the National Opinion Research Center's General Social Survey, with permission

Worked example 10B

? Petrol-driven cars cause pollution. They emit carbon dioxide, which builds up in the atmosphere causing global warming.

They also emit oxides of nitrogen, which affect breathing and can be fatal for people with respiratory problems.

Here are the figures for a popular medium-sized car as published by the UK Government:

Petrol consumption	5.8 litres per 100 km
Carbon dioxide	133 grams per kilometre
Oxides of nitrogen	15 milligrams per kilometre

A similar car of a different make recorded these results for a journey of 7 km under test conditions:

Petrol consumption	0.407 litres
Carbon dioxide	930 grams
Oxides of nitrogen	0.105 grams

How does the similar car's petrol consumption and pollution compare with the first car?

Solution

Quantities can only be compared if they have the same units.

Petrol consumption: For the first car the unit of petrol consumption is given as litres per 100 km. The second car uses 0.407 litres in 7 km.

So, it uses $\left(\frac{0.407}{7}\right)$ litres per km or $\left(\frac{0.407}{7}\right) \times 100 = 5.8$ litres per 100 km (to 1 dp).

This is exactly the same figure as for the first car.

Carbon dioxide: For the first car the unit of CO_2 emission given is grams per kilometre. The second car emits 930 grams in 7 kilometres.

So, it emits $\left(\frac{930}{7}\right) = 133$ grams per km (to the nearest whole number).

This too is the same as for the first car.

Oxides of nitrogen: For the first car the unit for these emissions is milligrams per kilometre. For the second car, the emissions are given as 0.105 grams per 7 kilometres.

This is $\left(\frac{0.105 \times 1000}{7}\right) = 15$ milligrams per km.

So, this also is the same as for the first car.

It is now clear that the two cars have the same fuel consumption and emission levels.

Exercise 10B

1 There are two measurements for heavy loads: the tonne and the ton.
A tonne is 1000 kg and a ton is 2240 pounds. A pound is 0.454 kg.
Find which is heavier: a 7-tonne truck or a 7-ton truck.

2 Anil keeps a record of how much petrol his car uses.
It works out at 0.148 litres per 1 mile.
He needs to convert this into litres consumed per 100 km.
He knows these approximate conversions:

1 gallon = 4.55 litres	1 litre = 0.22 gallons
1 mile = 1.6 km	1 km = 0.63 miles.

He performs this calculation to work out litres of petrol consumed per 100 km:

0.148 litres to travel 1 mile
which is 0.148 litres to travel 1.6 km
$\frac{0.148}{0.148}$ *litres to travel* $\frac{1.6}{0.148}$ *km*
= 10.8 (rounded to 1 dp).

(a) What has Anil actually calculated?
(b) Calculate the correct answer.

10C Scale drawings

It is often helpful to draw a scale drawing, particularly when you are working with lengths and areas in 2 dimensions. This is often the case when there are regulations to be met.

Activity 10C Placing a carbon monoxide detector

A carbon monoxide detector must be fitted on a wall and be at least 1 metre away from any fuel-burning appliance, such as a boiler, cooker or fire.

Where could a carbon monoxide detector be placed in this kitchen?

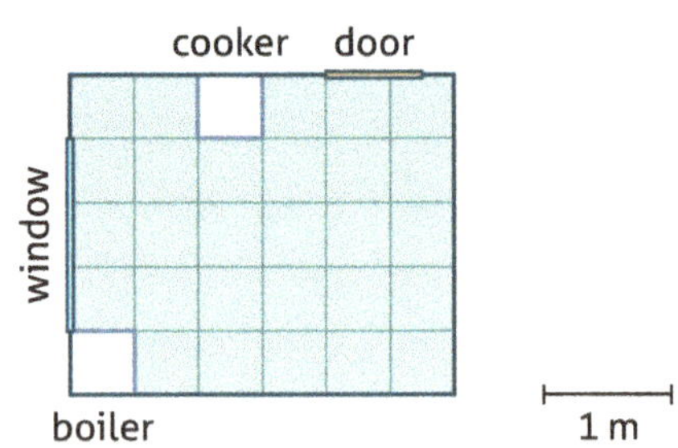

Worked example 10C

There are restrictions on where trees can be planted near houses:

- medium trees, > 30 feet high, must be more than 15 feet from the house
- small trees, ≤ 30 feet high must be more than 8 feet from the house.

The trees must also be more than 10 feet away from water pipes, and 20 inches from the boundary.

The diagram shows a house, measuring 28 feet by 15 feet, within a plot measuring 75 feet by 25 feet. The dotted lines represent a water pipe.

The owner wishes to plant three medium trees. Identify suitable positions for the trees.

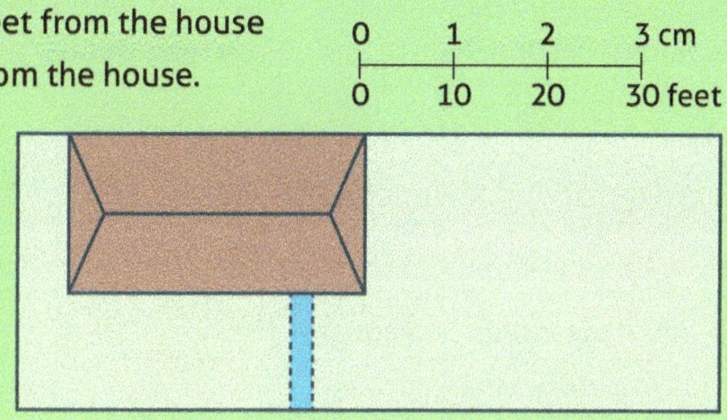

Solution

Medium trees must be more than 15 feet from the house.

None can be placed alongside the house.

The space at the side is only 10 feet wide.

There must be at least 20 inches between the trees and the boundary, so allow 2 feet (24 inches).

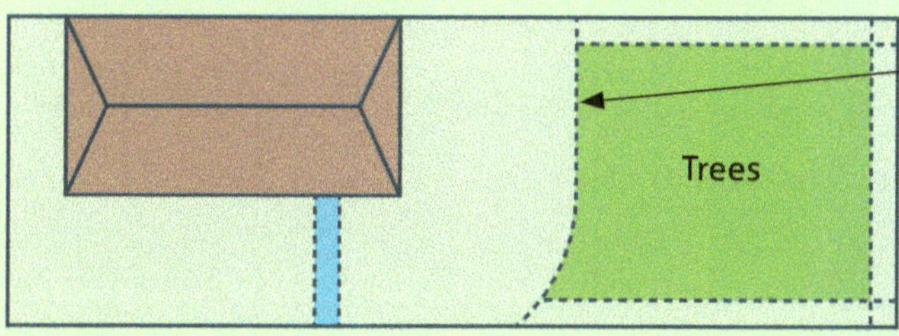

Each point on this line is 15 feet away from the house.

The pipe is within 15 feet of the house, so does not affect where the trees can be planted.

The trees can be placed anywhere in the shaded area.

Exercise 10C

1 Kevin is designing his garden. He draws a scale drawing with a scale of 1:100. The area around the vegetable plot is covered with gravel.

(a) Write down the dimensions of the shed.

(b) Write down the dimensions of the patio.

(c) The vegetable plot is a raised bed measuring 1.5 cm by 2.5 cm on the scale drawing. It has a depth of 30 cm in real life. How much soil is needed to fill it?

(d) What is the area of gravel in the garden?

(e) Lawn fertiliser is applied at a rate of 35 g per m^2. What weight of fertiliser is needed for one treatment of the lawn?

(f) Both the border and vegetable plot need weeding. What is the longest distance that Kevin will have to reach without treading on the soil?

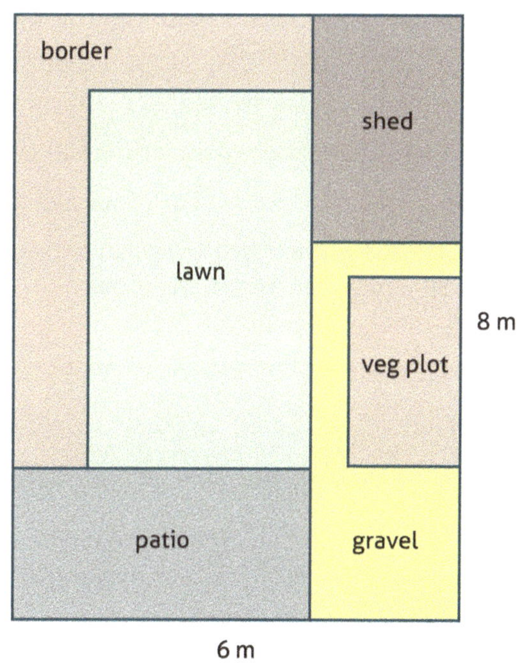

10D Health and safety

You have just encountered two situations where regulations had to be met. The theme is continued in this section with some health and safety regulations involving distances, areas and volumes.

Activity 10D Fit for a wheelchair

Workplaces are obliged to make their spaces accessible.

An office measures 3 m square.

It has one door and a window.

It requires furniture:
work desk; filing cabinet; cupboard; chair.

A typical wheelchair measures 710 mm wide and 1065 mm long.

Make a scale drawing and arrange the office so that the wheelchair can access all the necessary areas and navigate through the door.

Make assumptions, where needed, to solve the problem.

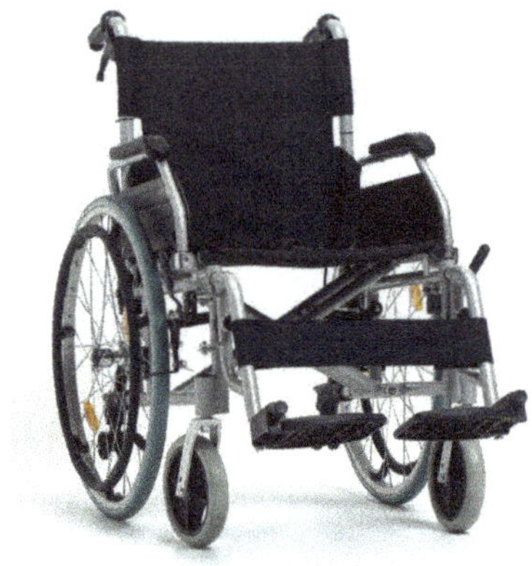

Worked example 10D (1)

A small charity is looking for office accommodation. They find a room that measures 6 m by 3.5 m. It is 2.7 m high.

Regulations stipulate that each worker requires 11 m^3 of space and that a height above 3 m must be counted as 3 m.

Is there enough room for the five members of staff, according to the regulations?

Would the five members of staff feel there was enough room?

Solution

Volume of room = $3.5 \times 6 \times 2.7$
$= 56.7\ m^3$

Volume of a cuboid is length × width × height.

Five members of staff need $5 \times 11 = 55\ m^3$.

So, there is just enough room for the five members of staff within the regulations.

The floor area = $6 \times 3.5 = 21\ m^2$. Each member of staff has $4.2\ m^2$.

There are desks, cupboards and chairs to accommodate, plus there needs to be enough space for the door to open, so it might be a little cramped!

Worked example 10D (2)

Gradients of roads in quarries for rigid vehicles should not exceed 1 in 10.

In this example 1 in 10 means 1 unit up (or down) for 10 units horizontally. You may come across other situations where it means 1 unit up for 10 units travelled, which is slightly different.

A road in a quarry rises $1\frac{1}{2}$ feet in 4 yards. Does it meet the safety standard?

Solution

4 yards = 12 feet

The Imperial unit of yards is converted to feet. There are 3 feet in a yard.

Gradient = $\frac{1.5}{12} = \frac{1}{8}$
= 1 in 8

The same unit of length must be used in a ratio.

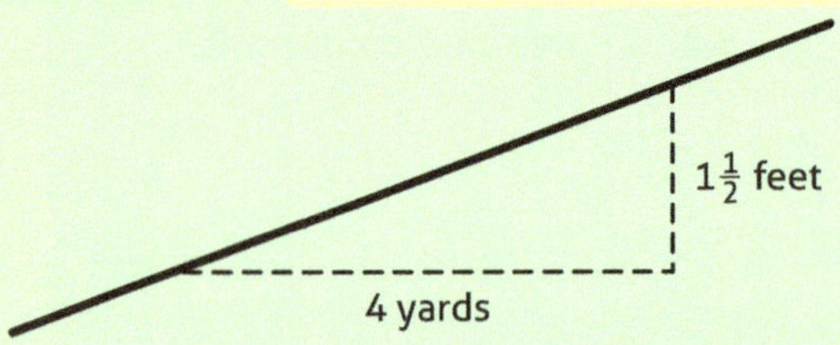

This is steeper than 1 in 10, so does not meet the standard.

Exercise 10D

1 The slope of a ramp must be less than 1 in 12 to conform with health and safety regulations.

> 1 foot = 12 inches = 30 cm

(a) A ramp is 30 feet long and rises 20 inches. Does it satisfy the regulations?

(b) A ramp is 10 m long and rises 3 feet. Does it satisfy the regulations?

(c) Bob is constructing a ramp. It has to rise by 40 cm. What is the minimum length it can be to satisfy the regulations?

2 A small company rents office space consisting of 2 rooms. One room measures 4.6 m by 2.8 m and is 3 m high. The other room measures 1.8 m by 7.6 m and is 3.5 m high.

Regulations stipulate that each worker requires 11 m^3 of space and that a height above 3 m must be counted as 3 m.

How many staff can the space accommodate?

10E Maps

Maps are a familiar example of very accurate scale drawings. This section looks at some of the calculations you may have to do when working with them.

Activity 10E What's in a map?

The map shows part of Alton Towers.

Alton Towers is a theme park and the places marked are rides or other attractions.

The distance from Cloud Cuckoo Land to Oblivion is 200 m.

(a) Estimate the distances between:

(i) Oblivion and The Smiler

(ii) Dark Forest and Wicker Man.

(b) What do you think the green lines represent?

(c) What do you think the blue areas represent?

(d) Estimate the compass bearing of:

(i) Dark Forest from Oblivion

(ii) Oblivion from Dark Forest.

Worked example 10E (1)

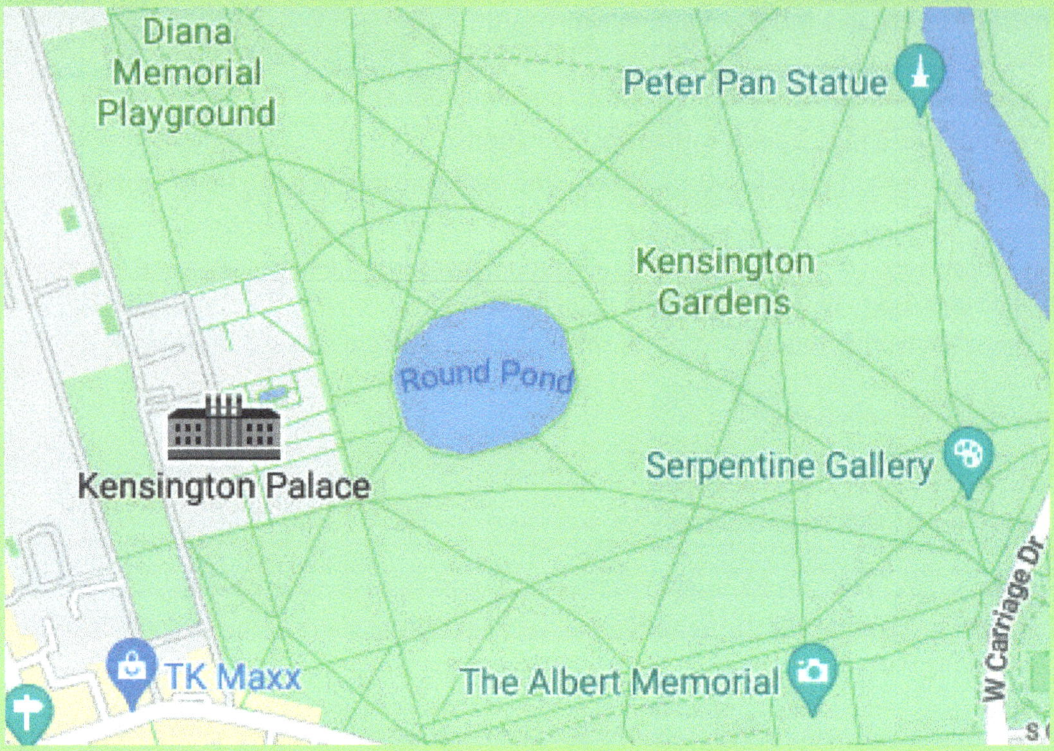

? The map shows Hyde Park, in London. The scale is 1:10 000.

The area of the Round Pond on the map is 3 cm^2.

(a) What is the distance between the Peter Pan Statue and The Albert Memorial in real life?

(b) The width of the Round Pond in real life is 230 m. What is its width on the map?

(c) What is the area of the Round Pond in real life?

Solution

(a) The distance on the map is 8 cm.
The distance in real life = 8 × 10 000 cm
= 80 000 cm
= 800 m.

The real-life distance is larger, so multiply by 10 000.

(b) The width of Round Pond in real life is 230 m.
The width on the map = 230 ÷ 10 000 m
= 0.023 m
= 2.3 cm.

The distance on the map is smaller, so divide by 10 000.

(c) The area of Round Pond in real life = 3 × 10 000 × 10 000 cm^2.
= 300 000 000 cm^2
= 30 000 m^2.

The area scale factor is the square of the length scale factor.

You will often use a map when you are travelling and sometimes you need to distinguish between the amount of ground you have covered and how far you are from your starting point.

- **Distance** is how far you actually travel. If you are walking, it can be measured by the number of steps you take.
- **Displacement** is how far away you are from your starting point. It is 'as the crow flies'.

There are similar distinctions between the words speed and velocity. Both are measures of how fast you are travelling.

- **Speed** is the rate of change of distance; it is how fast you travel over the ground.
- **Velocity** is the rate of change of displacement, so is how quickly you are approaching your final destination.

Worked example 10E (2)

On the right is a picture of a map. The scale of the real map is 2 cm to 5 km.

The distance from Falmouth to St Mawes, by road, is 18 cm on the map. It takes 56 minutes to drive.

The displacement of St Mawes from Falmouth is 1 cm on a bearing of 088° on the map.

(a) Calculate the average speed of the car.

(b) Calculate the average velocity of the car.

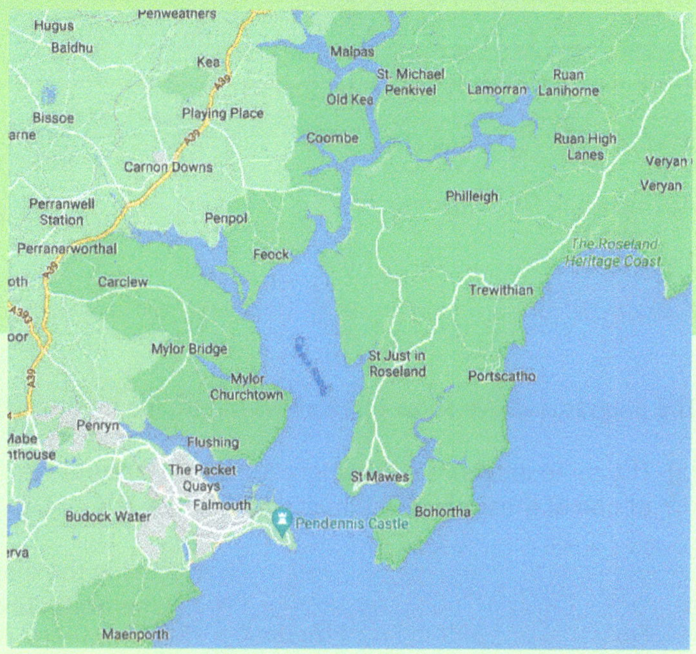

Solution

(a) Total distance = $\frac{18}{2} \times 5 = 45$ km

Average speed = $\frac{45}{56} \times 60$ km per hour

= 48.2 km per hour

(b) Total displacement = $\frac{1}{2} \times 5 = 2.5$ km

Average velocity = $\frac{2.5}{56} \times 60 = 2.7$ km per hour on a bearing of 088°.

Displacement is the **directed** distance from a starting point.

$$\text{Average speed} = \frac{\text{total distance}}{\text{time}}$$

$$\text{Average velocity} = \frac{\text{total displacement}}{\text{time}}$$

Exercise 10E

1 The scale on a map is 1:100 000.

(a) What does 1 cm on the map represent in real life?

(b) What distance in real life is represented by 3.2 cm on the map?

(c) What distance on the map represents 30 km in real life?

2 The scale on a map is 1:50 000.

(a) The perimeter of an area of woodland is 4.5 km in real life. What is the perimeter of the woodland on the map?

(b) The area of the woodland on the map is 5.5 cm². What is the area of the woodland in reality?

3 A robot moves around the edge of a rectangular room. It nearly completes three circuits of the room, stopping 75 cm from its starting point, along the 3.5 m edge.

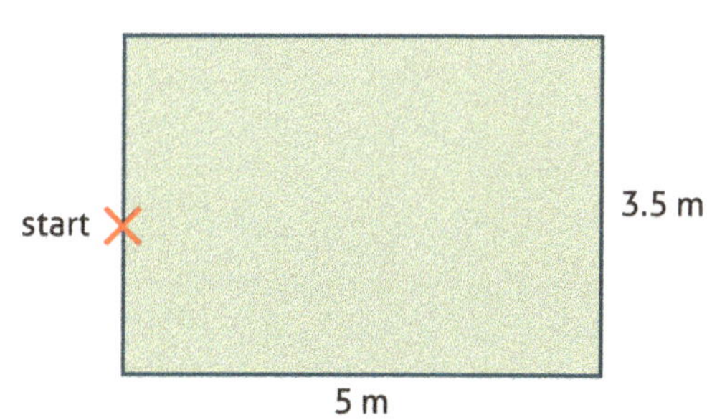

(a) What distance has it travelled?

(b) What is its total displacement?

(c) The journey takes 9.5 minutes. What is the average speed of the robot?

(d) What is the average velocity of the robot?

10F Drawing 3-dimensional objects

The last section of this chapter is about how you can draw a 3-dimensional object on a flat piece of paper. The section covers plans and elevations, and then isometric drawings.

Plans and elevations

One method is to do 3 scale drawings of different views of the object: from the front, one side and above. So, you end up with 3 true shape diagrams. They are called the **front elevation**, **side elevation** and **plan**.

Activity 10F (1) What shape?

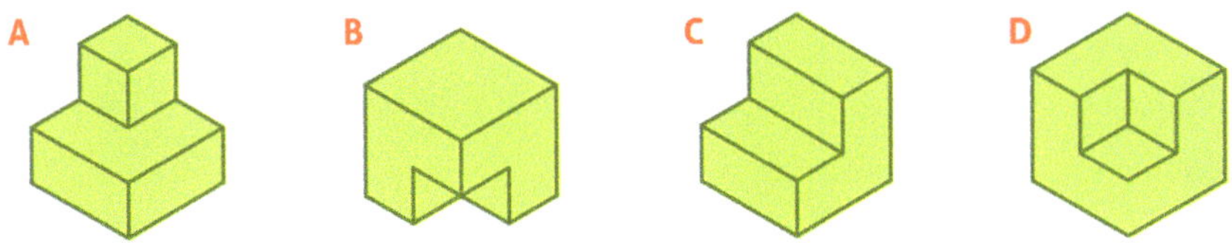

The diagrams above show three views of a shape made of cubes.

(a) The three diagrams on the right are three different views of one of these shapes. Which shape is it: A, B, C or D?

(b) Using the directions shown on the left, one of the three views is the front elevation of the object, another the side elevation and the third is the plan. Say which is which.

Worked example 10F

? Draw the plan, front elevation and side elevation of the triangular prism shown.

The triangle is equilateral.

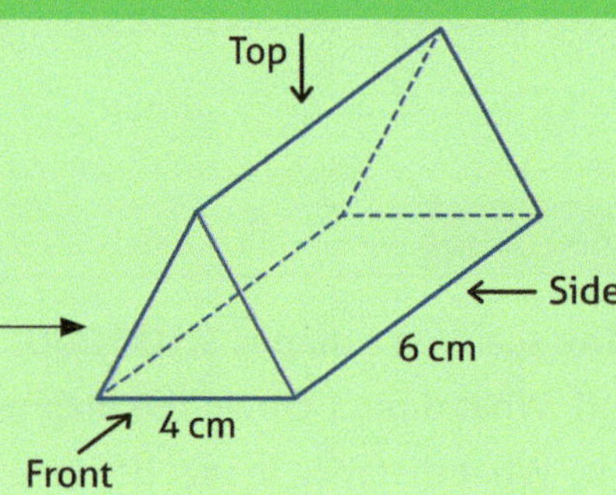

Not all of the lengths are true to scale in this view.

Solution

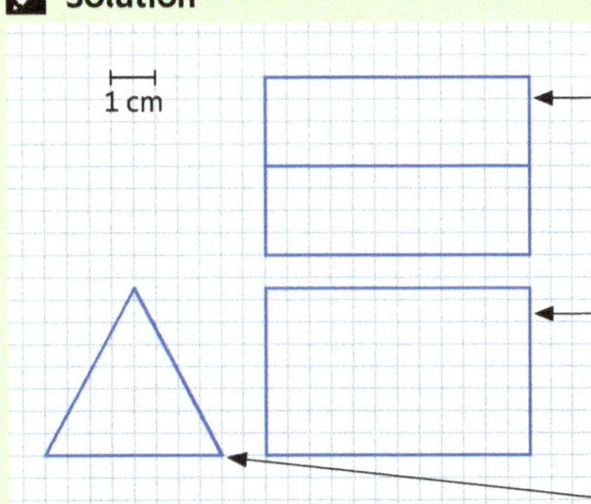

Plan

The outer rectangle measures 4 cm by 6 cm, the same as the base.

You cannot measure the height of the sloping side on this view.

Side elevation

The perpendicular height and length are true to scale.

You cannot measure the length of the sloping side on this view.

Front elevation

The triangle is true to scale.

You can measure the length of the sloping side on this view.

In these diagrams the lengths have been drawn to scale.

Isometric drawings

Another way of drawing a 3-dimensional object is to use isometric paper, as shown below. Notice that although the diagrams may look right, their faces are not true shape: right angles are not 90° on the page.

Activity 10F (2) A puzzle

Four cubes are placed together to make the shape shown in the isometric drawing.

When two shapes can be slid or turned onto each other count them as being the same, even though they may look different.

How many different shapes can you make with 4 cubes?

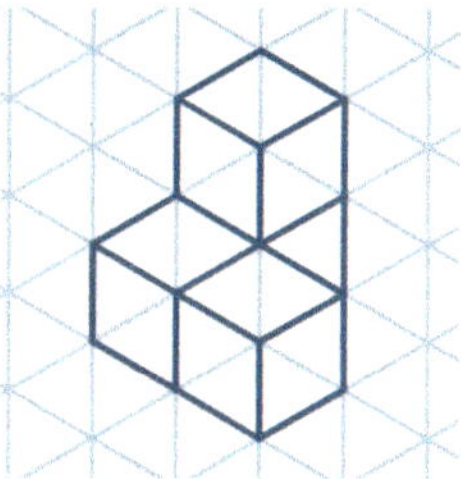

Exercise 10F

1 Draw an accurate plan, front and side elevations of this beaker.

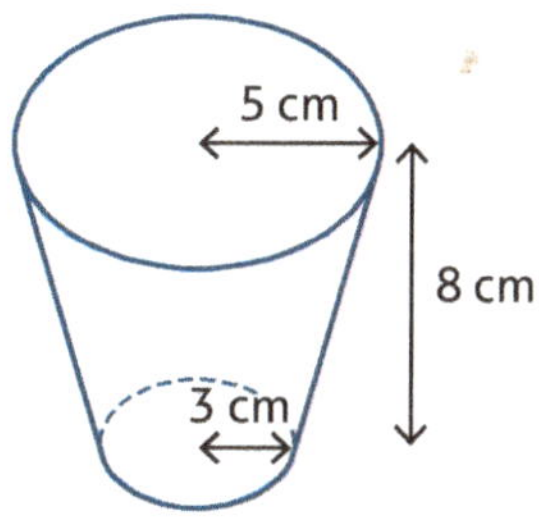

2 Draw an accurate plan, front and side elevations of this piece of a puzzle.

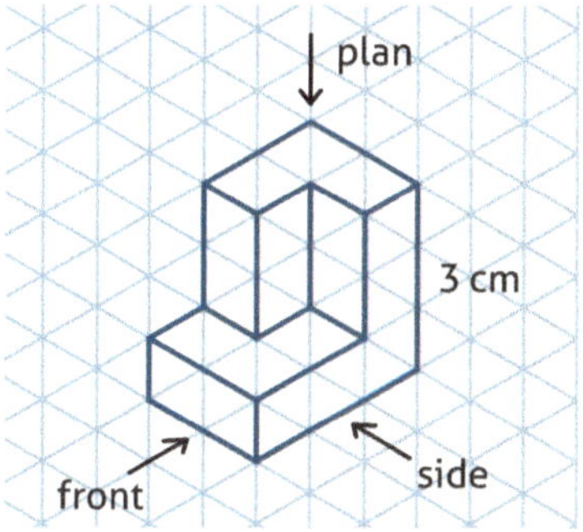

3 This shape is made out of 1 cm cubes.

(a) What is the volume of the shape?

(b) What is its surface area?

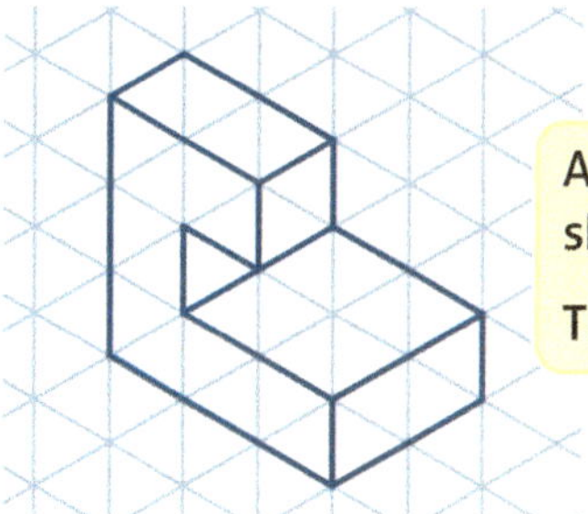

All the cubes used to make the shape are visible.

There are no 'hidden' cubes.

Test yourself 10G

1 A model of an aircraft is made to a scale of 1:72. The length of the model aircraft is 21.4 cm.

(a) What is the length of the original aircraft? Give your answer in metres.

(b) The surface area of the wings on the model aircraft is 140 cm^2. What is the wing area on the aircraft? Give your answer in square metres.

2 The film *Downsizing* (2017) is about a process that shrinks adult humans from, on average, 160 cm to a height of 12 cm. Currently, the 7 billion inhabitants of the planet generate 2 billion tonnes of waste. Estimate how much waste would be created if the whole population were downsized.

3 Health and safety regulations state that an electric socket in a bathroom must be at least 3 m away from any bath or shower.

The bath is 700 mm wide and 1700 mm long.

A quadrant shower tray goes in a corner and has radius at least 800 mm. A walk-in shower goes against a wall. Its rectangular tray is at least 170 cm long and 70 cm wide.

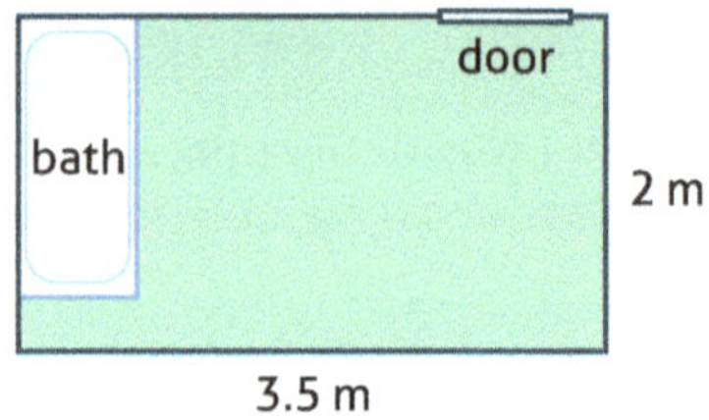

Noah and his family are redesigning their bathroom. Noah wants to include an electric socket and a shower in the room, in addition to the bath.

Is it possible to have either an electric socket, or a shower, or both? Justify your answer.

4 The plan, front elevation and side elevation of a piece of a 3D packing puzzle are shown on the right. The shorter sides are 1 cm long and the longer sides 2 cm.

(a) Draw the shape of the puzzle piece on isometric paper.

(b) What is the volume of the shape?

(c) What is the surface area of the shape?

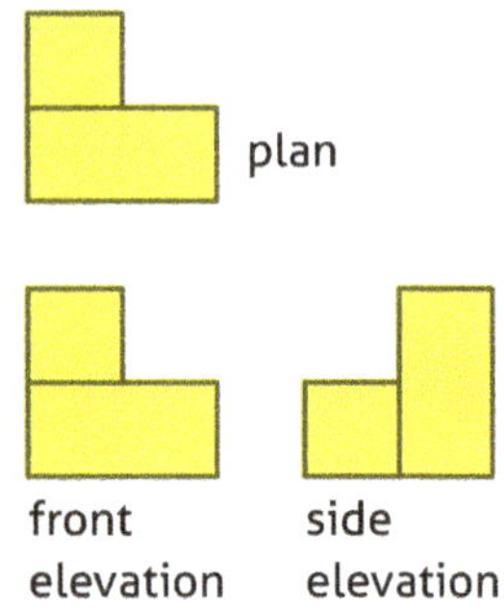

5 Marina is studying undersea fauna from her boat. She sails on a bearing of 060° for 50 km and then on a bearing of 150° for 80 km to get to a site of interest.

(a) Draw a scale diagram to show her journey. Choose a suitable scale.

(b) The average speed for the journey is 30 km/h. How long does the trip take?

(c) What is the boat's displacement from its starting point?

(d) Work out the boat's average velocity.

(e) Describe the course that Marina must set to travel directly back to her starting point.

(f) The boat takes two and a half hours to get back to its starting point. What is its average speed for the homeward journey?

(g) What is the average speed for the whole trip?

Practice question

A popular theme in horror films is giant insects.

A typical ant is about 4 mm long and 1 mm wide. It weighs 0.004 g. Its legs are 0.1 mm in diameter.

(a) The largest species of ant is 5 cm long. What does it weigh?

(b) How much would a 1 metre long ant weigh?

(c) Estimate the diameter of the legs of a 1 metre long ant.

The strength of a limb depends on the area of its cross-section.

(d) Estimate how many times stronger a human arm is than an ant's leg.

(e) Do you think that your answer to part (d) means that we need not worry about giant ants? Give one reason to justify your answer.

Now that you have studied this chapter, you should know:

- How to use relationships between lengths, areas and volumes of similar figures to model real world situations.
- How to work with time, length, area and volume to meet given regulations.
- How to calculate with commonly used units, including converting to the same units.
- How to interpret scale diagrams and maps and use them in problem solving.
- How to solve problems using 2D representations of 3D objects.
- How to solve problems involving distance, displacement, speed, velocity and acceleration.

CHAPTER 11

Fermi estimates

There are occasions when you want to make estimates of quantities when you have little or no idea about their size. A moment's reflection will help you to identify some facts that you know and are relevant to the problem. For the rest you have to make reasonable assumptions.

The first two sections of this chapter are about the facts you can expect to know and remember. In Section 11C you move on to making assumptions in a process known as Fermi estimation. This is named after Enrico Fermi, a physicist, who specialised in estimates of this kind.

11A About yourself

What do you know about yourself as a representative of the human species?
You may know your height but is it in centimetres or feet and inches?
Do you know your weight?
Are you particularly tall or short?
What size of shoe do you take?
How typical are you?

Activity 11A Which is closest?

These questions are about measurements of a 'typical' adult human being.

(a) Select the most appropriate measurement from the box on the right to go in each of the blank cells of the table on the left.

Height	
Weight	
Heart rate	
Length of foot	
Width of hand	
Circumference of head	

1.7 m	5 cm
40 cm	15 cm
120 beats/min	70 kg
10 cm	30 cm
2.1 m	100 kg
90 beats/min	20 cm
40 kg	50 cm
1.9 m	60 beats/min

(b) Which was the hardest to choose? Why was that?

Some things are familiar to you because of your experiences or interests, so you know facts that others may not. Other people, with different experiences and interests, will know other facts.

There are some facts that you would expect to know, such as time and calendar facts and relationships between commonly used measures of length, weight and capacity.

There are others, like those highlighted in Activity 11A above and Worked example 11A on the following page, that you should commit to memory.

Worked example 11A

? Estimate the number of times a human's heart beats during their lifetime.

Solution

A human heart beats roughly 70 times per minute.
A human lifetime is on average 80 years.

> To make this estimate you need to know that:
> a typical human heart rate is between 60 and 80 bpm (beats per minute), and
> life expectancy in the UK is about 80 years.
> These are useful figures to remember.

Known facts:

- 60 minutes = 1 hour
- 24 hours = 1 day
- 365 days = 1 year.

So, the number of heart beats in a lifetime

$= 70 \times 60 \times 24 \times 365 \times 80$

$= 2.943 \times 10^9$ or 3 billion.

> The answer is rounded to 1 significant figure as the measurements used are rounded.

So, a human heart beats roughly 3 billion times during a lifetime.

Exercise 11A

1 An A380 Airbus has a payload of 576 000 kg.
Estimate the maximum number of human beings that could be carried on an A380 Airbus.

> People's weight depends on their height and build. Typical values are 60 kg for women and 70 kg for men. These are figures to remember.

2 Roger walks from Exeter to Exmouth, a distance of 8 miles.
Estimate how long he takes.

> An adult human walks at a speed of about 3 to 4 mph, another figure to remember.

11B About the world

You often need to estimate the values of quantities relating to the world around you.

You may already be familiar with those values in Activity 11B, at least to an accuracy of 1 significant figure. Others are highlighted in the examples and exercises in the rest of this chapter. Some of them may be less familiar to you

Activity 11B Find the match

Select the most appropriate units and number from the box on the right to go in each of the blank cells in the table on the left.

Description	Number	'Units'
World population		
UK population		
Population of London		
Capacity of a coach		
Capacity of a train carriage		
UK motorway speed limit		

billion people	80
7.8	mph
people	70
9.3	million people
million people	66.3
56	people

Which match are you least confident with? Why is that?

Worked example 11B

? Estimate how many teenagers live in Liverpool.

A big city in the UK has a population of about a million.

Solution

The teenage years are from 13 to 19, so 7 years.

A human lifetime is about 80 years.

The teenage years are about $\frac{1}{10}$ of a lifetime.

The population of Liverpool is approximately 1 million.

There are approximately $\frac{1}{10} \times 1\,000\,000 = 100\,000$ teenagers in Liverpool.

The actual population of Liverpool at the time of writing is 0.9 million. Using 1 million is good enough for a rough estimate of the number of teenagers.

Exercise 11B

1 Estimate the number of pensioners in the UK. Pensioners are those aged 66 and older.

2 Estimate the number of pupils taking GCSEs in the summer.

3 Estimate how many coaches would be needed to evacuate the population of a small town.

11C Fermi estimation

The first step in carrying out a Fermi estimate is deciding on the method you are going to use. Tied in with this is knowing the quantities you need to use and their values.

Activity 11C How many pieces of fruit?

Look at the tray of fruit on the right.

(a) How can you work out the number of pieces of fruit?

(b) What do you need to know to work out the number of pieces of fruit?

(c) How can you find out the information you need in part (b)?

Some of the values used may be estimates based on reasoning about what you can see.

Some of the values may be commonly known facts or estimates such as those in sections 11A and 11B.

You may have to make assumptions about the rest of the values.

Worked example 11C

Estimate how long it would take to cycle from Lands End to John O'Groats.

Solution

The method:
Work out how much of the journey can be completed each day and calculate the number of days.

You need information about:

- the distance from Lands End to John O'Groats
- how many hours a day you can cycle
- the average speed you can maintain.

The distance depends on the route taken and any diversions to find accommodation and meals.
Assumption: 1000 miles

The hours spent cycling per day depend on fitness and the pressure to do the trip quickly.
Assumption: 12 hours

The average speed will depend on the quality of the bike, fitness of the rider, choices about the terrain covered and levels of tiredness. It will depend to some extent on the previous answer.
Assumption: 20 miles per hour

The calculation:
Time to cycle 1000 miles = 1000 ÷ 20 = 50 hours.
Number of days = 50 ÷ 12 = $4\frac{1}{6}$ days.

Between 4 and 5 days seems a sensible estimate from the assumptions made.

This seems quite a short time and it may be that the speed should be less and the number of hours per day could also reduce.

The process of making a Fermi estimate can be summarised by the following steps:

- Decide on your method for making the estimate.
- Summarise your calculation in words or using algebra.
- Substitute the values you know.
- Make reasonable assumptions about those you don't know.
- Carry out the calculation.
- Decide whether your estimate is good enough.

Exercise 11C

1 A Sociology student wants to estimate how many hours are spent on the internet each year by teenagers in the UK.

(a) Suggest a way to work this out.

(b) What information does she need to make the estimate?

(c) What assumptions must she make?

(d) Work out an estimate for the number of hours spent on the internet each year by teenagers in the UK.

2 Estimate the number of hotel rooms available in the UK.
List your assumptions and give details of the calculation.

3 Estimate how long it would take to walk from London to York.
List your assumptions and give details of the calculation.

11D Estimating probability

Estimating the likelihood, or risk, of an event is another situation where you can use this approach.

Activity 11D Is it likely?

For each of these events, estimate how likely you think it is.

Write your answer as a percentage.

(a) You will break your leg next week.
(b) There will be coastal flooding somewhere in the UK next year.
(c) It will rain on your next birthday.
(d) You will be struck by lightning some time in your life.

Worked example 11D

? Estimate how likely it is that a person you meet in Manchester is a vet.

Solution

First estimate the proportion of people who are vets.

Estimate 1 in 2000 are vets.

Since the proportion is estimated it does not matter how big Manchester is.

So, probability is $\approx \frac{1}{2000}$.

Roughly 0.1 % of a school year go on to study to become vets each year. In the past the numbers were smaller.

Exercise 11D

1 Estimate the probability that a random person you meet has the same birth month as you.

2 Estimate whether you are more likely to be injured in a road accident than win money on a scratch card.

11E Units

All measurements require the use of units. Some of them, such as metres and days, are very familiar and you have a sense of how large they are, but that may not be true of all the units you may have to use.

Activity 11E Choose a unit

Choose a suitable unit from the box on the right to measure:

(a) the weight of a lorry
(b) the time taken to eat breakfast
(c) a dose of medicine
(d) the length of a sports field
(e) the weight of a full suitcase
(f) the weight of a sugar cube.

kilogram	tonne
gram	mile
millimetre	metre
millilitre	litre
minute	hour

Worked example 11E

? Suggest a method of estimating each of the following quantities.

State a suitable unit for the answer. Then use your method to suggest a possible value.

(a) The length of a car.

(b) The time it takes to write one page of words.

(c) The weight of a cup of coffee.

1 tonne is 1000 litres

1 litre is 1000 ml

1 ml is 1 cm^3

Solution

(a) Think about how many people lying down would match it. A suitable unit is metres.
1 person from the front to the back of the driver's seat.
1 person from the back of the driver's seat to the end of the car.
So, the car is 2 people long.
Each person is roughly 2 metres tall and $2 \times 2 = 4$.
The length of the car is about 4 metres.

(b) Think about how many words there are on a page and estimate how long you take to write a certain number of words. Then scale up. A suitable unit is minutes.
You write about 10 words on a line.
There are about 30 lines on a page.
So, there are 300 words on a page.
You write about 15 words in a minute $\frac{300}{15} = 20$.
It takes about 20 minutes to write a page.

(c) Estimate the volume of coffee in a mug and use the answer to give its weight.
Both grams and kilograms could be suitable units.
The mug is 10 cm tall and 8 cm diameter (so 4 cm radius).
The mug is a cylinder, so its volume is given by the formula $\pi r^2 h$.
This is approximately $3 \times 16 \times 10$ or about 500 cubic centimetres (cm^3).
1 cm^3 of water weighs 1 gram and coffee has the same density as water.
So, the weight of the coffee is about 500 g or $\frac{1}{2}$ kg.

Exercise 11E

1 Without taking any measurements, estimate the area of the green part of the shape on the right, given that the area of the white part is 2 cm^2.

2 Suggest a method of estimating each of the following quantities.
State a suitable unit for the answer.
Then use your method to suggest a possible value.
(a) The length of a double-decker bus.
(b) The weight of a packed suitcase.
(c) The speed of a marathon runner.

3 Estimate the weight of water contained in the tub when you are having a bath, using suitable units.

11F Powers of 10 estimates

The metric system was designed to be easy to use.

The units go up in powers of 10. There are 10 millimetres in a centimetre, 10 centimetres in a decimetre, 10 decimetres in a metre and so on.

This section is about working in powers of 10 when making estimates.

Such estimates are easy to make but obviously not very accurate.

However, for some purposes they are good enough.

Some metric units are not often used; for length, the most common are millimetre, centimetre, metre and kilometre.

Activity 11F Powers of 10

In each of (a)–(d), choose the closest number.

(a) Height of the Eiffel Tower in metres	10	100	1000	10 000	100 000
(b) Distance from London to New York in km	10	100	1000	10 000	100 000
(c) Number of schools in the UK	10	100	1000	10 000	100 000
(d) Number of cafés in Manchester	10	100	1000	10 000	100 000

(e) Describe how you decided on your answer for the number of schools in the UK.

Using only powers of 10 often makes it easier to decide what number to use.

Worked example 11F

? Using powers of 10, estimate the cost of supplying a laptop to every child of school age in the UK.

Comment on your answer.

Solution

Method 1

Number of schools in the UK ≈ 10 000 or 100 000.
Number of students per school ≈ 100 or 1000.

Using the lower of one of the estimates with the higher of the other makes it likely that errors cancel each other out.

Total number of students ≈ 10 000 × 1000 = 10 000 000
or ≈ 100 000 × 100 = 10 000 000.

Cost of a laptop ≈ £100.

Both £10 and £1000 are unlikely for the cost of the laptop, so £100 has been used.

The cost of providing every school-age child with a laptop would be roughly £1 billion.

Method 2

Proportion of the population that are of school age ≈ 10 %.
Population of the UK ≈ 100 million.

You almost certainly know that the population of the UK is between 60 and 70 million. However, for this estimate only powers of 10 are being used and the nearest is 100 million.

Number of people of school age ≈ 10% of 100 000 000 = 10 000 000.

The proportion is an underestimate, and the UK population is an overestimate so 10 million students seems about right.

The estimated cost of a laptop is £100.
Total cost ≈ 10,000,000 × 100 = £1,000,000,000.
The estimated cost is £1 billion.
This is the same figure as that given by Method 1.

Exercise 11F

1 Nick estimates how many hours he will spend playing games online during two years of study at A level.

(a) Suggest a method for making this estimate.

Nick decides to use only powers of 10.

(b) Carry out the estimation.

2 It is about 4×10^5 kilometres to the moon. Estimate, to the nearest power of 10, how many marathons you would need to run to travel that far.

A One hundred B One thousand C Ten thousand D A hundred thousand E One million

11G Is the estimate good enough?

A 'back of the envelope' estimate such as you have been doing in this chapter is good enough for some purposes. Is it good enough for all purposes?

Activity 11G Good enough?

A government advisor wants to know how much carbon dioxide is being put into the atmosphere by diesel cars in the UK, and so the effect of phasing out such cars.

(a) Make a powers of 10 Fermi estimate.

(b) Do you think it is good enough to influence policy? Justify your answer.

Worked example 11G

? Melania wants to set up a dog-walking business.

She wonders whether it would provide enough money for her to live on.

(a) What information does she need to estimate her potential income from the business?

Melania carries out the estimate for her dog-walking business:

(b) By making suitable assumptions, estimate whether the business can generate enough income to live on.

(c) Does she need to refine her estimate?

Solution

(a) Some things Melania can decide:

The amount of time each day she can spend.
The number of walks each day.
The number of dogs she can walk together.

She has to make an assumption about how much she can charge.

(b) The amount of time each day she can spend = 8 hours.
The number of walks each day = 6.
The number of dogs she can walk together = 4.

Assumption: She charges £10 per dog per walk.

Daily income = £6 × 4 × 10 = £240.

(c) This looks promising; however, she is planning to act on this evidence, so should examine the numbers very carefully.
The time collecting the dogs could be substantial and limit the number of walks per day.
Some dogs may not be happy walking with other dogs, so she may have to limit the number of dogs.
The owners may think that £10 is a lot to pay for the likely 45-minute walk that the dogs will get.
There may not be that much business in the area, so she cannot fill her schedule.
The estimate is good enough to make it worth looking in more detail at the proposition.

Exercise 11G

1 (a) Estimate the number of texts sent in the UK each day.
(b) Suggest a likely use for this information.
(c) Is the estimate good enough for its likely use? Justify your answer.

2 (a) Estimate how many turkeys will be eaten at Christmas in the UK.
(b) Suggest a likely use for this information.
(c) Is the estimate good enough for its likely use? Justify your answer.

Test yourself 11H

1 Queues of cars that are 20 miles long are reported on the M6 motorway. The motorway has 3 lanes.
Estimate how many cars are in the traffic jam.

2 Estimate the weight of the water in an Olympic-sized swimming pool.

An Olympic-sized pool is 50 m long, 25 m wide and at least 2 m deep.

3 The film, *Snowpiercer*, is about a train that has the whole of the remaining population of the world on board.
Estimate how many carriages would be needed for the population of Birmingham.

4 Estimate how many take-away cups of coffee are drunk in the UK in a year.

5 Estimate how many office blocks there are in London.
List your assumtions and explain why someone may find this useful to know.

6 Estimate how many medals are needed for the Olympic Games.

7 Use powers of 10 to estimate the area of land, in square metres, needed to sustain a family of four for a year.

Practice question

A new attraction is to be built near Liverpool.
The projected cost is £23 million.

(a) Estimate how many people live within 2 hours of Liverpool.
List your assumptions and show your calculations.

(b) Assume that people will visit once every two years and be prepared to pay £20 each.
Assume that the running costs are £1.5 million each year.
How long will it take to recoup the initial investment?

Now that you have finished this chapter, you should know:

- How to set up a method for making a Fermi estimate of a quantity or probability.
- How to summarise the method as a calculation.
- How to distinguish between those values you know and those that require assumptions.
- How to make reasonable assumptions where necessary.
- How to carry out the calculation using sensible units.
- How to decide whether your estimate is good enough.
- How to make a powers of 10 estimate.

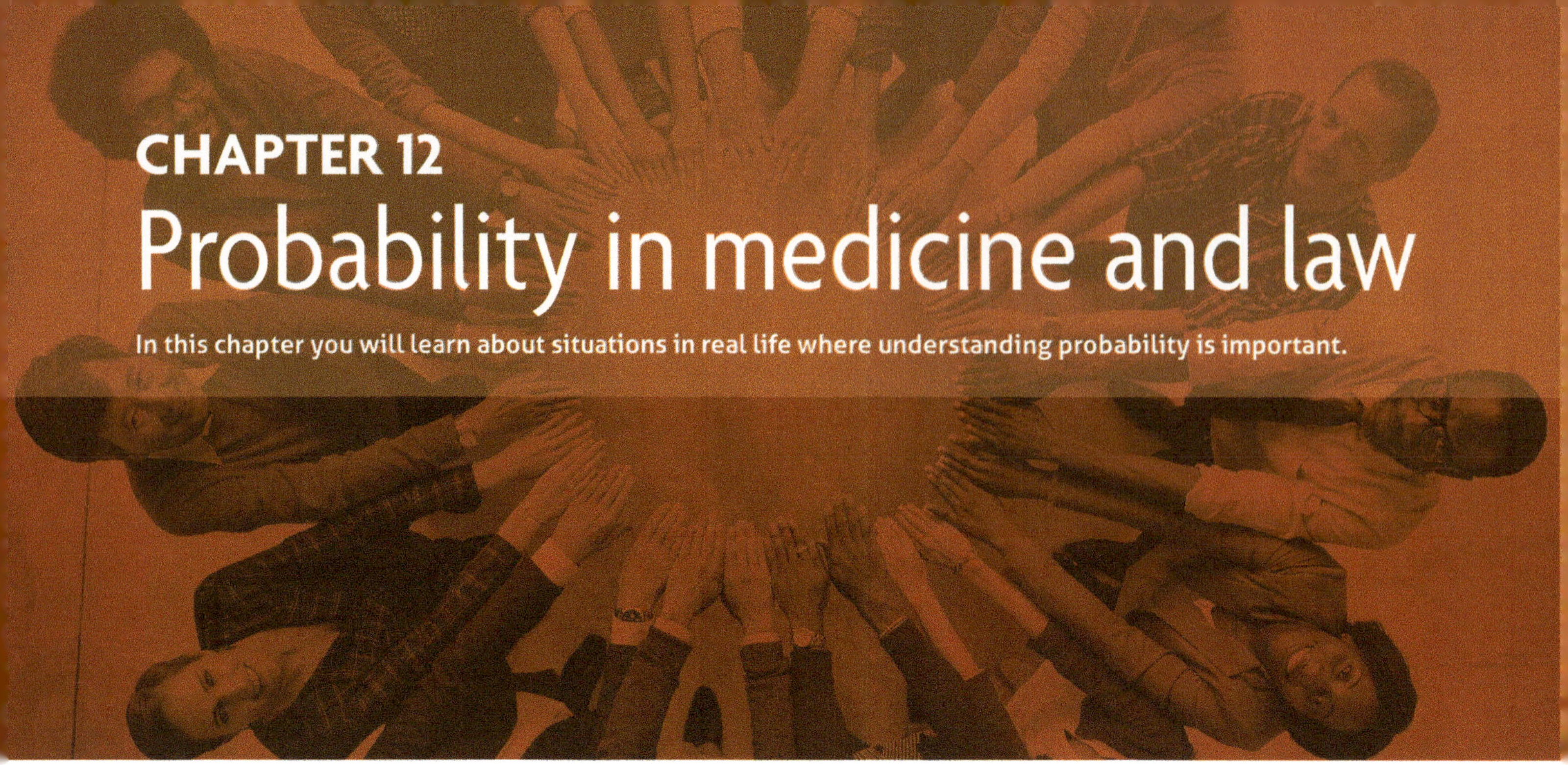

CHAPTER 12

Probability in medicine and law

In this chapter you will learn about situations in real life where understanding probability is important.

12A Conditional probability

Different people may have different risks of suffering from a particular illness – you came across this idea in Chapter 1.

The probability that a person over 60 has high blood pressure is a **conditional probability**. It is sometimes written as the probability that a person has high blood pressure **given that** he/she is over 60. The extra information that the person is over 60 affects the probability that he/she has high blood pressure. The events 'has high blood pressure' and 'is over 60' are said to be **dependent**.

Activity 12A Does it make a difference?

The *Titanic* was a ship that had its first voyage in 1912. It sank after hitting an iceberg.

The table below shows the number of survivors from each class of passengers.

Class	Survived	Died	Total
1st	201	123	324
2nd	111	173	284
3rd	172	537	709
Total	**484**	**833**	**1317**

Is the probability of a randomly chosen 1st class passenger surviving the same as the probability that a randomly chosen surviving passenger was first class?

Worked example 12A

? Use the information from the table in Activity 12A to find the following probabilities:

(a) The probability that a randomly chosen 3rd class passenger survived.

(b) The probability that a randomly chosen surviving passenger was 3rd class.

Solution

(a) For a 3rd class passenger, the probability of surviving is $\frac{172}{709} \approx 0.24$.

(b) For a surviving passenger, the probability of being 3rd class is $\frac{172}{484} \approx 0.36$.

Out of the 709 3rd class passengers, 172 survived.

When working with frequencies, probabilities can be calculated by working out how many out of the relevant total.

Exercise 12A

1 In a survey, 1000 adults are asked, 'Do you know what your blood group is?' The results are summarised in the table below.

Age	18–24	25–49	50–64	65 and over	Total
Know blood group	20	166	137	129	**452**
Do not know blood group	86	249	107	106	**548**
Total	**106**	**415**	**244**	**235**	**1000**

Assume that the survey is representative of the population.

(a) What is the probability that a randomly chosen person knows their blood group?

(b) What is the probability that a randomly chosen person is aged 65 or over and knows their blood group?

(c) A person is chosen at random from the population aged 65 and over, what is the probability that the person knows their blood group?

(d) A person chosen at random from the population knows their blood group. What is the probability that the person is aged 65 and over?

(e) Jim says that most people who know their blood group are 65 or over. Comment on whether Jim is right.

(f) What is the probability that a randomly chosen person, aged 50 or over, knows their blood group?

12B Using representative frequencies

In Question 1 of Exercise 12A, you have information from a survey. If the survey is representative of the population, you can use the frequencies to find probabilities.

If you have probabilities, you can change them to representative frequencies to make working with them easier. This is like assuming that you have a representative sample of the whole population. You will usually have to choose a sensible size for the sample.

Activity 12B Probability to frequency

In a biscuit factory, 0.5% of all biscuits are faulty.

A quality controller checks whether the biscuits meet quality standards.

She correctly identifies 96% of biscuits that are faulty. She also wrongly identifies 1% of biscuits that are OK as faulty.

(a) Using the total representative frequency of 20 000 in the square shaded grey in the two-way table below, complete the numbers in the rest of the table.

	Identified as faulty	Identified as OK	Total
Faulty	96		**100**
OK			
Total			**20 000**

(b) Use your representative frequencies to find the proportion of biscuits identified as faulty that are actually OK. You may want to look back at Chapter 1 to help you with this.

(c) Explain why 20 000 is a better starting total representative frequency than 1000 in this case.

You can work with representative frequencies in either a two-way table (as in Activity 12B above) or in a frequency tree (as in Worked example 12B below). You will get the same answers for probabilities with either method.

Worked example 12B

For the situation in Activity 12B, draw and complete a frequency tree for a total representative frequency of 20 000 biscuits.

Hence, find the proportion of biscuits identified as faulty that are actually OK.

Solution

A frequency tree for a total representative frequency of 20 000 biscuits is given below.

0.5% of biscuits are faulty.

0.5% of 20 000 is 100, so 19 900 are OK.

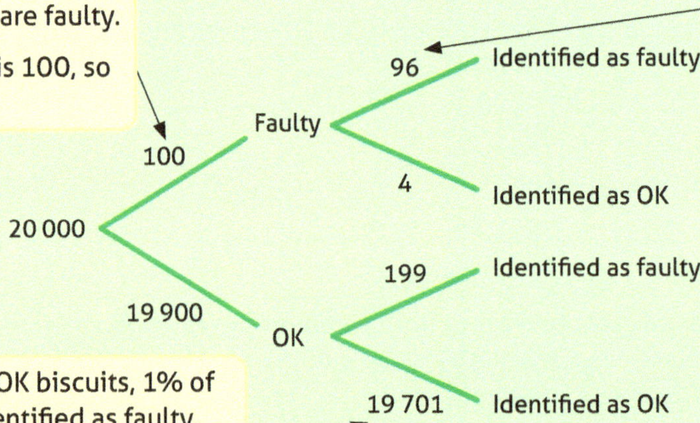

For the 100 faulty biscuits, 96% of them will be identified as faulty.

96% of 100 = 96

This leaves 100 − 96 = 4, which will be identified as OK.

For the 19 900 OK biscuits, 1% of them will be identified as faulty.

1% of 19 900 is 199.

This leaves 19 900 − 199 = 19 701, which will be identified as OK.

96 + 199 = 295 biscuits are identified as faulty.

Proportion of biscuits identified as faulty that are actually OK = $\frac{199}{295} \approx 0.6746$.

False positive and false negative

If a biscuit is OK but is identified as faulty, this is a **false positive**.

If a faulty biscuit is identified as OK, this is a **false negative**.

False positives and false negatives occur with all kinds of testing.

Exercise 12B

1 A school has two boxes of 20 calculators to lend to students. All the calculators in one box work. Only half the calculators in the other box work. Both boxes look the same and all the calculators look the same.

(a) Use a copy of the blue tree diagram to complete the representative frequencies for an imaginary 40 times where a teacher chooses a box at random and tests one calculator.

(b) A teacher chooses a box at random and tests one calculator. The calculator works.

What is the probability that she has the box with all the calculators working?

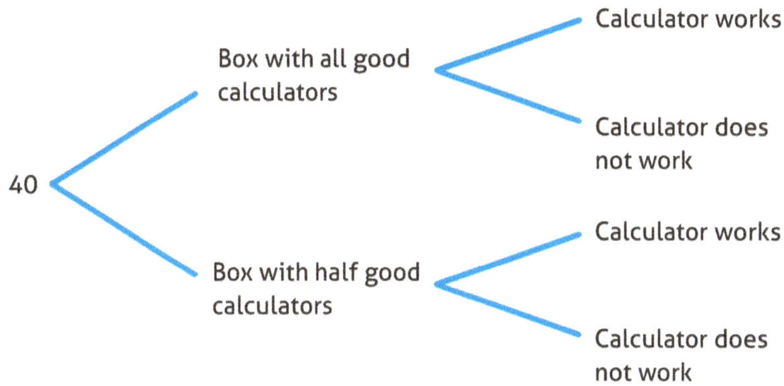

2 Unwanted emails, often used for advertising, are referred to as spam. It is estimated that 85% of all emails are spam. A spam filter is software that checks each email to see whether it is spam or legitimate. It stops the emails identified as spam getting through to the person they are addressed to.

A particular spam filter identifies 5% of spam email as legitimate. It also identifies 2% of legitimate email as spam.

(a) Fill in the representative frequencies on the orange frequency diagram, starting with a total frequency of 10 000 emails.

(b) What proportion of emails identified as legitimate are actually spam?

(c) What proportion of emails that are identified as spam are actually legitimate?

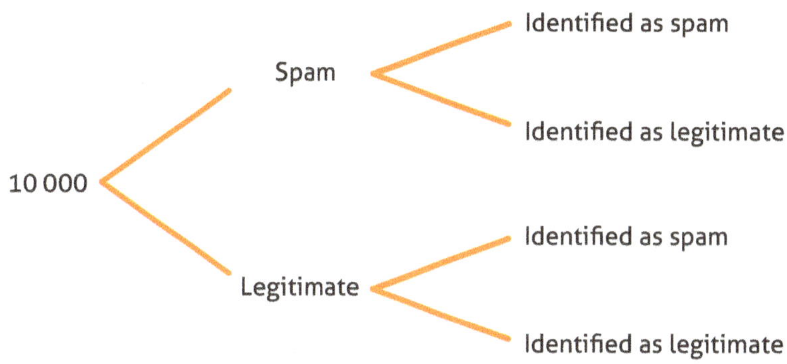

A new spam filter is better at identifying spam email. It will only identify 1% of spam email as legitimate. However, it will identify 4% of legitimate email as spam.

(d) Would you recommend the new spam filter to a business? Explain your reasoning.

12C Medical testing and probability

In this section you will be introduced to the use of probability in interpreting the results of medical tests.

Activity 12C Should everyone be tested?

A new disease can make people ill for weeks. People who have the disease don't have symptoms for the first few days but can still spread it.

A particular country has 1 in 1000 people with the disease.

The UK government is deciding whether to test all people travelling from that country when they arrive in the UK.

What factors should the government consider when making the decision?

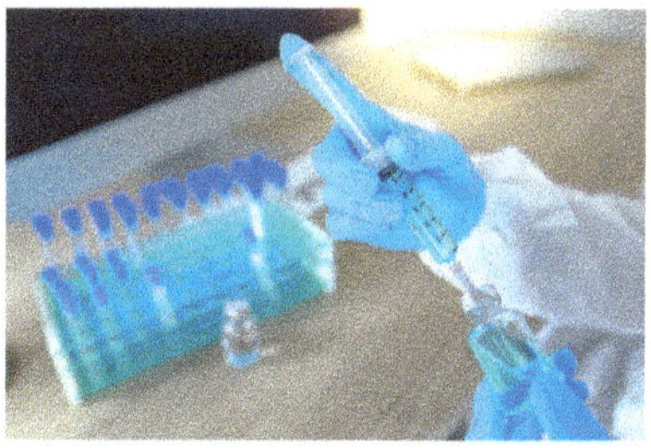

Worked example 12C

? On average, 1 in 1000 people have a particular illness. A test for the illness will correctly identify 90% of those who have the illness. It will also correctly identify 95% of those who do not have the illness.

A positive result in the test means that someone is identified as having the illness.

(a) Represent this information on a frequency tree, using a suitable total representative frequency.

(b) (i) What is the probability that someone with a negative test result does actually have the illness?

(ii) What is the probability that someone who gets a positive test result does actually have the illness?

(c) Which is the bigger problem for this test: false negatives or false positives? Justify your answer.

Solution

(a) To represent the information on a frequency tree, imagine 100 000 representative people.

The probability of having the illness is out of 1000, the other probabilities are percentages which are out of 100. So, starting with 100 000 should work well.

1 in 1000 have the disease, so 100 out of 100 000.

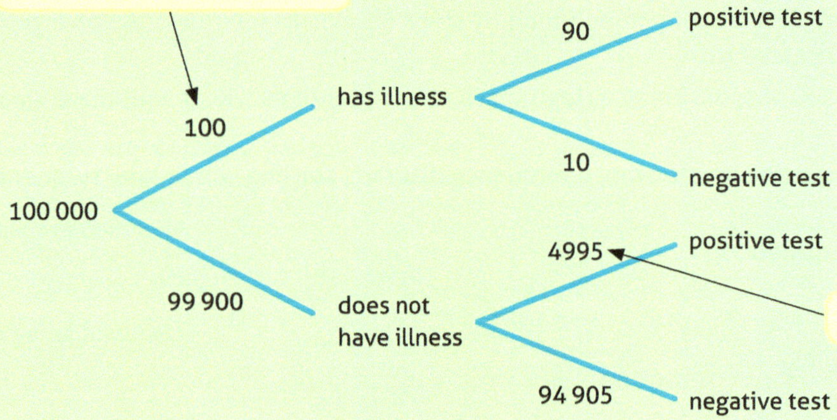

95% of 99 900 is 4995.

(b) (i) There are 10 + 94 905 = 94 915 people with a negative result.
10 of them have the illness.
Probability of someone with a negative test result having the illness = $\frac{10}{94\,915} \approx 0.0001$.

(ii) There are 90 + 4995 = 5085 people with a positive result.
90 of them have the illness.
Probability of someone with a positive test result having the illness = $\frac{90}{5085} \approx 0.018$.

(c) To decide whether the bigger problem is false positives or false negatives, you can look at either representative frequencies or probabilities.

Out of the 100 000 representative people tested, 10 are false negatives and 4995 are false positives, so the bigger problem is false positives.

Reducing false positives

No test is 100% accurate. The possibility of a false positive when conducting a medical test is well known; false positives can be very problematic when the thing being tested for is rare. There are a number of ways that are used to reduce the possibility of a false positive for any particular test. These include the following.

- Only testing people who have an increased likelihood of having the illness.
- Testing a second time when someone has a positive test and only reporting the test as positive when both show positive. This works when results of tests are independent of each other.

Exercise 12C

1 Half of people with a high temperature have the illness referred to in Worked example 12C. People with a high temperature are tested for the illness. The test will correctly identify 90% of those who have the illness. It will also correctly identify 95% of those who do not have the illness.

(a) Represent the information in the following two-way table for a total 100 000 people tested.

	Test positive	Test negative	Total
Has illness			50 000
Not ill			
Total			**100 000**

(b) What is the probability that someone with a negative test result does actually have the illness?

(c) What is the probability that someone who gets a positive test result does actually have the illness?

2 A test for a particular illness is 90% accurate when used in ideal conditions. It will correctly identify 90% of those who have the illness. It will also correctly identify 90% of those who do not have the illness.

The test is used on people with a 75% chance of having the illness.

(a) Represent the situation on the frequency tree for a total 1000 people tested.

(b) What is the probability that someone with a negative test result has the illness?

(c) What is the probability that someone who gets a positive test result does not have the illness?

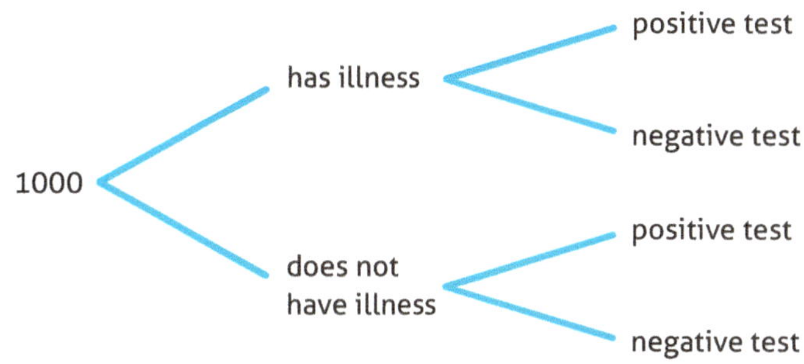

12D Do events with small probability happen?

Some events are very unlikely but nonetheless they do occur. It is fair to say that if something can happen, sooner or later it will happen.

Events with small probability are likely to happen a small number of times in a large population. You are very unlikely to win the lottery, but nearly every week someone does!

Activity 12D Is that surprising?

The probability of being struck by lightning in a year in England is about one in two million, so is it surprising that this sometimes happens?

Worked example 12D

The population of England is 56 million.

(a) Using the probability of one in two million from Activity 12D, how many people in England would you expect to be struck by lightning each year?

(b) About half of people who are struck by lightning have only minor injuries. What is the probability that Ben is struck by lightning next year and has only minor injuries?

Solution

(a) If one in two million is struck by lightning, the expected number struck is $56 \div 2 = 28$.

(b) Out of the 28 struck by lightning, 14 will have minor injuries.

Probability $= \dfrac{14}{56\,000\,000} = \dfrac{1}{4\,000\,000}$

Exercise 12D

1 Two in ten thousand light bulbs produced by a factory are faulty.

(a) A particular chain of shops sells 50 of the light bulbs on an average day.

How many faulty light bulbs does the chain of shops sell in an average year?

(b) A typical customer buys two of the light bulbs in an average year.

How many customers would it take for them to expect to get one faulty light bulb between them in an average year?

2 Sandy has just started at university. One of the students she meets has the same birthday as her.

Sandy tells her mum that this is an amazing coincidence and that she doesn't think it will have happened to anyone else in her year at university (apart from the person she has just met).

(a) A student meets another student. What is the probability they share the same birthday? State any assumptions you make.

(b) Comment on whether there are likely to be other students in Sandy's year who have met someone who shares their birthday.

12E The prosecutor's fallacy and the defendant's fallacy

In this section you will think about how to interpret conditional probabilities that come up when trying to decide whether someone is guilty of a crime.

Activity 12E How strong is the evidence?

A person has been accused of committing a crime. DNA evidence has been found at the scene of the crime.

Only one in 3 million people has DNA that would match the evidence.

Does this mean that there is only a one in 3 million chance that the accused person is innocent?

The prosecutor's fallacy

The one in 3 million probability in Activity 12E is the probability that DNA matches the evidence, given that the person is innocent.

The claim that there is a one in 3 million chance that the accused person is innocent is about the probability that the person is innocent given that the DNA matches.

The **prosecutor's fallacy** assumes that these two probabilities are the same.

Each of the probabilities is a conditional probability, but in the first case we know that the person is innocent and in the second case we know that the person matches. These probabilities are not usually the same. will see why in Worked example 12E.

Either the person is innocent or not, but we don't know which it is. Probability is being used to measure degree of belief in the person's innocence.

Worked example 12E

A person has been accused of committing a crime. DNA evidence has been found at the scene of the crime.

Only one in 3 million people has DNA that would match the evidence.

Assume that there is no other evidence.

There are potentially 27 million suspects from the population; these are all the people whose sex matches the DNA and who are old enough to have committed the crime.

What is the probability that the accused person is guilty?

Solution

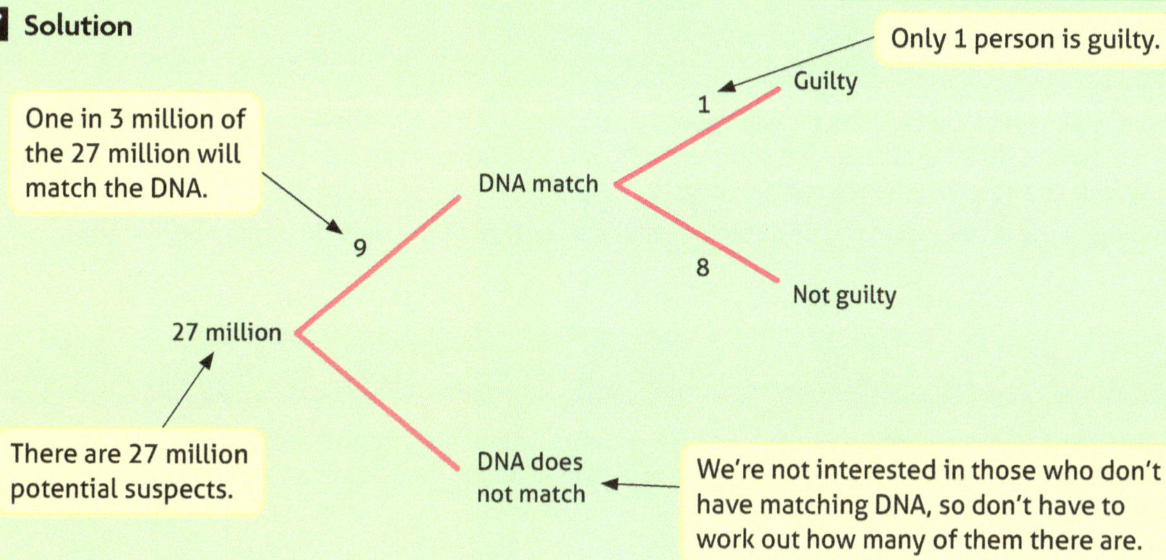

The accused person is one out of the 9 matches.

So, if there is no other evidence, the probability that the person is guilty is $\frac{1}{9}$.

- Additional evidence is needed before the accused can be found guilty.
- The term **prosecutor's fallacy** is sometimes used more generally for confusing conditional probabilities that sound similar but are the other way around, such as the probability of a positive test result given that you are ill and the probability of being ill given that you have a positive test result.
- Saying that there is a $\frac{8}{9}$ chance that the accused person is innocent is an example of the **defendant's fallacy**. There may be other evidence which suggests that the accused could be guilty. In that case they are not as likely to be innocent as other people who would match the DNA.

Exercise 12E

1 A robbery is committed. The robber leaves hair at the scene of the crime. Only 2% of the population have hair that matches the hair at the scene.

The police are fairly certain the robbery was committed by an adult who lives in the local area.

(a) There are 250 000 adults who live in the local area. The hair evidence narrows the number of potential suspects.

How many potential suspects are there?

(b) The police visit someone who has a criminal record and find that he has something in his house that matches one of the stolen objects. He was seen close to the scene of the robbery just after it happened. Which of the following is the probability that this person is guilty of the crime?

Less than $\frac{1}{5000}$ $\qquad$ $\frac{1}{5000}$ $\qquad$ More than $\frac{1}{5000}$ $\qquad$ Close to certain

Justify your answer.

2 A lie detector is a machine that identifies when someone is lying by measuring signs of stress while answering questions. A lie detector correctly identifies 80% of lies and 70% of true answers.

(a) Assume that people lie 25% of the time. Fill in the two-way table for 1000 representative responses checked by the lie detector.

	Identified as lie (positive result)	**Identified as true (negative result)**	**Total**
Lie			
Truth			
Total			**1000**

(b) A lawyer claims that the lie detector will show a positive result 80% of the time when someone is lying, so there is an 80% chance that someone who got a positive result was telling a lie. Use suitable calculations to show that the lawyer is wrong.

(c) According to the results from the lie detector, what percentage of the time do people lie?

12F The gambler's fallacy

Imagine you are playing a game that involves tossing a coin. You are hoping for the coin to land heads. The last five times you played the game, the coin landed tails. Is it more likely to land heads next time?

Activity 12F Does it even out?

The average debt per household in the UK (including mortgages) is £60,000.

A researcher is taking a sample of 10 randomly selected households.

(a) What would be a reasonable estimate for the mean debt for the sample?

The first household in the sample has £500,000 debt.

(b) Which is the best estimate of the mean debt for the sample:

(i) less than £60,000

(ii) around £60,000

(iii) more than £60,000?

Worked example 12F

? The average debt per household in the UK (including mortgages) is £60,000.

A researcher is taking a sample of 10 randomly selected households. The first household in the sample has £500,000 debt.

(a) What would be a reasonable estimate for the mean debt of the remaining 9 households in the sample? Justify your reasoning.

(b) What would be a reasonable estimate for the whole sample of 10 households?

Solution

(a) The sample mean is likely to be close to the population mean, so £60,000 is a reasonable estimate.

(b) One of the households has a debt of £500 000. Assume the other 9 have a mean debt of £60 000.

The total debt of all 10 households is £500,000 + 9 × £60,000 = £1,040,000.
Estimated mean = £1,040,000 ÷ 10 = £104,000.

> If you took lots of different samples of size 9, the population means would be normally distributed with mean £60,000. There is more about this in Chapter 16.

The gambler's fallacy

The name of this fallacy comes from the games of chance where a particular result sometimes occurs.

For example, a fair coin has an equal chance of landing heads or tails. If you are tossing the coin ten times, you would expect roughly equal numbers of heads and tails to occur. Suppose you get a run of 5 heads; the gambler's fallacy suggests that the next tosses will be more likely to be tails to balance things out, but the coin has no memory and is equally likely to land heads or tails each time it is tossed.

In Activity 12F people may expect the sample mean to be close to the population mean and assume that if the first household in the sample has a large debt then the rest of the sample will have small debts to make things even out, but each member of the sample is chosen independently.

Exercise 12F

1 In a national lottery game, people choose 6 numbers from 1 to 59. Each week, numbers are drawn at random from 1 to 59, so that any set of six numbers is equally likely to come up. If the player's numbers match the drawn numbers, the player wins a prize.

(a) Which of the three statements below is an example of the gambler's fallacy?

(i) Don't choose 6 consecutive numbers.

(ii) Number 7 has come up a lot. It must be lucky.

(iii) Number 47 hasn't come up as often as other numbers. It must be due soon.

(b) A player in the national lottery sees the statements above and chooses 7, 47 and four other numbers. Does this give the player a better chance of winning a prize? Justify your answer.

Test yourself 12G

1 Facial recognition software is being tested on a standard set of photographs. The table shows how accurate the software was in identifying whether the person in the photograph was male or female.

Use the information in the table to find the following probabilities.

(a) A photograph of a male is identified as a female.

(b) A photograph identified as female was actually a male.

	Identified as male	Identified as female	Total
Photo of male	343	7	**350**
Photo of female	21	279	**300**
Total	**364**	**286**	**650**

2 Some sportspeople take drugs to boost their performance. They are tested for these drugs. A test for a particular drug correctly identifies 99% of people who have taken the drug. However, among those who have not taken the drug, it incorrectly identifies 2% as having done so.

(a) Fill in the frequencies in the two-way table for a representative 10 000 sportspeople tested. Assume that 50% of sportspeople take the drug.

	Identified as drug taker (positive result)	Identified as not taking drug (negative result)	Total
Takes drug			
Does not take drug			
Total			10 000

(i) What percentage of sportspeople test positive for the drug?

(ii) What is the probability of a positive test being incorrect?

(iii) What is the probability of a negative test being incorrect?

(b) Repeat part (a) but this time assume that only 10% of sportspeople take the drug.

(c) At a particular sports event, 6% of the sportspeople test positive for the drug. Which of the following statements is most likely to be true? Justify your answer.

- Less than 6% of sportspeople at the event took the drug.
- 6% of sportspeople at the event took the drug.
- More than 6% of sportspeople at the event took the drug.

3 A test for a particular illness is used on people who have a 2% chance of having the illness. The test will correctly identify 99% of those who have the illness. It will also correctly identify 90% of those who do not have the illness.

(a) Complete the frequency tree, for a representative 10 000 people tested.

Use it to show that the probability someone with a positive test result does actually have the illness is approximately 16.8%.

(b) Find the probability that someone with a negative test result does not have the illness.

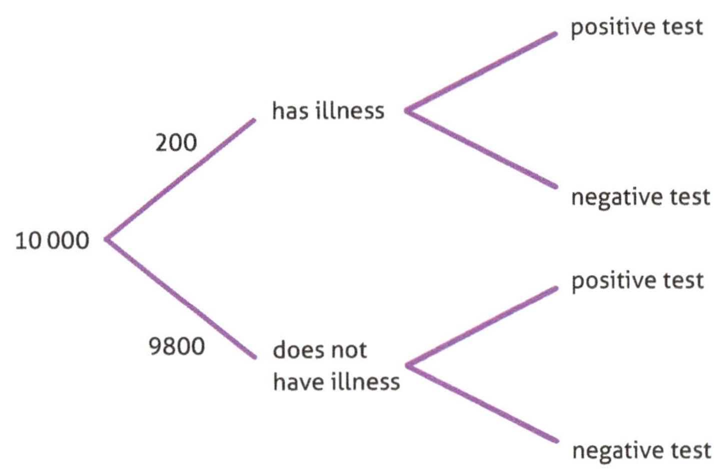

4 After medical testing for a rare illness, one sixth of those who tested positive actually have the illness. Those who have a positive result are given a second test. The second test will correctly identify 98% of those who have the illness and will also correctly identify 95% of those who do not have the illness. Assume that the result of the second test is independent of the first test.

(a) The frequency tree is for the second test. Explain why a starting frequency of 600 is a good one to use.

(b) Complete the frequency tree for the second test.

(c) What is the probability that someone who gets a positive result on the second test has the illness?

(d) Comment on the effectiveness of testing those with a positive result a second time as a way of preventing false positives.

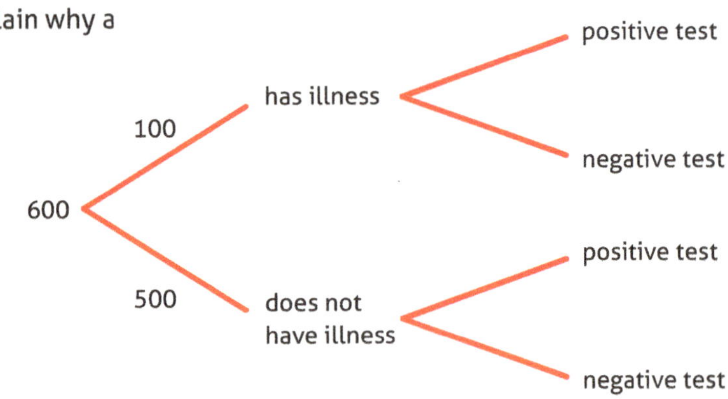

5 A new test for an illness is being developed. It will be used on people who have a 4% chance of having the illness. There are two options: Option A and Option B.

Option A	Option B
The test will correctly identify 99% of those who have the illness.	The test will correctly identify 95% of those who have the illness.
The test will correctly identify 95% of those who do not have the illness.	The test will correctly identify 99% of those who do not have the illness

Investigate which option is better. You must show relevant calculations to support your decision.

Practice question

A lecturer uses plagiarism detection software to check whether students' essays have been copied.

The software correctly identifies 96% of essays that are copied. It also wrongly identifies 1% of original essays as copied.

Previous experience suggests that 10% of essays are copied.

(a) Fill in the frequencies for 10 000 essays on the following frequency tree.

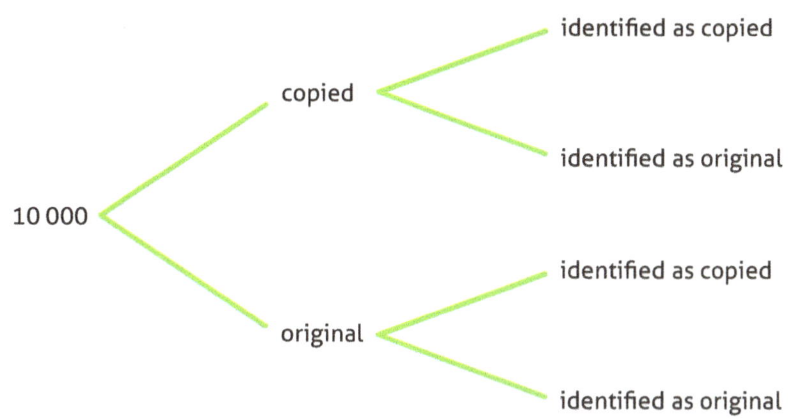

(b) Find the probability that an essay identified as copied is actually original.

(c) The lecturer does not want to make a false accusation against a student. She is considering two possible ways to follow up when an essay is identified as copied:

(i) Use the same software on the essay again and assume it has been copied if the software identifies it as copied both times.

(ii) Interview the student and only assume the essay has been copied if the student cannot talk sensibly about the content.

Which of these would you advise the lecturer to do? Explain your reasoning.

Now that you have completed this chapter, you should know:

- That the probabilities of events which sound similar can be quite different.
- How to use representative frequencies to work out conditional probabilities.
- Some applications of probability to medicine and law.
- How to recognise some examples of incorrect reasoning in probability.

CHAPTER 13

Is it for real?

Have you ever read something that you know is meant to deceive you? Or seen a diagram that is deliberately misleading? This chapter is about some of the tricks you should look out for.

13A Percentage points

There are two different senses in which the word percentage is used and this can be confusing.

What is a scam? Do all scams involve numbers? Money?

Activity 13A The meaning of the word percentage

A pharmaceutical company is launching a new drug for treating heart disease. Here are two possible public statements about it:

- Our new drug reduces heart attack risk by 50 percent.
- In clinical trials, it reduces heart attacks by 0.1 percentage points, from 0.2% of people to 0.1%.

(a) Explain why, in fact, both are saying the same thing.

(b) Which statement sounds better for the drug?

Worked example 13A

? In 2020 the UK planned to cut foreign aid from 0.7% to 0.5% of Gross National Income.

Some newspapers reported this as a drop of only 0.2%.

(a) Why is this figure misleading?

(b) What would have been a more accurate statement?

Solution

(a) A change from 0.7% to 0.5%.

It is correct to describe this as 0.2 **percentage points** rather than 0.2%.

(b) The percentage drop from 0.7% to 0.5% is

$\frac{0.7 - 0.5}{0.7} \times 100 = \frac{20}{0.7} = 29\%$ to the nearest whole number.

So, either '0.2 percentage point drop'
or '29% drop'
would be a more accurate statement.

Confusing **percentage change** and **percentage points change** is a common trick. Good journalists will quote both if necessary.

Exercise 13A

1 Here are the male and female unemployment rates in the EU for April and May 2020.

Month	Male unemployment (%)	Female unemployment (%)
April	6.9	6.4
May	7.2	6.4

Eurostat EU

(a) By how many percentage points had the male unemployment rate increased from April to May?

(b) What is the percentage increase in male unemployment from April to May?

(c) By what percentage was female unemployment less than male unemployment in May?

2 Here are some shock media headlines together with some facts.

Use these to decide if the shock nature of the headlines is justified.

(a) **Child's risk of brain cancer triples after just two CT scans (*Daily Mail*, 7 June 2012)**

Fact: The normal risk of children developing brain cancer in later life is 6 per 100 000 children. The original research on brain cancer and CT scans was thoroughly checked and found valid. (Cancer Research)

(b) **Dying for a tan (Sun online, 7 July 2020)**

Fact: Normal risk from skin cancer is 25 per 100 000. Recent research indicates that sunbed use before age 30 increases this by 75%. (Cancer Research)

(c) **Half a glass of wine every day increases breast cancer risk (BBC News, 23 May 2017)**

Fact: Research over several years has shown that half a glass of wine a day increases the risk of breast cancer by 5%. Out of 1000 women who do not drink, 125 may be expected to contract breast cancer. (Cancer Research)

13B First impressions

When reading or interpreting any graph or chart, ask yourself:

- What is the source of the data? Is it reliable?
- Is the type of display used suitable for the data?
- How have the axes been labelled? Do the scales start at zero?
- What is the message or story?
- Is the sample size large enough for you to draw reliable conclusions?

Activity 13B First impressions

The two charts below show the same monthly percentage unemployment rates for one year.

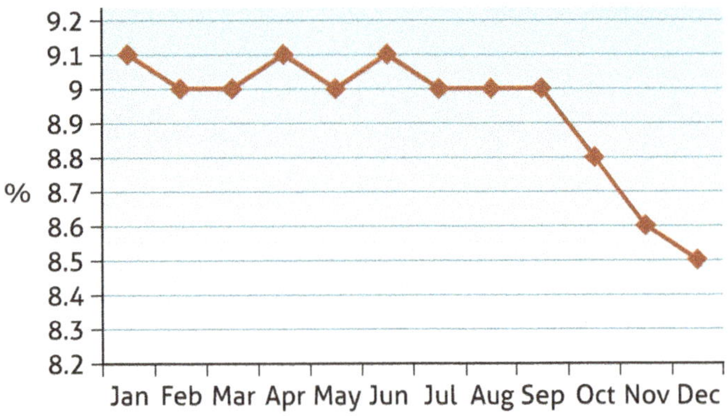

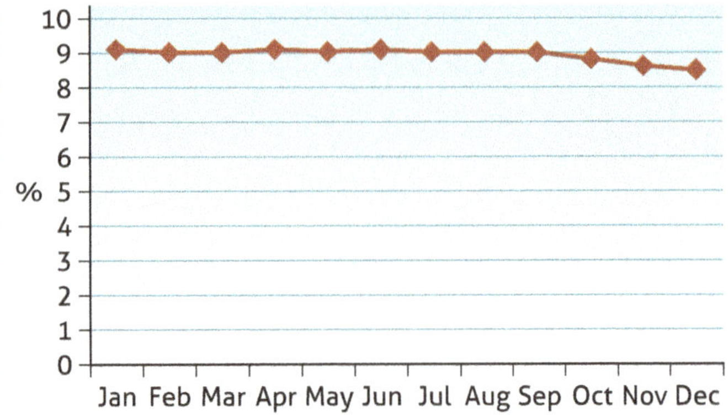

Give single statements to make a case for and against each chart.

Worked example 13B

? (a) Here are four charts, (i)–(iv). Look at each one, for no more than about five seconds, then jot down, in one or two sentences, your first impressions of the story the chart is telling.

(i) **Gun deaths in Florida**

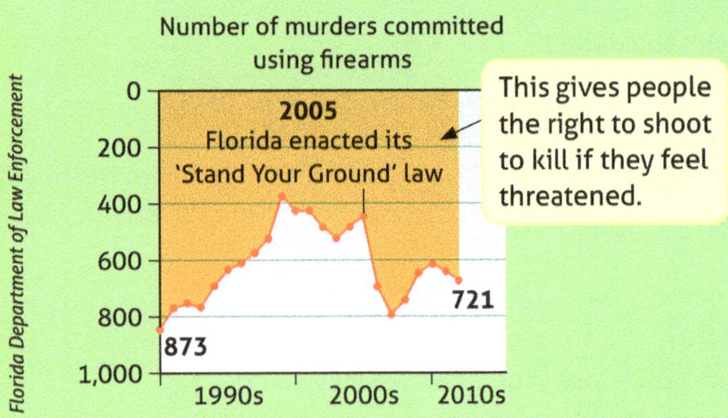

(ii) **UK driving accidents**

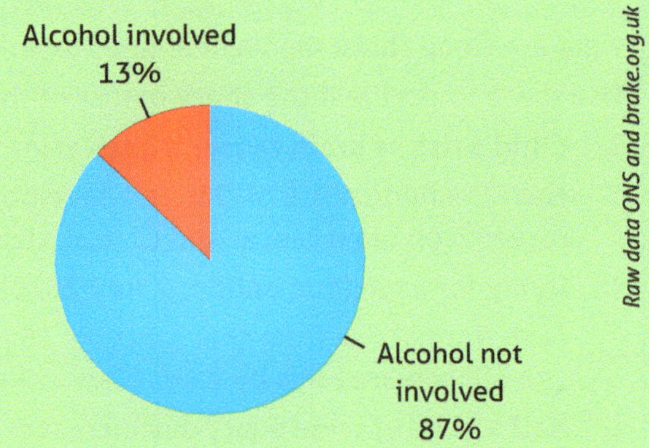

(iii) **Working hours**

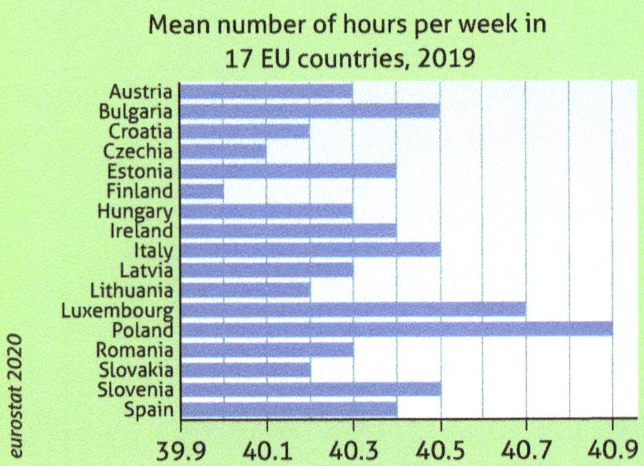

(iv) **Changing world temperatures**

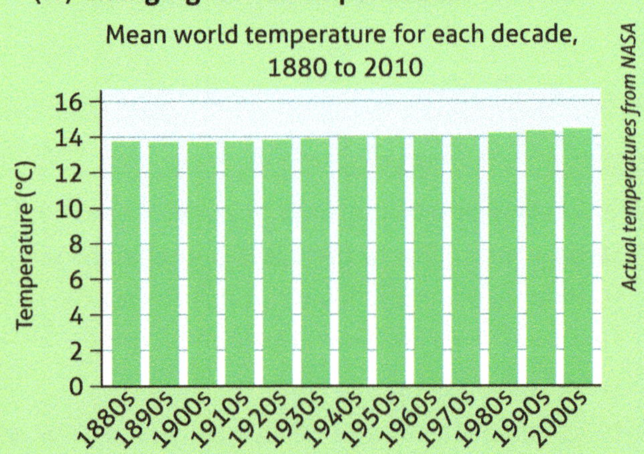

(b) Now look more closely at each chart and give your considered view of what each one is saying.

If this is not the same as your first impressions explain why.

Solution

	First impressions	Considered view
(i)	The new law reduced the number of murders involving firearms, which went up slightly and then decreased.	The exact reverse is true. The number of murders involving firearms increased after the law, went down then increased again. The vertical axis is increasing in the downwards direction rather than normal up.
(ii)	The number of accidents involving alcohol is a lot less than those where alcohol is not involved. Drinking is not an important factor in road accidents.	To comment on the effect of alcohol on the likelihood of a driver having an accident you need to compare the proportions of those drinking who have accidents with the proportion of those not drinking who have accidents. The pie chart shows the proportions of accidents not the proportions of drivers having accidents. So, it is not possible to draw a conclusion.
(iii)	Some countries, for example Poland, work a lot longer hours in manufacturing than, say, Finland.	There are differences in the numbers of hours worked but these are at most about an hour. The differences are exaggerated by having the horizontal axis from 39.9 hours to 40.9 hours.
(iv)	The temperature rise is very small – almost insignificant – if at all.	There is a rise in temperature overall. The chart makes it look very small but the effects of such a rise in temperature need to be known before its significance can be judged.

Exercise 13B

1 Is the chart on the right a fair way of comparing the five broadband providers?

What other information might be useful?

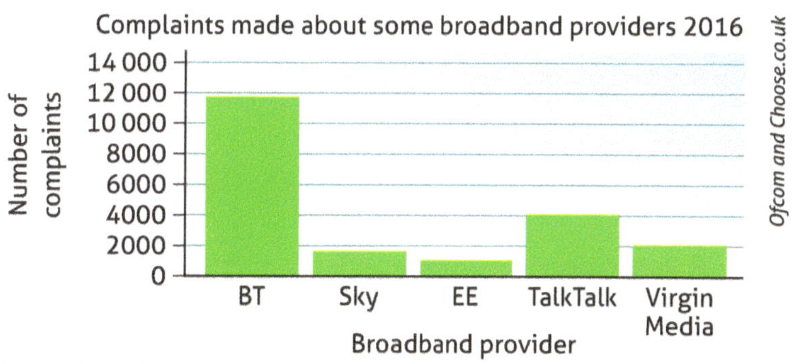

2 Here are some percentage support figures for five political parties A, B, C, D and E, for three, two and one months before an election.

Party	Months before election		
	3	2	1
A	17	20	22
B	19	19	20
C	22	21	22
D	20	25	19
E	22	15	17
Total	100	100	100

Show how the people in each of the following situations could use accurate bar charts to suggest that their views are correct.

(a) A journalist showing that the five parties have about the same support.

(b) A candidate for party A who says her party is making large and consistent gains compared with the others.

(c) A candidate for party C who says that his party is about twice as popular as either party D or E.

13C Correlation does not imply causation

When two variables change together they are associated; you can see association on a scatter diagram. If the points on the scatter diagram lie roughly on a straight line, the association is described as correlation.

The scatter diagrams below show data with positive and negative correlations.

They include lines of best fit: the stronger the correlation the closer the points are to the line.

You can use a spreadsheet to draw a scatter diagram and a line of best fit.

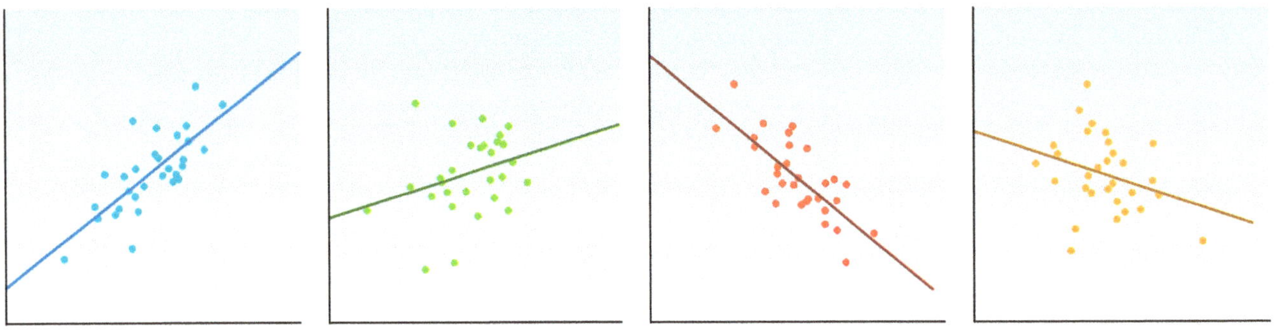

Strong positive | Weak positive | Strong negative | Weak negative

A common misunderstanding is to think that because two variables are correlated, one must cause the other. That is not so. Correlation does not imply causation.

Activity 13C An unexpected correlation

A famous, and at first sight weird, correlation is that between chocolate consumption per year and the number of Nobel prizes won by that country.

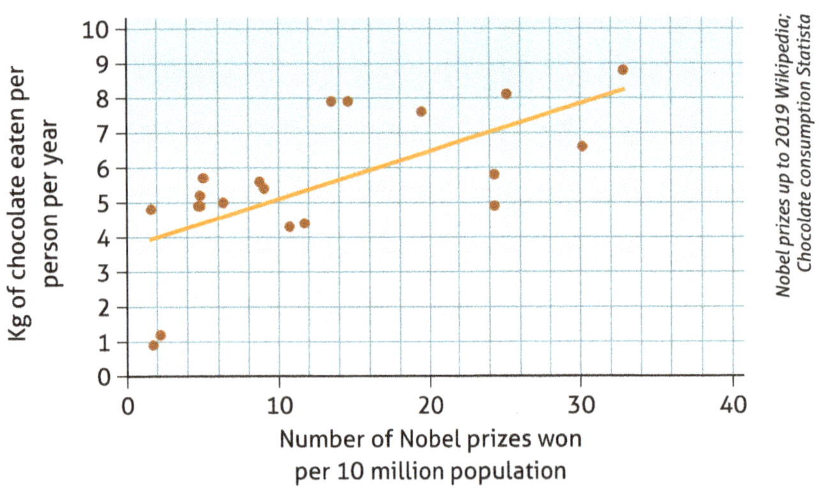

The data in the scatter diagram are up to 2019. The original work was published in a world-famous medical journal. The author claimed that eating more chocolate would **cause** an increase in the number of Nobel prizes won by a country.

Explain why the author was wrong and make your own suggestions as to how the correlation arises.

Worked example 13C

The effect of playing violent video games on aggressive behaviour has been investigated many times. Conclusions have ranged from a strong correlation to no correlation.

Here are the essential details of one experiment.

- 648 school students aged 12 to 19 were involved.
- The subjects were asked to select their 5 favourite video games and to score each of them on a scale of 1 to 7 for violence.
- The mean of these 5 scores was used as a measure of each student's exposure to violent video games.
- The subjects completed a recognised and respected questionnaire from which their level of aggression was calculated.
- The two numbers, exposure to video game violence and aggression level, showed a fairly strong positive correlation.
- The researcher claimed this proved that playing violent video games caused aggression in the players.

A **subject** is a person who takes part in an experiment. They are observed by the **researchers**.

(a) Comment on each of the points about this experiment in the table below.

(b) Suggest a possible reason for the correlation.

Solution

(a)

Point	Comment
The sample size was satisfactory.	Yes, 648 is a large sample.
The sample composition was satisfactory.	The information is not given but there were young adults who may well have had different views from the younger subjects.
The way the games were rated for violence was satisfactory.	Subjects rated the video games for violence themselves, so the score was a matter of opinion. Subjects selected their 5 favourite games but there may well have been others that affected them.
The way subjects were rated for their level of violence was satisfactory.	The aggression questionnaire is widely recognised to be valid, which is good.
The level of correlation proved that playing video games caused aggression.	This was a completely false claim by those running the experiment, correlation does not imply causation.

(b) It may well be that a third factor is involved. For example, certain personality types may enjoy playing violent video games and are given to aggression.

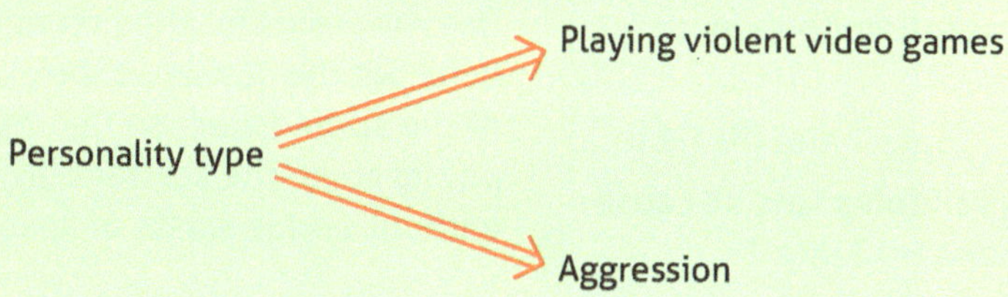

⚠ Many statisticians are wary about using correlation in cases where the variables are values from two time series. Many time series have either an upward trend or a downward trend. So, any two time series are likely to be correlated just because each variable changes with time, even though there is no meaningful relationship between them.

Bear this in mind when answering Question 2 in Exercise 13C.

Exercise 13C

1 Use this scatter graph to decide if there is any correlation between rain and sunshine in Lerwick. Describe any other trends shown in the graph.

Rain and sunshine, Lerwick in Shetland in June (from 1931 to 2017)

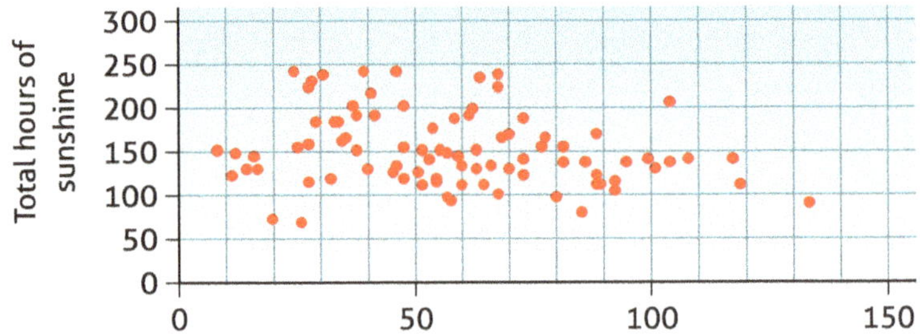

2 Table 1 shows data on UK chocolate sales (x) and PC sales (y), and Table 2 shows data on thefts from persons (x) and unemployment rate (y).

Year*	2008	2009	2010	2011	2013	2014	2015	2016	2017	2018
Chocolate sales (£ million)	473	421	245	159	135	119	132	178	183	188
PC sales (£ million)	115	94	80	32	30	15	13	16	20	31

Statista 2020

Table 1

* Figures for 2012 unavailable

Year	2010	2011	2012	2013	2014	2015	2016	2017	2018
Thefts from person (thousands)*	542	605	543	567	448	363	370	385	467
Unemployment rate (percentage)	7.9	8.1	8.0	7.6	6.2	5.4	4.9	4.4	4.1

ONS

Table 2

* Excludes fraud, shop lifting, bank robberies, etc.

(a) Enter the (x, y) data into a spreadsheet and use them to draw a scatter diagram.

(b) Describe any correlation that is present in (i) Table 1 and (ii) Table 2.

(c) Suggest a single cause that may affect both variables, and so cause any correlation between them in (i) Table 1 and (ii) Table 2.

Use the x-axis for the first row of data. Set the scales on the axes of the spreadsheet so that the points cover the scatter diagram without undue waste of space.

13D Bias

The word **bias** describes a systematic error in a particular direction when data are collected. There are many ways in which bias can occur. For example, in a survey:

- subjects tend to respond in the way they think the researcher wants or would be pleased by
- leading questions influence subjects to answer a certain way
- a subject may have a personal interest in the result.

This is sometimes called a halo effect.

Bias may be intentional, but it can also be unconscious.

Activity 13D Seeing and removing bias

Describe why these situations are biased. Suggest how the bias might be removed.

(a) A radio station polls listeners on a controversial issue by talking to people who phone into the station.

(b) A landline telephone survey about a new mobile phone.

Worked example 13D

? Explain why each of these experiment/survey situations may produce biased results.

	Situation	Explanation
(a)	A study investigating medication treating pneumonia selects subjects from pneumonia patients in a hospital.	The study omits patients who did not go to hospital but recovered at home.
(b)	Volunteers are asked to take part in a long-term study of eating habits.	Those who are interested in healthy eating are more likely to volunteer. Healthier people are more likely to complete the long-term study.
(c)	Workers in a building demolition company are asked to report any mental health issues they may have.	Some people are reluctant to report mental health issues to employers for fear of losing their jobs. (This also applies to any other sensitive medical issues.)
(d)	A random selection of patients are emailed and asked to attend the clinic within the next week for special blood and blood pressure monitoring.	Some groups (e.g. the elderly) do not use email as much as others. Some people will be nervous at being called in, so blood pressure might be higher than normal. (This has happened.)

Online social media surveys are very popular; they are cheap and can contact thousands of subjects. Suggest three reasons as to why they are biased.

Exercise 13D

1 Local councils in the UK need to know the number of rough sleepers in their areas.

(a) This number is usually found by simply going out onto the streets and counting the number of rough sleepers on a particular day and time. A local council does its count of rough sleepers once a year: between 10 pm and 11 pm on 1 August.

Explain why this probably gives a biased result.

Suggest ways bias could be reduced.

(b) Another method is to collect details from hospital admission records.

Why might this also be biased?

13E Regression to the mean

Thirty students on a forest management course were shown how to identify a particular fungus that attacks trees. They were then given a test in which they had to state whether or not each of 20 trees had the fungus. Unfortunately, none of them had understood the lecture so they all guessed the answers. Their scores were as follows.

Correct	< 4	4	5	6	7	8	9	10	11	12	13	14	15	> 15
Students	0	1	0	1	3	3	4	6	4	3	2	1	2	0

The lecturer says that the 5 students with the lowest scores must take a second test as a re-sit. The 5 with the highest scores will be entered for a competition and for practice they must also take the second test.

The 10 students taking the second test still guessed the answers.

Their scores were:

re-sit students: 9, 10, 10,10, 11
competition students: 8, 9, 10, 10, 12.

So, the re-sit students improved.
The lecturer concluded they had done some study.

The competition students did less well; the lecturer thought they must have been up late playing video games the night before.
Neither of the lecturer's conclusions were correct. The scores were just the typical chance outcomes when people are equally likely to be right or wrong at each guess.

The fact that the weak seemed to get better and the good seemed to get worse is an example of **regression to the mean**.

In this case the probability of a right answer was $\frac{1}{2}$

so, the mean score was 10 out of 20.

Particularly large or small values of a random (or partly random) quantity tend to be followed by others that are closer to the mean of the whole set.

Some uncertainty or randomness is often involved in measurements.

Activity 13E Chance events

Regression to the mean is likely to occur when a value is wholly or partly random. State whether each of the following is:

completely random
not at all random, or
partly random and partly not.

(a) The score of the home team in a football match.
(b) The total when two dice are rolled.
(c) A competitor's score in a general knowledge competition.
(d) The temperature of the water in a boiling kettle.

Sporting performance is a combination of skill and luck (randomness). This is because of weather conditions, referee decisions, etc. If the skill levels of teams are roughly the same, then chance will be an important element in the result. This is why it is difficult to predict football match results, especially when teams are close in position in their league.

Worked example 13E

Amber is a netball player. Her team plays in a league, so teams are at more or less the same skill level. The chart shows the goals she scored for her last 22 matches.

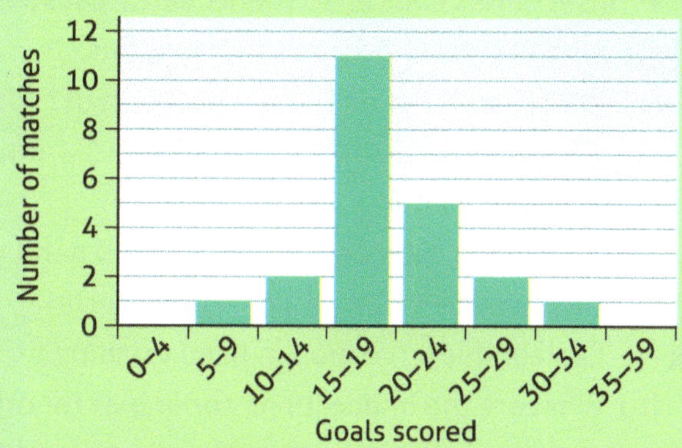

She scored a mean of 19 goals for these matches.

In one match she scores 11 goals.

The coach is not pleased and talks to her sternly.

In the next match Amber scores 17 goals. The coach says that players always do better when he has told them they must improve their game.

(a) Explain why this might not necessarily be true, particularly in Amber's case.

The team's best player, Jess, has a mean goal score of 26. Last game she scored 34. After the game the coach does not speak to her at all. In the next game she only scores 25. The coach thinks she is sulking and has not tried hard.

(b) Why is this not necessarily the case?

Solution

Scores in netball, as in all sporting events, depend on a combination of chance and skill, so regression to the mean is likely to occur.

(a) Amber's score of 11 was a lot lower than her mean score of 19.

Regression to the mean suggests that Amber's score in the next match would likely be closer to her mean.

So, her score of 17 could well have been pure chance and not the result of the coach's stern talk.

(b) Jess's mean score is 26. Her score of 34 is much higher than this, so it is likely that her score in the next match will be closer to the mean and hence lower.

The explanation of her score of 25 could be regression to the mean rather than sulking.

Exercise 13E

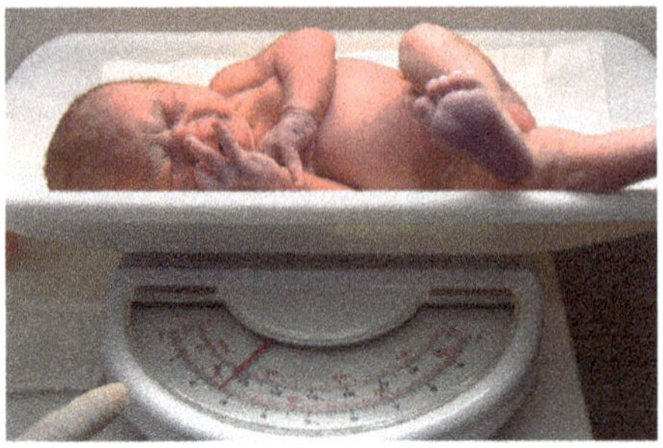

1 Explain how regression to the mean might explain these results.

(a) Women whose first-born babies weigh more or less than average weight, may expect that a second baby will tend to be lighter or heavier, respectively, than their first child.

(b) Amy is practising the multi-choice part of her driving test. There are 50 questions. The pass mark is 43. These are her scores for her last 10 practice tests.

44 43 40 42 41 36 47 45 44 38

She does another practice test and scores 47. She thinks she is certain to pass but doesn't.

Test yourself 13F

1 A study finds a correlation between an increase in the number of insect bites, B, in a month, and the number of car accidents, A, in that month.

Here are three possible explanations:

- Insect bites cause car accidents, $B \Rightarrow A$.
- Car accidents cause insect bites, $A \Rightarrow B$.
- Insect bites and car accidents partly cause each other, $A \Leftrightarrow B$.

(a) Find two other possible explanations.

(b) Which explanation is the most obvious? Justify your choice.

2 In an experiment about antiseptic hand washing in a hospital, staff were informed of the test and introduced to the researcher.

The researcher recorded an average of 5 hand washes per hour.

The next week a CCTV, part of normal security, recorded an average of 2 hand washes per hour.

(a) Explain this large drop in hand washing.

(b) What are the implications about bias for other experiments involving the observation of people?

3 Reporting on a piece of medical research, a newspaper reported that eating a high fibre diet reduced the risk of breast cancer by 6%.

A close look at the primary data showed that the risk of a woman getting breast cancer dropped from 3.3% to 3.1% by eating a high fibre diet.

(a) How did the newspaper obtain the figure of 6%?

(b) Comment on whether this was the best way of presenting the information.

4 A survey involving A and E (accident and emergency) patients was carried out in a Canadian hospital. Its purpose was to gain overall information about how patients were treated on arrival and their general initial impressions about admission.

A full-time researcher was employed during working hours. One of his first tasks was to obtain written permission from the patients to access their medical records; only then could they be given the survey. Several hundred detailed survey sheets were generated over two months.

Why would the results run the risk of being biased?

5 Several years ago a positive correlation was found between coffee drinking and heart disease. This was based on a large random sample. Subjects' coffee drinking (or not) habits were compared with their heart health. The researchers claimed that coffee causes heart disease.

(a) Explain why the researchers' conclusion was not justified.

(b) Suggest a different possible conclusion. You may choose to use the fact that smoking is now known to contribute significantly to a large number of diseases.

Practice question

Amber buys a window cleaning round.

According to the books, which have been checked, the mean profit a week is £295.

This can vary randomly over the year, although weekly profits outside the £245 to £345 bracket are rare.

(a) Amber's profit after the first week was only £250.

Should she be worried? Justify your answer.

(b) Amber wonders if the weather might affect business. She looks at the books again to find the number of window cleaning jobs each month. Amber plots these on a scatter diagram against total monthly rainfall and hours of sunshine, with the help of data from the Met Office.

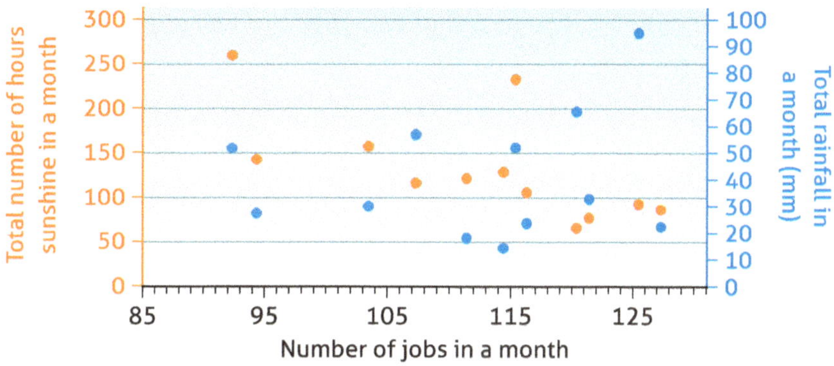

Which seems to have more effect, if any, on business: rain or sun?
Use the scatter diagram to justify your answer.

(c) According to a publication for window cleaners, they have an estimated risk of 0.0125% of getting seriously injured in a year. A new safety system claims a 20% risk reduction. Amber cannot believe this as 0.0125% − 20% = − 19.9875, which is obviously wrong.

What error has Amber made in understanding the '20% risk reduction'?
Calculate the correct value of the risk assuming the claim is true.

(d) After two years Amber decides to sell the window cleaning business.
She produces the chart below to show the profit over the two years.

Amber used the correct figures and there had been freak storms.
However, she made errors in how she presented the data.

Write down two errors she has made and explain how these might give a false impression.

Now that you have finished this chapter, you should know:

- That many scams are based on numbers and money.
- How to distinguish between 'percentage' and 'percentage points'.
- How to identify misleading statistical diagrams.
- How to recognise bias in published data.
- How to interpret scatter diagrams.
- How to recognise the difference between correlation and causing an effect.
- How to recognise regression to the mean and its possible implications or effects.

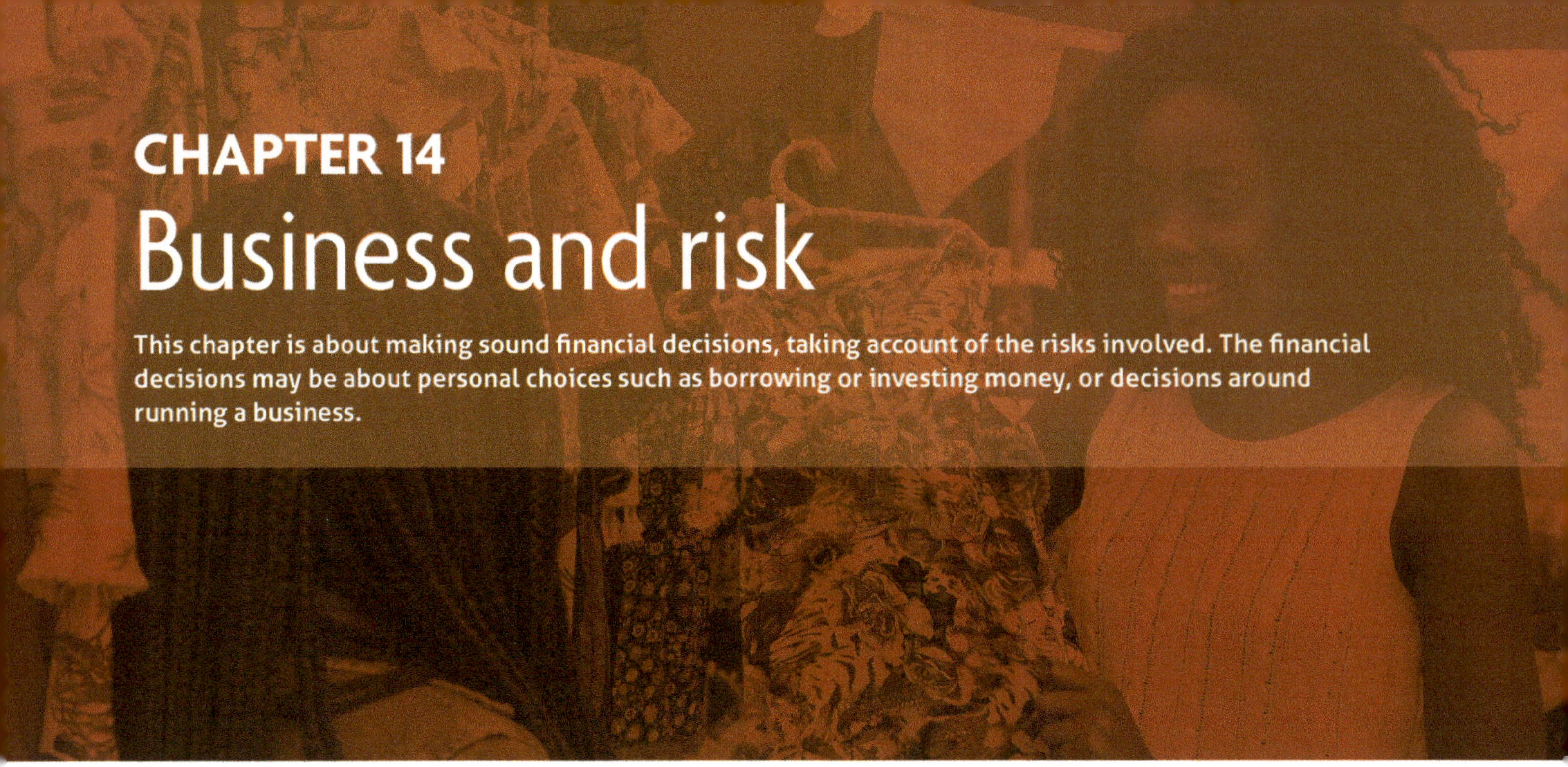

CHAPTER 14
Business and risk

This chapter is about making sound financial decisions, taking account of the risks involved. The financial decisions may be about personal choices such as borrowing or investing money, or decisions around running a business.

14A Which is cheaper?

Many shops and organisations offer cheap deals. However, you need to be wary: some may seem very attractive but can be deceptive.

Activity 14A Buying jam doughnuts

Debbie is buying jam doughnuts for her class.
She needs 12 doughnuts.

She looks at the offers:

- 5 for 65p
- 4 for 55p
- 20p each

Find the cheapest way she can buy 12 doughnuts.

What if she needed 13 doughnuts? Or 14? Or 11?

Worked example 14A

? Gordon buys a new laptop. The price is £499.

He has a choice of two ways to pay:

- 36 instalments of £19.17
- use a credit card that charges an annual rate of 21.8%. He will pay £250 a year.

(a) Work out how much he would pay in each way.

(b) What advice would you give Gordon? Give your reasons.

Solution

(a) **Instalments**
This option costs 36 × £19.17 = £690.12.

Credit card
After 1 year:
The interest is 21.8% of £499 = £108.78.
So, Gordon's debt is £499 + £108.78 = £607.78.
He repays £250, so his remaining debt is £607.78 − £250 =£357.78

> You can calculate this as 1.218 × £499.

After 2 years
Gordon's debt is 1.218 × £357.78= £435.78
He pays £250 so the remaining debt is £435.78 − £250 = £185.78.

After 3 years:
The debt is 1.218 × £185.78 = £226.28.
So, he makes a final payment of £226.28.
The total cost is £250 + £250 + £226.28 = £726.28.

> The amounts are rounded to the nearest penny.

(b) Using a credit card is more expensive by £726.28 − £690.12 = £36.16.

A problem with the credit card method is that £250 is a lot of money to find all at once, so it is riskier.

Paying by instalments is the cheaper option. The monthly payments are quite small, so he should be able to afford them, making that option less risky.

Larger packs usually work out cheaper than smaller ones.
Suggest some reasons why someone may prefer to buy a smaller pack.

Exercise 14A

1 (a) What percentage discount is equivalent to 'Buy one, get another for half price'?

(b) How much would 5 items priced at £17.50 each cost under the offer in part (a)?

(c) Suggest a reason why retailers decide to use 'Buy one, get another for half price' rather than just discounting by the percentage found in (a).

2 Marie buys a TV for £399.

(a) She uses her credit card. It charges interest with APR (i.e. an equivalent rate) of 26% per year. Assuming she pays £180 at the end of each year, work out her final payment for the TV.

(b) The store offers the opportunity to pay 36 monthly instalments of £18.70. Should Marie take this option? Give reasons for your advice.

14B Profit and loss

This section covers the basic calculations in profit and loss. Business decisions are very often based on the likely profit from a particular model; however, the risk involved must always be considered as well.

Activity 14B Bottles of water

Jaya buys 200 bottles of water for £180.
She plans to sell them at £1.20 each.

Her friend says she should sell them at 80p each because she would sell more of them.

Should Jaya take her friend's advice? Explain your decision.

Worked example 14B

? Ivan buys 300 packs of tennis balls for £270. He does some market research that suggests £2.50 per pack is a reasonable price to charge.

There is a risk that he will not sell all of the balls by the end of the summer and so needs either to sell them more cheaply or pay to store them through the winter months.

To make a profit you have to take more money in than you pay out.

(a) How much should he charge per pack to make 35% profit?

(b) Ivan calculates that he has spent 12 hours on buying and selling the tennis balls. The minimum wage is £8.20 per hour.

How much should he charge per pack of tennis balls to cover his time at minimum wage, assuming he makes no further profit?

(c) Ivan wants to cover his costs, including labour, during the summer season. He reckons that he can only be certain of selling half of the packs, and that it will still take 12 hours of his time. Will a charge of £2.50 per pack enable him to do this?

(d) After selling half of the packs he drops the price to £1.50 per pack and sells 90 more packs within the 12 hours he has budgeted.

What is his effective rate of pay for those 12 hours of work?

Solution

(a) 35% of £270 is £94.50.
£270 + £94.50 = £364.50

So, the cost per pack $\frac{£364.50}{300} = £1.22$.

The answer to (a) is £1.215, so the selling price is rounded to the nearest penny. Similarly, the answer to (b) is also rounded to the nearest penny.

(b) Ivan's pay is 12 × £8.20 = £98.40.
So, the total cost is £270 + £98.40 = £368.40.

The charge per pack is $\frac{£368.40}{300} = £1.23$.

(c) Ivan sells 150 at £2.50 making 150 × £2.50 = £375.00.
His pay and costs are £368.40 .
Yes, a charge of £2.50 enables him to cover his costs and pay himself minimum wage.

(d) Income from 90 packs at £1.50 each is 90 × £1.50 = £135.

Total income minus expenditure is £375 + £135 − £270 = £240.

His hourly pay is $\frac{£240}{12} = £20$.

Exercise 14B

1 Anya buys 250 boxes of candles for £600. She estimates her costs are £210 altogether. She sells the candles for £4 per box.

(a) How much profit does she make?

(b) There is a risk that Anya may not sell all of the candles. What could she do to avoid losing money?

2 Yuri makes 120 wooden bowls and sells them at a local market. His materials cost £1.40 per bowl. In an hour he can make 2 bowls and aims to pay himself £12 per hour for his time. Any surplus goes towards paying for the equipment he has bought.

He is able to pay £72 towards his equipment.

What does he charge for the bowls?

3 Delia bakes 80 cakes for sale at the market. She charges £4.50 for each cake. She estimates that the cakes cost £1.48 each to make, and she takes 10 hours altogether to make them. The profit provides her wages.

(a) How much does she earn per hour if she sells all of the cakes?

(b) Delia has sold 55 cakes by the final hour of the market. She does not want to risk having any cakes left, so drops the price to £1 per cake for the last hour.

Comment on her strategy.

14C Adjusting to the market

Constructing a business model will involve the likely profit and loss from the enterprise but it will also include consideration of the risks involved.

Activity 14C What to do?

From Activity 14B, Jaya sticks with her plan to sell her 200 bottles of water at £1.20 each.

She sells 150 of them at that price.

(a) Has she covered her cost of £180?

(b) What do you think she should do with the remaining 50 bottles?

Worked example 14C

? Alan constructs a business model for selling mobile phone cases on his market stall:

- He will spend £4500 on 400 mobile phone cases.
- He will sell them at £20 each.
- He calculates that, if he sells enough of them, he will make a profit.
- There is a risk that he will not sell all of them.
- He plans to reduce the price if demand drops.
- His time is not accounted for as he is running the stall anyway.

(a) How many does he need to sell to cover his initial spend?

He sells 150 but then demand drops and he decides to reduce the price to £15.

(b) How many more does he need to sell at this price to cover his initial spend?

(c) Alan decides to sell the remaining phone cases below cost price. What advantages are there in doing that?

Solution

(a) To break even selling them at £20 he needs to sell $\frac{4500}{20} = 225$.

He needs to sell 225 mobile phone cases to cover his initial spend.

(b) Income from sales of the first 150 is £20 × 150 = £3000.

He needs a further £1500 to recover his money.

Number to sell at £15 is $\frac{1500}{15} = 100$.

He needs to sell 100 phone cases at £15 to recover his money.

(c) He increases his income from the mobile phone cases.
The additional income offsets any eventual loss or increases any profit.
He does not have to store the unsold product.

In fact, he would prefer to sell more to cover his initial outlay because his time and other costs are not covered. If he sells other products too, so much the better.

Exercise 14C

1 Sajid constructs a business model:

He pays £300 for 80 battery chargers.

He plans to sell them for £7.99 each.

He sells 35 at that price.

(a) What is the income from the 35 battery chargers?

(b) His business model includes making a profit of 25%.

What should he charge for the remaining battery chargers?

14D Planning the product and price

There are many reasons why demand for a product may be lower than expected. Sometimes the price is too high. It may also be that the market is saturated and no one wants to buy that product, whatever the price.

Activity 14D What product? What charge?

Ranjit plans to start a business delivering hot meals to people who are housebound.

(a) List some of the issues he must take account of.

(b) How many people can he manage on one round?

(c) What would be a sensible price range?

(d) Should he offer a choice of meals?

(e) Give reasons for your answers.

Worked example 14D

A new start-up plans to sell sandwiches.

It will deliver to local businesses.

It is constructing a business model, and this includes deciding what to charge for a pack of sandwiches.

Question

(a) List the **preliminary** cost items for their business.

Solution

- Transport to fetch supplies and deliver the sandwiches.
- Cold storage for sandwiches in transit.
- Utensils needed to make sandwiches.

(b) List the cost items for **each pack** of sandwiches.

- Bread, spread, filling
- Packaging
- Time to make it

(c) Estimate the **time** that two people would need to work per day.

- 10 am until 1:30 pm delivering
- 2 hours making the sandwiches
- 2 hours shopping, cleaning, and paperwork

(d) Set, and justify, a **price** for the product.

- Sandwiches in supermarkets cost around £3–£4 a pack.
- The convenience for a customer of having the sandwiches delivered would offset a higher charge but £5 is probably a maximum.
- Suggest £4.25 to start with.

(e) Suggest any **other factors** they should they consider.

- The number that can be sold in the time window.
- The proximity of offices where the sandwiches can be sold.
- What to do with unsold sandwiches.
- Any legal costs or health and safety considerations.

The business model will also involve the costs. The cost per sandwich is likely to be much less than £4.25.

Investigate whether it is worth using butter and home-baked bread.
Would the price mark-up cover the additional costs? Is there a demand for these?

Exercise 14D

1 (a) Give a reason why it is not a good idea to set the price of a product too high.
(b) Give a reason why it is not a good idea to set the price of a product too low.

2 Sarah sets up a mobile hairdressing business.
She constructs a business model.
She plans to visit clients' houses.
Should she charge more for a cut than a hairdressing salon?
Justify your answer.

14E Supply and demand

It is often the case that the higher the price of an article the lower the sales are likely to be. The relationship can be illustrated by a demand curve. Demand curves can be very helpful in constructing business models.

Activity 14E Price versus quantity

An item costs 20p to produce.

Use the demand curve to find:

(a) the selling price if 100 are sold

(b) the number sold if the selling price is 41p

(c) what happens with a selling price of 20p

(d) the profit if the selling price is 30p.

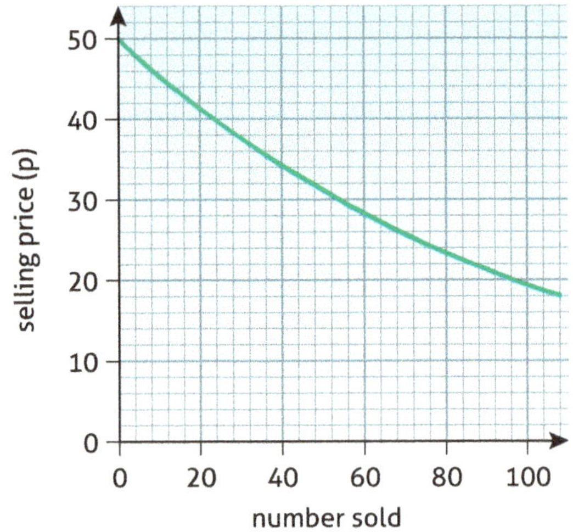

Continue this spreadsheet for larger numbers sold, taking readings from the graph for column B.

	A	B	C
1	Number sold	Selling price (p)	Profit
2	10	45	£2.50
3	15	43	£3.45
4	20	41	£4.20
5	25	39	£4.75

(e) Find the selling price that gives the most profit.

Worked example 14E

? A range of trainers, priced at £50, is only selling 200 pairs per month.

Market research suggests that every price reduction of £5 will increase monthly sales by 100.

The trainers cost £32 to produce.

(a) Set up a spreadsheet to model the likely profit for each price point.

(b) What price will generate the most profit?

Solution

(a)

	A	B	C
1	Price	Number sold	Profit
2	£50	200	£3,600
3	£45	300	£3,900
4	£40	400	£3,200
5	£35	500	£1,500
6			

=(A2−32)*B2

(b) The model suggests that £45 is the optimal price to charge as it yields the most profit.

Note that it is neither the highest nor the lowest price.

£3900 is the largest profit.

Exercise 14E

1 A company sells mugs.
Market research provides this demand curve.
The mugs cost £2.50 to produce.
Find the optimal selling price.

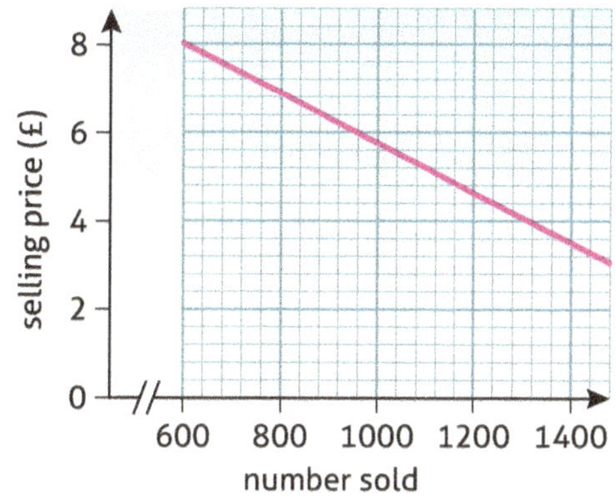

14F Predicting the future

Successful sales are often built on market research. Finding out what sells well and what people want enables you to remove some of the risk from your business venture. You can never remove it all though.

Activity 14F What's your sandwich filling?

(a) Predict the percentage of people who prefer each filling.
(b) Now hold a vote among your friends or classmates to see which filling they prefer.
(c) Were you right?
(d) Are your results in (b) a good basis for deciding how many of each type of sandwich for the canteen to stock?

Ham and cheese
Cheese and onion
Tuna mayonnaise
Tomato salad

Worked example 14F (1)

? A stall-holder selling cupcakes looks at last week's data to decide how many of each type of cake to stock each day.

The spreadsheet shows last week's sales:

	A	B	C	D	E	F	G	H
1	Cake	Monday	Tuesday	Wednesday	Thursday	Friday	Saturday	Totals
2	Vanilla	5	3	7	12	21	29	77
3	Red velvet	8	7	3	8	15	18	59
4	Cappuchino	7	6	8	9	23	19	72
5	Chocolate	12	15	8	11	31	25	102
6	Lemon	8	4	7	5	28	22	74
7	Totals	40	35	33	45	118	113	384

(a) (i) What formula is in cell H2? (ii) What formula is in cell E7?

The stall-holder plans to bake nearly 500 cakes; they must be in batches of 12 of the same type of cake.

(b) Use the data to suggest how many of each type of cake she should bake.

(c) What are the risks of this method?

(d) Suggest an improved basis for her decision.

Solution

(a) (i) =SUM(B2:G2) (ii) =SUM(E2:E6)

(b)

	A	H	I	J
1	Cake	Totals		Next week's batches
2	Vanilla	77		8
3	Red velvet	59		6
4	Cappuchino	72		8
5	Chocolate	102		11
6	Lemon	74		8
7	Totals	384		41

=INT(H2/384*500/12+0.5)

INT means 'integer part of'.

+0.5 means the answer is rounded to the nearest whole number, not rounded down.

(c) People buying cakes are not necessarily going to be consistent from week to week; her customers are likely to vary each week.

(d) Use a rolling average of the weekly sales of each type of cake.

Playing a game of chance also involves risk. The probabilities are easier to determine but you are more likely to lose, since that is what the provider of the game plans to happen.

Worked example 14F (2)

In a raffle there are 500 tickets, each costing £2.

There are 3 prizes of £10, 2 prizes of £50 and 1 prize of £100.

(a) Find the organisers' expected profit from the sale of all of the tickets.

(b) Find your expected profit if you buy 1 ticket.

(c) What does your answer to (b) tell you?

Solution

(a) The raffle organisers expect to receive £2 × 500 = £1000.
They expect to pay out £100 + £50 × 2 + £10 × 3 = £230 in prizes.
Their expected profit is £1000 − £230 = £770.

(b) This table shows calculations for your expected winnings from 1 ticket.

Prize	Probability	Expected winnings
£100	$\frac{1}{500}$	$\frac{1}{500}$ × £100 = £0.20
£50	$\frac{2}{500}$	$\frac{2}{500}$ × £50 = £0.20
£10	$\frac{3}{500}$	$\frac{3}{500}$ × £10 = £0.06
	Total	£0.46

Notice that your expected profit is what you expect to get on average in the long run. It may well not be the same as what you are hoping for.

In the table the expected winnings from the various prizes are added together. Then you must subtract your costs, in this case £2 for your ticket.

Your expected profit, by subtracting the ticket cost, is £0.46 − £2 =−£1.54.

(c) On average, you will lose.

Exercise 14F

1 A game is played with a six-sided die with faces numbered 1 to 6. You pay £1 per go and win £5 if you roll a 6. How much money do you expect to gain if you play 60 times?

2 The charge to join an outdoor Christmas lights event is £10. The organisers expect 200 people if the weather is fine; if the weather is poor the event will be cancelled. The organisers are planning 5 events. They have fixed costs of £4000 whether or not the events are held.

Using data from previous years, the organisers estimate the probabilities of different numbers of events taking place to be as follows.

Number of events	0	1	2	3	4	5
Probability	0.05	0.15	0.25	0.3	0.15	0.1

(a) What is the greatest possible loss for the organisers?

(b) What is the probability that the organisers make a loss?

(c) Calculate the organisers' expected profit.

The organisers decide to take out insurance to guard against possible losses if the weather is bad.

The premium for the policy is £1000.

It will pay out £2500 if more than 3 events are cancelled because of bad weather.

(d) What is the greatest possible loss for the organisers now?

(e) (i) What is the organisers' expected profit now?

(ii) What is the insurance company's expected profit?

14G Getting it right

It's not possible to predict the future exactly, so constructing a business model is an uncertain process. When determining quantities to manufacture or the price to charge for a product, the associated risk must always be considered.

Figures from previous years are often used as a starting point but they need to be refined by looking at trends that may affect the market.

Market research can also be very helpful. There may also be particular factors at work in the area of the market where the business operates.

Activity 14G How many items?

You are running a business and the price for your product is set.

You now have to decide how many items to produce on the next production run.

For each product below, choose one source of information from the list to the right that you feel is most appropriate, justifying your decision.

(a) Biker jackets

(b) Luxury soaps

(c) Luxury chocolates

(d) Speciality breads

(e) Jewellery

(i) last month's data from your company

(ii) last year's data from your company

(iii) last quarter's data from your company

(iv) market research

(v) national sales information

Worked example 14G

? A garden centre is working out what seeds to plant ready for selling in the Spring. They look at what they planted last year.

They look at the results of a survey of gardeners about their plans for their gardens next year. Many of them are planning to increase the growing of their own food.

Last year:	
Grow your own food	20%
Bedding plants	40%
Perennials	10%
Annuals	30%

Last year the annuals and bedding plants needed heavy discounting to sell them all.

(a) What are the advantages and disadvantages of repeating what they did last year?

(b) What other information would be useful?

(c) What are the risks of using the survey results as a basis for their planting?

(d) Make a recommendation on what seeds to sow this year, justifying your changes to last year.

Solution

(a) Advantage
They know how well it worked.

Disadvantages
It may not have been a good decision last year.
Circumstances may be different this year.

(b) Knowing how many of each type of plant was sold last year
Knowing different pressures that will affect the market this coming year.

(c) Respondents to the survey are unlikely to include less dedicated gardeners, who will impulse buy what is available.
It does not take account of unseasonable weather during Spring months.

(d)

Last year		**This year**
Grow your own food	20%	30% survey suggests these are popular.
Bedding plants	40%	30% these were hard to sell last year.
Perennials	10%	20% these can be sold for most of the year.
Annuals	30%	20% these were hard to sell last year.

Exercise 14G

1 An insurance company reviews its charges for flood insurance.
Which of the following factors should they take into account?

(a) Number of claims on flood insurance in the last 12 months.

(b) Frequency of storms in the last 12 months.

(c) Frequency of drought conditions over a 5-year period.

(d) Change in pattern of storms over a 5-year period.

(e) Location of the property requiring flood insurance.

(f) Age of the property requiring insurance.

Give reasons for your answers.

Test yourself 14H

1 Walt buys a scratchcard each day. It costs £1 and he has a 0.1% chance of winning £50. How much can he expect to win or lose over 10 years?

2 Tallis buys a cooker costing £349 using her credit card rather than choosing the store's plan of 24 monthly payments of £22.69. Her credit card charges APR interest of 25.2%. She plans to pay off £200 at the end of each year.
Is this better value than the store's option?

3 Dido constructs a business model:

She pays £7500 for 5000 bottles of a cold cure.
Her market research suggests a selling price point between £3.50 and £5.50.
She prices them at £5 each to start with.
She plans to drop the price by 50p if demand drops.
She thinks that a profit of 150% is reasonable.

She sells 2400 of them for £5.

She sells a further 600 at the next price point by the beginning of March.

It is now past the peak selling time for her product.

Dido risks being left with 2000 bottles of cold cure that will not keep until next winter.

(a) Advise her, giving your reasons.

(b) Will she make her planned 150% profit?

4 Nadine parks her car in a no parking zone each weekday while she is at work. She reckons that the probability that she gets a parking fine, of £50, on any given day is 10%.

The nearby car park charges £0.80 per hour.

How many hours does she need to park each day to make the fine better value?

5 A company sells frozen pizzas. Their present business model is:

Sell at £7.50 each.
Pizzas cost £2.25 to produce.
They sell 1000 pizzas per month.

They conduct market research and obtain this demand curve.

It goes through the points (1000, 7.5), (2000, 5) and (3000, 2.5).

Would you advise them to change their selling price? If so what should it be?

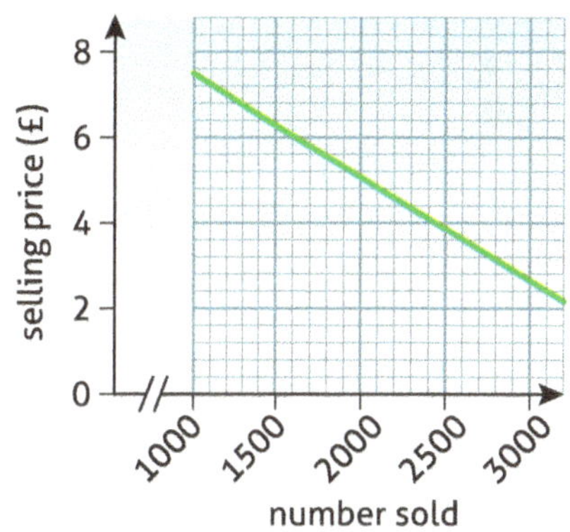

Practice question

A store has an offer that if you buy 3 items, the cheapest is free. Nick buys 6 items costing £3.99, £12, £4.75, £11.99, £7.50 and £15.

(a) How much does he pay if the items are taken in the order they are listed?

(b) What is the least that Nick could pay?

(c) What is the most that Nick could pay?

Now that you have completed this chapter, you should know:

- How to make sound financial decisions.
- How to construct a business model for a small business.
- How to set an optimal price for a product.
- How to use data to mitigate risk.
- How to use probability in making business decisions involving risk

CHAPTER 15

Making decisions

This chapter is about using the available information when making decisions. It includes realising that in some situations more information is needed before an informed decision can be made and that sometimes assumptions are required.

15A Selecting a winner

In a single race with individuals taking part there is no doubt who is the winner: the person first across the finish line. It isn't as clear when the race is one of a series or a team event.

Activity 15A Who has won?

- A typical Formula 1 season has 21 races.
- In each race, points are awarded to the first 10 drivers to finish: 25, 18, 15, 12, 10, 8, 6, 4, 2, 1.
- A driver in the top 10 can gain 1 more point for the fastest lap of the race.
- The winner of the annual championship is the driver with the most points in total.
- If two drivers end the season with the same number of points, the driver with more wins is the winner. If they have the same number of wins, it is decided on second places and so on.

When dealing with complicated rules it can help to set them out as bullet points like this.

What is the least number of races a driver needs to win to be certain of winning the annual championship? Assume that each time they win they get a bonus point for the fastest lap.

Worked example 15A

? In a cross-country team-race, each team consists of 5 runners.

All runners complete the race.

The runner who finishes first is awarded 1 point, the runner who finishes second is awarded 2 points and so on.

A team's score is the sum of the points awarded to all of its 5 members. The team with the lowest total score wins.

This table shows finishing positions in a race between two teams the iPlods (IP) and the Happy Feet (HF).

Position	1	2	3	4	5	6	7	8	9	10
Team	IP	HF	HF	IP	IP	HF	IP	HF	IP	HF

The iPlods scored 1 + 4 + 5 + 7 + 9 = 26.

(a) Which team won the competition?

(b) With this scoring system, is it ever possible for there to be a draw in a two-team race?

(c) What is the highest possible winning score? Give two ways in which it can be achieved.

Solution

(a) Happy Feet scored 2 + 3 + 6 + 8 + 10 = 29 points, so the iPlods won with 26.

(b) The total number of points available in a race is 1 + 2 + 3 + 4 ... + 10 = 55. This is an odd number of points, so the two teams cannot have the same score.

(c) The highest possible winning score is 27, with the other team scoring 55 − 27= 28.
There are many ways of scoring 27.
Here are two of them: 1 3 5 8 10 and 2 3 5 7 10.

Exercise 15A

1 Since about 1980, in UK football, a win gains a team 3 points, a draw 1 point and a lost game zero. This is called the 3–1–0 system. Before this, the 2–1–0 system was used.

Below are tables showing the top 5 teams, in the top league, arranged in position (rank order), using the old 2–1–0 system.

What would the order have been using the 3–1–0 system?

1975/1976 Season	**Position based on the 2–1–0 system**				
Position	**Team**	**Won**	**Drawn**	**Lost**	**Points**
1	Liverpool	23	14	5	60
2	Queens Park Rangers	24	11	7	59
3	Manchester United	23	10	9	56
4	Derby County	21	11	10	53
5	Leeds United	21	9	12	51

In the following Questions 2 and 3, teams are ranked using the method in Worked example 15A.

2 In a cross-country race there are 3 teams, each of 4 runners.

All 12 runners finish.

(a) What is the highest points score possible for a team?

(b) Explain why it is possible for all 3 teams to get the same points score.

3 In the cross-country race in Question 2, one of the teams finishes 10th, 11th and 12th. Could the other runner in the team have saved the team finishing last overall?

15B Ranking and voting

This section involves ranking or putting in order. For example, ranking TV programmes in order of preference or gymnasts in order of their artistic interpretation. Elections provide another example of ordering.

Activity 15B In no particular order ...

This table shows the rank order decided by four judges P, Q, R and S, for the six contestants in a gymnastics competition, A, B, C, D, E and F.

So, Judge P ranked C as 1st, E as 2nd, A as 3rd and so on.

Devise two methods of combining the scores that give different rank orders for the six contestants.

Contestant	Judge P	Judge Q	Judge R	Judge S
A	3	4	5	6
B	4	1	2	5
C	1	2	6	3
D	5	5	4	4
E	2	6	1	1
F	6	3	3	2

Some countries, such as the UK, use the 'first-past-the-post' system to elect their members of parliament: in each constituency the candidate who gains the most votes is the winner.

Many other countries use proportional representation; for each party, the numbers of members elected are in the same proportion as the votes cast. There are several systems for proportional representation.

Worked example 15B

The table shows the votes cast in a fictional election.

- There are 7 constituencies, labelled 1 to 7.
- Each constituency has 100 voters and they all vote.
- There are 4 political parties A, B, C and D.

In the UK there are 650 constituencies. Typically, they each constituency has about 80 000 voters but not all of them vote.

	Constituency							
Party	1	2	3	4	5	6	7	Total votes
A	33	12	12	41	11	26	14	149
B	32	47	37	39	32	24	45	256
C	15	19	32	5	22	25	35	153
D	20	22	19	15	35	25	6	142
Total	100	100	100	100	100	100	100	700

(a) Find the number of members elected for each party if the first-past-the-post system is used.

(b) Calculate the number of members elected for each party if proportional representation is used.

(c) Comment on the effects of the different voting systems.

Solution

(a) Under the first-past-the post system the winning party for each constituency is as follows.

Constituency	1	2	3	4	5	6	7
Winner	A	B	B	A	D	A	B

So, Party A gets 3 members, Party B gets 3 members, Party C none and Party D 1 member.

(b) Here are the total numbers of votes for each party and the percentages of the overall vote, to 2 decimal places.

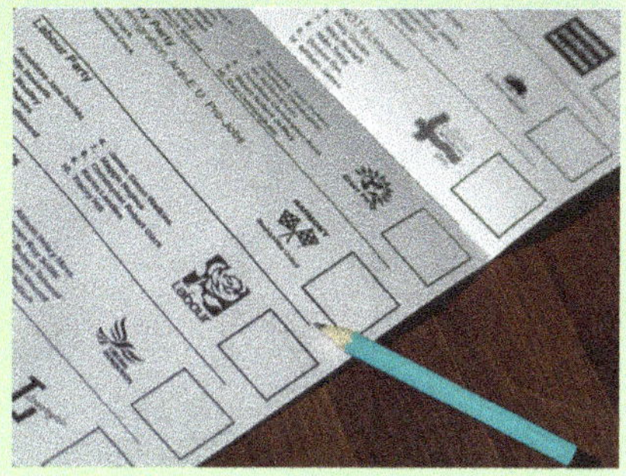

Party	A	B	C	D
Total votes	149	256	153	142
% of vote	21.29	36.57	21.86	20.29

Party A gained 21.29% of the votes. So, under proportional representation (PR), Party A would get 21.29% of the 7 seats available.

$\frac{21.29}{100} \times 7 = 1.49.$

Rounded down to the nearest whole number this is 1.

Doing the same calculation for the other parties gives the figures in the table below.

Party	A	B	C	D
Number of seats under PR	1	3	2	1

(c) The outcomes are different under the two systems. Under PR, Party A gets fewer seats and Party C gets 2 instead of 0. There is a more even distribution of members between the parties using PR.

Exercise 15B

1 In some countries, for example Argentina, Israel and Turkey:

- people only vote for a particular party on election day, not a person
- the number of seats a party wins is proportional to their number of votes
- however, who takes up a seat is decided by a rank order list published by each party.

So, for example, if party A gains 40% of the vote in a 360-seat parliament, they will win 144 seats. Only the first 144 people on the party's list will become members of parliament.

(a) In a 180-seat parliament a party gains 20 million votes out of a possible 60 million votes. Catalina is 50th in that party's list. State with reasons whether she will become an MP.

(b) In the same election a party gains 200 000 votes.

How many seats should they win? Comment on your answer.

2 Some countries vote for a president by putting their preferences in rank order.

Imagine four candidates Dupont, Garcia, Lee and Sharma, are entering an election, with a total of 100 million voters. Here are the results, shown in millions, in the table. Dupont gained 33 million 1st preferences, 17 million 2nd preferences, 31 million 3rd preferences and 19 million 4th preferences, etc.

Preference \ Candidate	Dupont	Garcia	Lee	Sharma
1st	33	19	32	16
2nd	17	36	18	29
3rd	31	18	34	17
4th	19	27	16	38

(a) Devise two different ways of using voter preferences to decide who wins the election.

(b) In each case use your method to decide who will be president.

15C Making comparisons

Sometimes you want to compare two quantities.

A common technique for comparing values at different times or places involves setting one of them as an index value; typically, it is assigned the value 100. Index numbers were covered in Chapter 3.

Activity 15C Gender ratio

People planning social provision often need to know the proportion of males to females. They need a number which is easily understood and compared. The World Health Organisation (WHO) use the number of males per 100 females, usually rounded to the nearest whole number. This is called the gender ratio, even though it isn't actually written down as a ratio.

(a) From 2013 to 2017, in Great Britain (England, Scotland and Wales), 1 820 581 female and 1 918 893 male babies were born.

What is the gender ratio for babies born in Great Britain from 2013 to 2017?

(b) According to the World Bank, Namibia has the smallest gender ratio for babies. It is 101.

That year a total of 70 240 babies were born.

Estimate how many of them were boys.

Various measures are used to compare countries' development. This helps to ensure that support is provided where it is most needed. The Human Development Index (HDI) takes into account many features such as quality of life, breadth of education available, gender equality and life expectancy. Economic factors may seem less important if a country has, for example, an unequal justice system or a poor gender equality record.

There are situations when you want to compare two quantities where you don't know the value of either, but you can use careful reasoning to decide between them.

Worked example 15C

The diagrams show two garden ponds, A and B. They are drawn to the same scale. Which has the larger surface area?

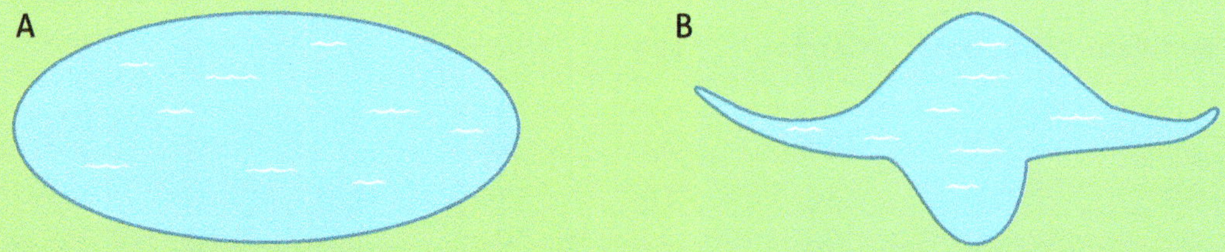

Solution

Both ponds have the same length and width.

However, the edge of pond A is convex whereas that of Pond B is largely concave.

So, Pond A has the larger surface area.

Exercise 15C

1 In each of these situations, no calculation is needed for the choice you have to make. Instead, you should give a reasoned answer.

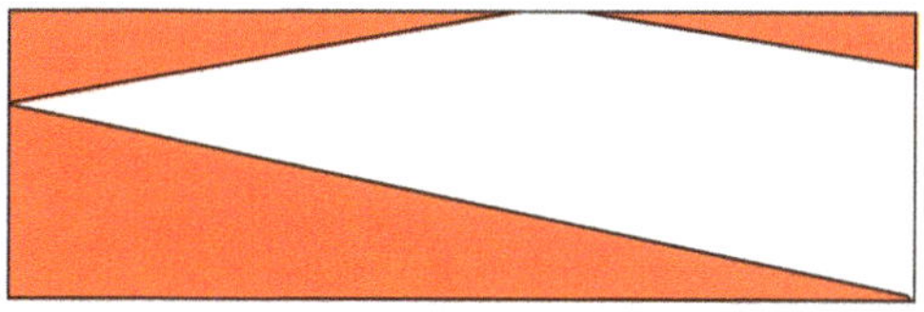

(a) Is it true that more of the area inside the rectangle is coloured red than is coloured white?

(b) Which is better value for money: a 5 kg bag of rice at £8 or a $2\frac{1}{2}$ kg bag at £5?

(c) Mark runs a 26 mile marathon in 3 hours. Oscar runs at 9 mph for long distances. Who is the faster runner?

(d) Richard and Henry are brothers. They know that they are descended from William the Conqueror but don't know how many generations ago that was. Richard says it was 40 and Henry says 20. Who is more likely to be right?

2 This table shows the gender ratios for four age ranges in China, Peru and the UK, based on UNO figures in 2020.

	At birth	1 to 14 years	15 to 65 years	66 years and above
China	113	115	106	88
Peru	105	101	99	89
UK	105	105	100	85

Describe and give likely reasons for the changes and differences between the gender ratios for the different age groups in the three countries.

15D Interesting but is it supported by the numbers?

Sometimes quite surprising claims are made but on closer examination and some calculations ...

Activity 15D Long-distance walking

Captain Barclay was a famous early 19th-century Scottish walker.

His most famous walking record was in 1809, when he walked 1000 miles in 1000 hours to win a bet of 1000 guineas.

1000 guineas is equivalent to over £80,000 in today's money.

1000 miles in 1000 hours is an average speed of 1 mile per hour. That sounds easy, but consider these facts:

- Average walking speed is about 3 to 4 miles per hour.
- The Olympic record time for the 20 km walk is 1 hour 18 min.
- The average adult needs between 6 to 9 hours sleep a day.

Draw up a realistic plan for walking 1000 miles in 1000 hours.

Do you think you could do it?

Assumptions always need noting so they can be checked.

Worked example 15D

? Trafalgar Square is often used for mass meetings and demonstrations.

It is roughly square, measuring 110 m by 110 m.

Health and safety experts say that for stationary crowds the maximum safe number of people is between 400 and 500 in each 100 m^2.

Is the News Flash shown on the mobile phone credible?

Solution

Assuming that the health and safety guidance is followed,
the maximum number of people per square metre is $\frac{500}{100} = 5$.

The area of the square is $110 \times 110 = 12\,100$ m^2.
Assuming all this space is available, the maximum crowd size is about,
$12\,100 \times 5 = 60\,500$, say roughly 60 thousand.
However, Trafalgar Square has fountains, so not all of its area is available to people.
So a crowd of 50 thousand is just about credible but possibly an upper limit.

Exercise 15D

In each of these questions there is a statement followed by some facts (which have been verified and checked). Decide which statements are valid – using the **facts** to justify your choice together with any **assumptions** you might make.

1 'We are about as far away now in time from the first moon landing, as I was to World War 1 when I saw them land on the moon.'
- World War 1 was 1914 to 1918.
- The first moon landing was 20 July 1969.

2 'Amy thinks that Matilda, her grandmother, could in theory have met somebody who had met Karl Marx.'
- Karl Marx (1818–1863).
- Matilda died 40 years ago when she was 96 years old.

3 'The world's population could in theory stand on the Isle of Wight.'
- The world's population is 8 billion.
- A billion is 1 000 000 000.
- 5 people can just about stand on a square metre of ground.
- The Isle of Wight has an area of 380 km^2.

Test yourself 15E

1 This table was produced to show the numbers of voters who made the three candidates for chairperson, Abebe, Bell and Cherner, their 1st, 2nd or 3rd choices.

However, the votes for one of the candidates are wrong.

Voters	Abebe	Points	Bell	Points	Cherner	Points
1st choice	41	123	17	51	51	153
2nd choice	14	28	32	64	16	32
3rd choice	35	35	41	41	23	23
Total points		186		156		208

(a) All those eligible to vote did so. How many of them were there?

(b) Explain how you can tell that an error has been made.

(c) The votes recorded for Abebe and Bell are correct but those for Cherner are wrong.
Give the correct results for Cherner.

(d) The candidates are given 3 points for a 1st place, 2 for a second place and 1 for a 3rd place.
Who was the winner?

2 According to a Government publication:

'In 6,000 miles a car will produce roughly its own weight in CO_2.'

Is this reasonable? The car quoted here has a weight of 1,400 kg.

Assume a petrol consumption of 5.8 litres per 100 km and a CO_2 emission rate of 133 grams per kilometre.

3 Use the information in this box to investigate the claims below. Mention any assumptions you might make.

(a) It is not possible to comfortably carry a million pounds in £20 notes.

(b) A kilo of gold is worth less than a kilo of £50 notes.

(c) A medium-sized rucksack can hold £1 million in £50 notes.

(d) It is just possible to comfortably carry £1 million worth of gold.

> £50 notes measure 156 mm by 85 mm by 0.113 mm and weigh 1.1 g.
>
> £20 notes measure 139 mm by 73 mm by 0.113 mm and weigh 0.85 g.
>
> The density of gold is 19.32 g cm^{-3} or 19.3×10^3 kg m^{-3}.
>
> Gold has a price of £48 per gram.
>
> The capacity of a medium-sized rucksack is about 40 litres.
>
> A comfortable load for the average person to carry is about 15 kg

4 Between February and May each year, gray whales migrate from Mexico to the Arctic.

Hiking from Lands End to John O'Groats is a recognised challenge for long-distance walkers. A typical time is between 10 days and a fortnight.

Without doing any detailed calculations, but giving your reasoning, decide which travels faster, a whale or a hiker?

5 In several sports, judges have to effectively rank competitors.

Make three comments about the ranking by these four judges (J_1, J_2, J_3 and J_4) of the six athletes A, B, C, D, E and F.

Describe 3 ways in which the scores could be used to give an overall order of the athletes. Give an advantage of each of your methods.

Judge	Athlete					
	A	B	C	D	E	F
J_1	5	4	1	6	2	3
J_2	2	3	6	1	5	4
J_3	5	4	6	1	2	3
J_4	3	6	1	2	5	4

Practice question

There are websites that rate films, TV programmes and music tracks.

People log in and give a particular film a star rating out of 10.
The mean star rating for all the people rating that film is then calculated.
These numbers are used to put the films in rank order.

To avoid a film going to the number one spot when only 100 people vote and all give it a 10, the mean star rating, 10 in this case, is multiplied by the correction number $= \frac{(\text{number of people voting for the film})}{(\text{number of people voting for the film} + 25\,000)}$.

If only 100 people vote, the correction number is $\frac{(100)}{(100 + 25\,000)} = 0.00004$.

For this case the rating for the film would be $0.00004 \times 10 = 0.0004$.
A rating of 0.0004 would not put the film into the top one thousand!
A rating of 0.8 only just gets a film into the top 250 on most sites.

This is the simplest method to calculate the correction number. Large commercial companies use much more complicated methods. These are kept secret as a film's ranking can have an excessively big effect on its sales.

This partially completed spreadsheet is used to calculate the mean stars, correction number and rating for the three films shown, using the correction equation above.

	A	B	C	D	E	F
1	Film	Voters	Total stars	Mean stars	Correction number	Rating
2	Hamilton	48 380	411 230	8.50	0.66	5.60
3	Platoon	89 096	721 677		0.78	6.33
4	Gladiator	1 326 284	11 273 414			

(a) Write down the formula in cell

(i) D4 (ii) E4 and (iii) F4.

(b) Calculate the numbers in the empty cells.

(c) Use your answers to part (b) to put the three films in rank order of rating.

(d) An online film rating site rates the three films as follows:

Hamilton 8.6 *Platoon* 8.1 and *Gladiator* 8.5.

How do these ratings and order of ratings compare to the ones ranked using the simple correction number in the above spreadsheet?

Now that you have finished this chapter, you should know:

- How to rank competitors in various sporting events and situations.
- How to investigate various methods of scoring in sporting events.
- How to consider the fairness of some simple methods of voting.
- How to calculate and interpret gender ratios.
- How to compare quantities.
- How to check and validate assertions using reliable data or assumptions.

CHAPTER 16

Trialling, sampling and bias

In medicine we often want to know whether a particular treatment works. In this chapter you will learn how to decide whether an effect is caused by an intervention.

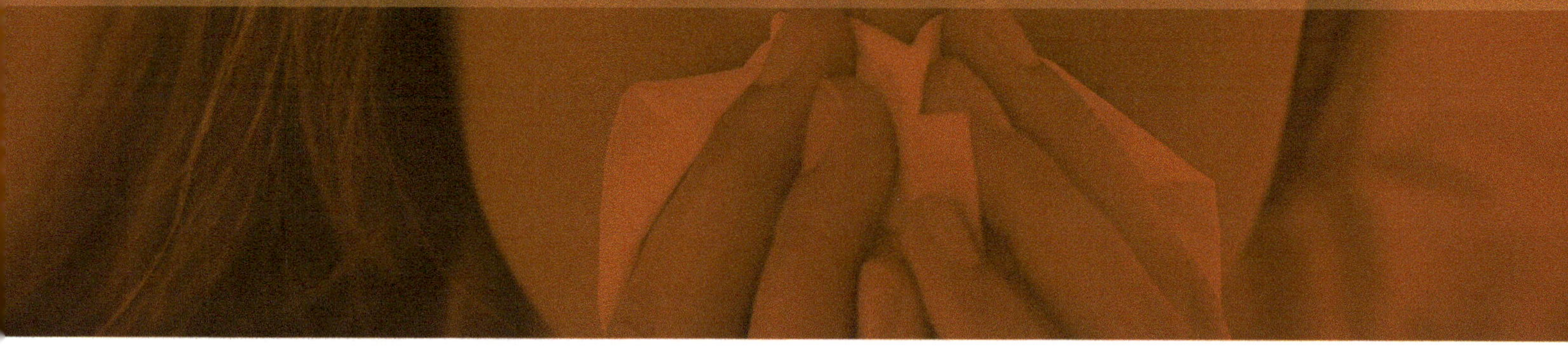

16A Using a control group

What is the cause?

In the 1930s and 1940s there was a large increase in the number of people dying of lung cancer in England and Wales.

Different possible causes were suggested, including the increase in pollution and an increase in cigarette smoking.

In the late 1940s, Doll and Hill conducted a study, interviewing hospital patients with lung cancer and similar patients without cancer. They found that the patients with lung cancer tended to smoke more. This helped to show that smoking was a factor in causing lung cancer.

The Doll and Hill study used a **control group** of patients who did not have cancer to compare with the patients who did have cancer.

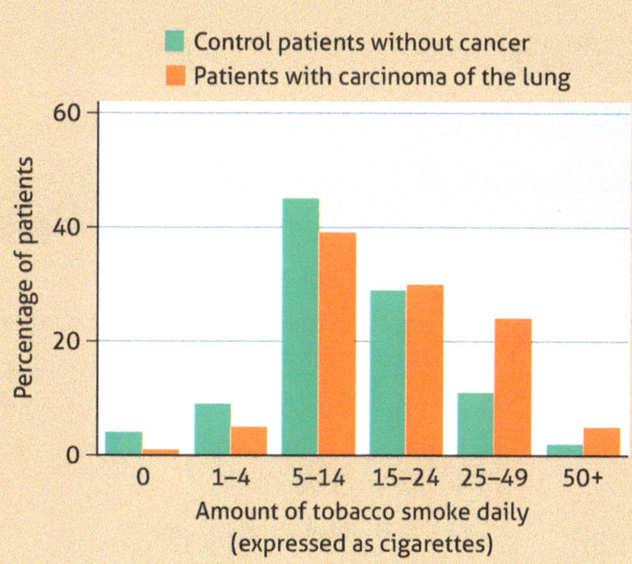

Doll, R. and Hill, A.B. (1950, 30 September) 'Smoking and Carcinoma of the Lung', British Medical Journal, p.742

Activity 16A How not to test something

Marina claims she has a herbal cure for the common cold.

She gives her cure to 50 people who have the common cold.

One week later, they are all better.

Marina claims her cure made them better.

What other explanations can you think of?

Worked example 16A

Marina tests her herbal cure in another way.

- She finds 50 people with the common cold.
- She asks for 25 volunteers to try the herbal cure. The other 25 people are given nothing (the control group).
- One week later she compares the two groups to see whether the cure makes a difference.

(a) Explain why this is a better test of the cure than the one described in Activity 16A.

(b) After a week, all the people who took the herbal cure are better but 5 of the control group still feel ill. Explain why this doesn't prove that the cure works.

Solution

(a) The test is using a control group and this allows the two groups to be compared. If there is no difference between the groups then the herbal cure does not have any effect.

Using a control group means that the effects of other variables, such as time, will not cause the difference between the groups.

(b) Marina asked for volunteers to take the herbal cure; maybe their belief in it has caused them to feel better.

The positive effect of believing that something will do you good when it has no real effect is called the **placebo effect**. It is well known that this can make people feel better. Also, Marina's sample size is fairly small.

There is more about the placebo effect and sample size later in the chapter.

Exercise 16A

1 Three groups of people take part in an experiment to test the effect of coffee on how well people sleep.

Group	A	B	C
Amount of coffee	3 cups of coffee a day	6 cups of coffee a day	No coffee

(a) Which is the control group?

(b) Lisa is conducting the experiment; she lets people taking part choose which of the groups they are in. Explain why this could lead to bias.

2 A café owner wants to know whether greeting customers with a smile causes them to spend more.

One day, some of the customers are greeted with a smile and others are not. The mean amount spent by each group is shown in the table below.

Group	Smile	No smile
Mean spend	£4.01	£3.53
Number in group	175	170

(a) What might the café owner conclude from the data?

(b) The café staff decided which of the customers to smile at. Explain why this might have led to bias and how it could have affected the results.

(c) A friend of the café owner suggests it would have been better to smile at all customers one day and none the next and compare the mean spend on the two days.

Give one reason in favour of the friend's design and one reason in favour of the café owner's original design.

16B Randomised controlled trials

Studies in medicine are often conducted using **randomised controlled trials**. These kinds of trial can also be used in other fields such as education. The flow chart below shows the process for a randomised controlled trial to see whether singing in a choir improves mental health for adults over the age of 65.

The thing that is being tested in the trial is often referred to as the **treatment** or the **intervention**.

The participants in a trial are often referred to as **subjects**.

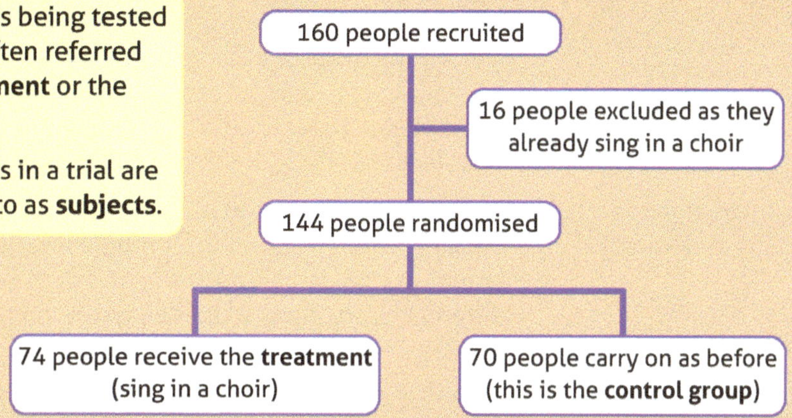

Notice that randomisation does not always result in exactly equal numbers in the groups: in this case 74 and 70 rather than 72 in each.

Before and after the **intervention**, the two groups were compared using standard health tests.

What is randomisation?

Participants in a study are allocated to groups at random, using a method which makes it equally likely that they are in any of the groups. The diagram shows a trial where there is one treatment group and one control group.

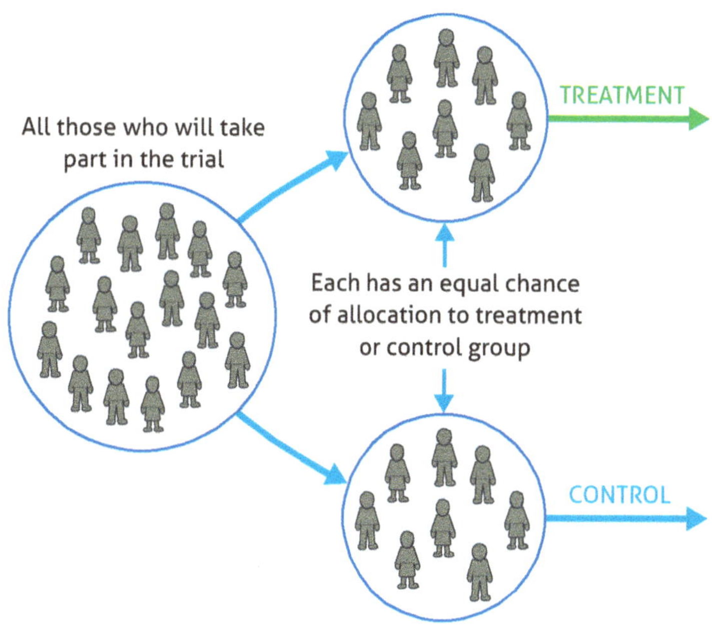

⚠ Sometimes it is not practical to randomise and sometimes it would cause ethical problems. For example, Doll and Hill (referred to at the start of this chapter) could not randomise a group of people and tell half of them to smoke and half not to smoke because it would take years before any difference would be seen and those who had been told to smoke would have been harmed if it was found that this did cause lung cancer.

Activity 16B (1) How and why to randomise?

Consider the trial described in the flow chart above.

(a) Think of some ways you could allocate the 144 people at random to the treatment and control groups.

(b) What are some benefits of allocating the people to the two groups at random?

Worked example 16B

144 people have been accepted to take part in a randomised controlled trial of whether singing improves mental health.

Solution

(a) Why is it helpful to have a control group when doing this trial?

(a) Having a control group allows a comparison between those who have sung in a choir and those who have not. If everyone in the trial sang in a choir, you could see whether they had improved their mental health, but this could be due to feeling positive due to taking part in a trial. You want to know if the singing itself has a positive effect.

(b) Why should subjects be allocated to the treatment and control groups at random?

(b) Allocating the people at random eliminates bias in deciding who goes in which group. It means that the two groups can be compared at the end of the study and any differences are likely to be due to the singing.

Bias is a systematic error. In this case, if you allowed people to volunteer to be in the treatment group, this could result in bias as they might be more enthusiastic and produce better results.

(c) Describe one way of using random numbers to allocate subjects to the treatment and control groups at random.

(c) Use a random number generator to allocate a random number from 1 to 10 to each person. If the number is even, the subject is in the treatment group, if odd in the control group. On a spreadsheet you can use the command =RANDBETWEEN(1,10).

(d) One randomisation of the subjects allocates 74 to the treatment group and 70 to the control group. Explain why it doesn't matter that there aren't exactly 72 in each group.

(d) The two numbers are close, so the two groups can still be compared. In any trial, there is a possibility that participants drop out before the end, so even if the two groups started out with equal numbers, they might well end up with one a little larger than the other.

There are other ways that you can use random numbers to allocate the subjects to the two groups. The important thing is to know one way that you can use when you need to.

Activity 16B (2) Random numbers on a calculator

Find out if your calculator has the facility to generate random numbers.

If it can, make sure you know how to use it.

Exercise 16B

1 A researcher wants to know whether having access to a dictionary improves year 9 school students' English results.

260 suitable schools volunteer to take part in the trial.

130 schools are assigned at random to the treatment group and told to provide all year 9 students with a dictionary, which they should have access to at all times.

The other schools are assigned to the control group and instructed not to provide dictionaries to year 9.

(a) For each of the following methods of assigning schools to the treatment group, decide:
- whether it will guarantee that there are exactly 130 schools in the treatment group
- whether each school is equally likely to be in the treatment or control groups.

Method	Will it guarantee 130 treated?	Equally likely?
Assign either 0 or 1, at random, to each of the 260 schools. Schools with 1 are in the treatment group.		
Number the schools 1 to 260. Choose 130 different random numbers from 1 to 260. The schools with those numbers go into the treatment group.		
Use a spreadsheet to give a random number to each school. Sort the schools in order of size by random number. The top 130 go into the treatment group.		
Put the schools in alphabetical order of name. The top 130 go into the treatment group.		

(b) One of the methods listed in part (a) is not a random method. Which one is it? Explain how using this method could introduce bias into the trial.

(c) Give a reason in favour of assigning whole schools to the trial and control groups rather than individual students.

16C 'Blind' and 'double blind' in statistical trials

Activity 16C Could there be another explanation?

Drinking a particular herbal tea is thought to help with concentration when studying.

- 100 people are randomised into two groups.
- 50 people in the treatment group drink the herbal tea before studying.
- 50 people in the control group don't drink herbal tea before studying.
- The treatment group report better concentration, on average.

Does this mean the herbal tea helps concentration or can you think of another explanation?

Worked example 16C

? The effect of the herbal tea on concentration is tested in another way.
- 100 people are randomised into two groups.
- They are all told that the study is to see what effect hot drinks have on concentration.
- 50 people in the treatment group drink the herbal tea before studying.
- 50 people in the control group drink a different herbal tea before studying (this tea is not believed to have any effect on concentration).
- They are all monitored to see how well they concentrate on the material they are studying.
- The results are recorded and the groups are compared at the end of the trial.

(a) Give two reasons why this is a better test of the herbal tea than the one described in Activity 16C.

(b) Is it better for the people doing the monitoring to know which individuals are in the treatment group or not? Give a reason for your answer.

Solution

(a) 1 The people taking part in the study don't expect the herbal tea to have a particular effect so this will avoid biasing the results.

2 They are monitored by someone else rather than reporting how well they concentrate for themselves; this will help with comparison of different people in a standard way.

(b) It is better for those doing the monitoring not to know, so that they cannot unintentionally influence the results, for example by their attitude to the participants.

Blind and double-blind trials

In a **blind** trial (or **blinded** trial), the participants do not know whether they are in the treatment group or the control group.

In a **double-blind** trial neither the participants nor the researchers know whether they are in the treatment group or control group.

Blinding is used to prevent bias due to participants or researchers expecting the treatment group to react in a particular way to the treatment.

To ensure that participants do not know whether they are receiving the treatment, a placebo may be used for the control group. This looks the same as the treatment but has no effect.

Some trials test a new treatment against an existing treatment, with the control group getting the existing treatment.

If the treatment being tested needs the participants to change their behaviour, a blind trial may not be possible.

Exercise 16C

1 A study is conducted to test whether copper bracelets help reduce hip pain in older people. 200 people take part in the study. 100 of them wear a copper bracelet. 100 wear a bracelet which looks the same but is made from a different metal.

After a month all participants answer a questionnaire about their pain levels.

(a) Identify the control group in this study. Choose the correct answer.

A Those who wear the copper bracelets.
B Those who wear the bracelets made from the different metal.
C There is no control group.

(b) Identify the placebo in this study.

(c) What would need to be done to make this a double-blind study?

16D The effect of sample size

A suitable size of sample is an important feature of a good trial.

Activity 16D How strong is the evidence?

Two randomised controlled double-blind trials are conducted to test the same medication. Previous experience suggests that people who are not treated for the condition are just as likely to improve as not.

Which trial provides stronger evidence of effectiveness?

Trial 1	Improvement	No improvement
Treatment	7	3
Control	5	5

Trial 2	Improvement	No improvement
Treatment	70	30
Control	50	50

Worked example 16D

The chart below illustrates the probabilities of different possible outcomes for the trials described in Activity 16D. It is based on the assumption that, in each group, participants are equally likely to improve as not; this is what happens without treatment, so the researcher will only conclude that the treatment is having an effect if the evidence is strong enough.

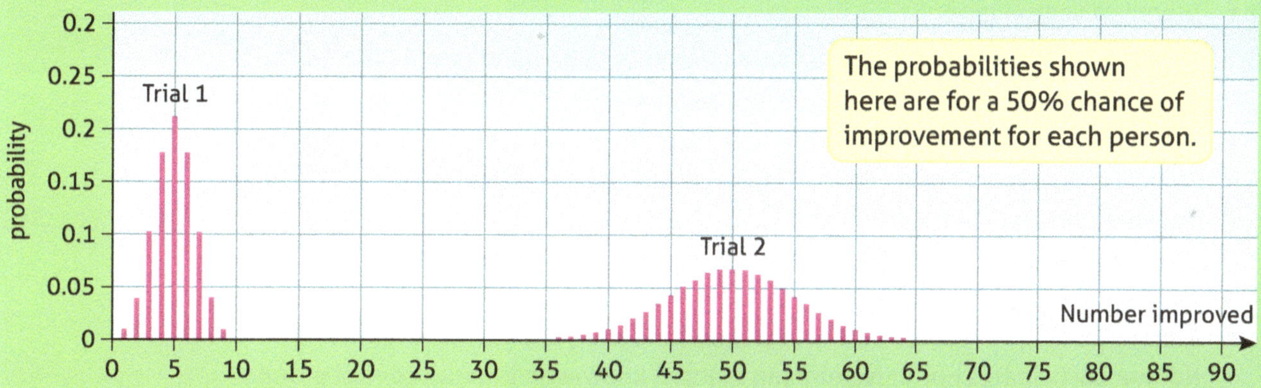

The percentages improving are the same for both trials. In both trials, 50% of the control group improved, which is what would have happened without treatment.

Consider the results for each group and comment on whether there is strong enough evidence that the treatment is having an effect.

Solution

(a) 7 improving is reasonably likely even without treatment, so the evidence that the treatment is effective is not strong enough.

(b) It is very unlikely that as many as 70 would improve without treatment, so there is evidence that the treatment is effective.

- 70 is in the **tail** of the probability distribution for 100 in a group.
- Larger samples provide stronger evidence but also increase costs. Researchers can use formulae to calculate suitable sample sizes.
- There is more about judging whether results are consistent with a 50% chance of success in Chapter 17.

Exercise 16D

1 Marie is planning a randomised controlled trial to test a new medication for back pain. She calculates that a sample size of 68 will be big enough to provide good evidence. A colleague suggests using a larger sample size.

(a) Give a reason in favour of using a larger sample size.

(b) Give a reason against using a larger sample size.

(c) The trial is started with a sample size of 68. There are 34 people in the treatment group and 34 in the control group. After a week of the trial, two people in the treatment group drop out. Marie knows two people with back pain. Should she ask them to join the trial in place of the two people who have dropped out? Justify your answer.

Test yourself 16E

1 Bob is planning a randomised controlled trial.

(a) Choose the best definition of randomised from the following list.

- It is not very well organised.
- Participants have an equal chance of being allocated to each group.
- There is no way of predicting the results.

(b) Choose the best definition of controlled from the following list.

- It is well organised.
- Strict records are kept of everything that happens.
- There is a control group.

2 Bob needs to decide whether the trial he is planning should be blind or double blind.

(a) Choose the best definition of blind from the following list.

- Participants don't know whether they are in the treatment group or control group.
- Participants have an equal chance of being allocated to each group.
- Researchers have no pre-conceived ideas about the results.

(b) Choose the best definition of double blind from the following list.

- It is kept completely secret.
- Nobody keeps any records of the results either during or after the trial.
- Neither participants nor those running the trial know who is in the control group.

3 Common features of a randomised controlled trial are listed on the left below. Reasons for their use are listed on the right.

Match each feature to the best reason for its use.

Feature	Reason for use
Randomised	Allows comparison
Control group	Ensures factors not being tested do not cause a difference between groups
Blind	Helps to ensure participants do not know whether they are in the treatment or control group
Placebo	Prevents distortion of results due to participants knowing what is expected

4 A manager is investigating whether sending a personal email with a money off voucher will encourage customers of a business to fill in an online survey.

60 customers are chosen at random. They are sent the link to the survey and asked to fill it in. 30 of them get a follow-up email from the manager offering 10% off the next time they shop and asking them to fill in the survey.

The results are shown in the table below.

	Follow-up	No follow-up
Number of surveys filled in	13	6

(a) Identify the control group for the study. Choose the correct answer.

A The customers with the follow-up email.
B The customers without the follow-up email.
C There is no control group.
D It's not possible to tell whether there is a control group.

(b) Decide whether each of the following statements is true or false. Justify each decision.

Those followed up were roughly twice as likely to return the survey as those who weren't.
This study proves that sending a follow-up email will double the rate of response when you are doing a survey.
A larger trial would provide better evidence.
The 60 customers involved should have been told about the trial and that some were getting a money-off voucher and others weren't.

5 A trial of a new hand cream compares it with an existing cream. The points below list the way the trial is conducted.

A 100 people who work outdoors and wash their hands frequently are enrolled.
B 50 people are chosen at random to use the new hand cream.
C The other 50 use the existing cream.
D The two hand creams are packaged in identical pots.
E Each participant has a code number and is given cream with the same code number.
F Researchers instruct the participants how to use the hand cream and monitor the condition of their hands.
G The researchers do not know which participants are using the new cream.

(a) For each feature of the trial listed below, list which of the points A to G describe that feature.

Some features may have more than one point.

(i) Randomisation
(ii) Control group
(iii) Double blind

(b) An alternative to the researchers monitoring the participants' hands would be to ask the participants to report on the condition of their own hands. Give a reason in favour of the researchers monitoring participants' hands.

(c) Jim says that a double-blind trial has to include a placebo. Is Jim right? Justify your answer.

Practice question

Aisha thinks that chamomile tea helps people to sleep better.

She says: 'When I drink chamomile tea before going to sleep, I sleep better. This proves it helps people sleep.'

(a) Explain why Aisha's experience does not prove that chamomile tea helps people to sleep better.

A randomised controlled trial was conducted on 300 volunteers. Some of them were asked to drink chamomile tea before going to bed. The others were asked to drink hot water. Neither group knew what the other group was asked. Their sleep was monitored using a device each volunteer wore on the wrist. The trial lasted two weeks.

Treatment	Chamomile tea	Hot water
Number of people	152	148
Total number of hours sleep	13406	12950

(b) The people taking part in the trial were allocated to the two groups at random.

Describe a way this allocation could be done.

(c) Which is the control group? Explain why a control group is needed in the trial.

(d) Discuss whether the data in the table provide convincing evidence that drinking chamomile tea helps people to sleep better.

Now that you have completed this chapter, you should know:

- What randomised controlled trials are and their importance in trying to establish causation.
- The concepts and importance of 'blind' and 'double blind' in statistical trials.
- That proportions from small samples are more variable than those from large samples.

CHAPTER 17*

Problem solving with the Normal distribution

In many situations there are variables with distributions that are Normal or sufficiently near Normal for it to be a good enough model to provide useful results. So, the Normal distribution is often useful as a model when solving problems.

* Chapter 17 covers content that is in both Part 2: Critical Maths and Part 3: Statistical Problem Solving.

17A Using the Normal distribution

This section builds on ideas that you met in Chapter 8.

Activity 17A Likely distributions

For each of the following variables, sketch its likely distribution and say whether you think the Normal distribution will provide a reasonable model.

(a) The weights of baby girls born at full term in a hospital's maternity ward.

(b) The number of heads when a fair coin is tossed 100 times.

(c) The number of minutes before or after its scheduled arrival time that a train arrives at its final destination.

(d) The different scores when fair dice with six faces are rolled.

(e) The number of 6s when 120 fair dice are rolled.

(f) The number of goals scored by a premiership football team in its matches during one season.

Notation

When a variable X has a Normal distribution, you can write $X \sim N(\mu, \sigma^2)$ where μ is the mean and σ is the standard deviation. The symbol $\sim$ means 'has the distribution'.

For example, $X \sim N(10, 3^2)$ means the X has a Normal distribution with mean 10 and standard deviation 3.

Standardisation

The number of standard deviations from the mean is called **the z-score** or the **standardised score**.

A z-score of 1 means that the data value is 1 standard deviation above the mean.
A z-score of −2 means that the data value is 2 standard deviations below the mean.

To calculate a z-score:

- subtract the mean
- then divide by the standard deviation.

For a data value, x then the z-score is $\frac{x-\mu}{\sigma}$.

When you convert values to a z-score it is called standardising. The standardised Normal distribution, N(0, 1), has mean 0 and standard deviation 1. You can convert any Normal distribution into the standardised Normal distribution.

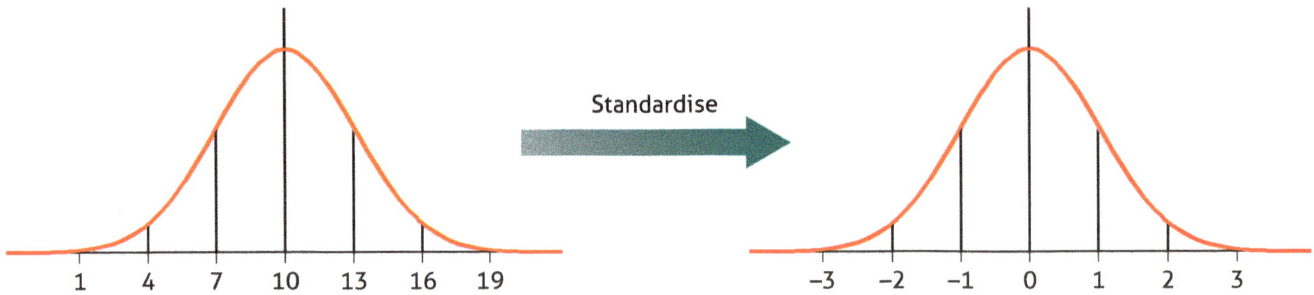

$X \sim N(10, 3^2)$

Normal distribution with $\mu = 10$ and $\sigma = 3$.

$Z \sim N(0, 1)$

Standardised Normal distribution with $\mu = 0$ and $\sigma = 1$.

For a Normal distribution:

- 68% of values (just over ⅔) lie within 1 standard deviation of the mean, $-1 \le z \le 1$.
- 95% of values of the variable lie within 2 standard deviation of the mean, $-2 \le z \le 2$.
- 99.7% of values (virtually all) lie within 3 standard deviation of the mean, $-3 \le z \le 3$.

Worked example 17A

The mean height of an adult male giraffe is 5 metres and the standard deviation is 0.9 metres.

(a) Calculate the probability that a randomly chosen adult male giraffe has a height of over 6.8 metres.

(b) Calculate the z-score of a giraffe with height 3.65 metres.

What does the z-score tell you? Does this mean the giraffe is not an adult male?

Solution

(a) $z = \frac{6.8 - 5}{0.9} = 2$ ← z-score is $\frac{x-\mu}{\sigma}$.

95% of values are 2 standard deviations from the mean.

So, the probability that a randomly chosen giraffe has a height of more than 6.8 m is 2.5% or 0.025.

(b) $z = \frac{3.65 - 5}{0.9} = -1.5$

This giraffe has a height that is 1.5 sd below the mean.

This is not far below the mean, so the giraffe could be an adult male.

Exercise 17A

1 Each diagram shows the Normal distribution where $X \sim N(0, 1)$.
Work out the percentage of values which fall within each of the shaded regions below.

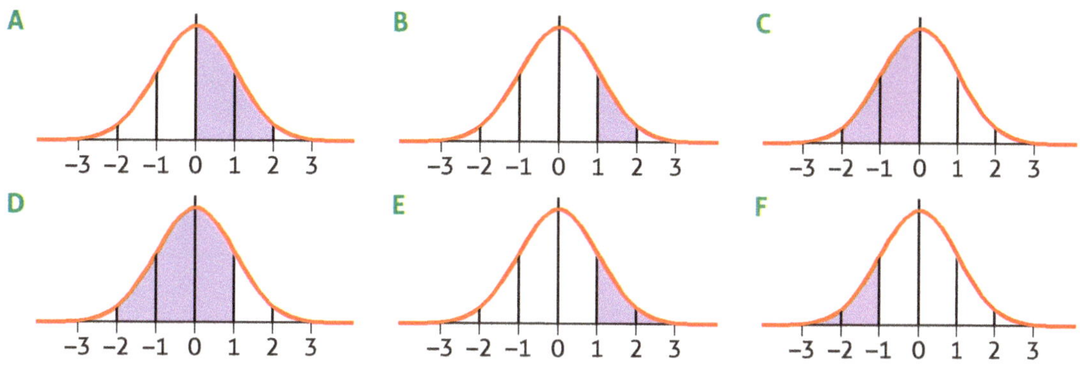

2 It is found that the distribution of the times, T minutes, for adults to solve a puzzle is given by $T \sim N(20, 5^2)$.

(a) Find the probability that a randomly selected adult takes:

(i) longer than 25 minutes (ii) less than 10 minutes (iii) between 15 and 25 minutes.

(b) Three friends do the puzzle. Sam takes 22 minutes, Amy 13 minutes and Leon 18 minutes. Work out their z-scores and say who among them is the most unusual.

17B Rolling dice

Two of the situations in Activity 17B involved rolling dice. In this section the ideas are taken further.
How can rolling dice set up variables for which the Normal distribution provides a good model?

Section B is core content for Critical Maths students but extension content for students studying Statistical Problem Solving.

Activity 17B Rolling a single die many times

(a) How many times would you expect each of the numbers 1, 2 ... 6 to come up when you roll a die 60 times?

(b) Follow these instructions to set up a spreadsheet to simulate 60 rolls of a die and to sort them out.

- Insert the formula =RANDBETWEEN(1,6) into cell A1 of a new spreadsheet.
- Copy the formula down the column to A60.
- Now try to sort the column using the sort command: What happens?

To stop the numbers from changing you need to highlight the column and select Copy, then highlight another column and use Paste values from the Paste menu. You can now sort the numbers from smallest to largest.

(c) Make a frequency table of the results of 60 rolls.
Are the frequencies as you expected?
Draw a vertical line graph to show the distribution.
Would you expect the distribution to be more or less even if you did 600 rolls? Try it out on your spreadsheet.
What about 6000 rolls?

Here are two conclusions you can draw from Activity 17B:

- The distribution of the different numbers is not Normal; it is a **uniform** distribution.
- The proportions from large samples are more variable than those from small samples.

Rolling 2 dice many times

When you roll 2 dice, the possible total scores are between 2 and 12.

Are they all equally likely?

Which of the vertical line graphs A–C shows the distribution of the total scores when two dice are rolled many times?

A

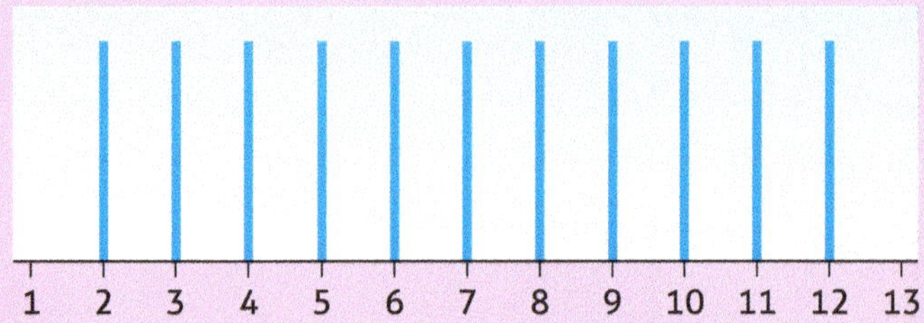

B

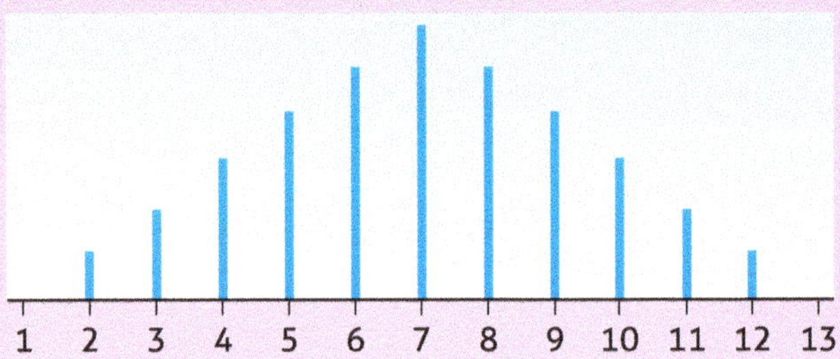

C

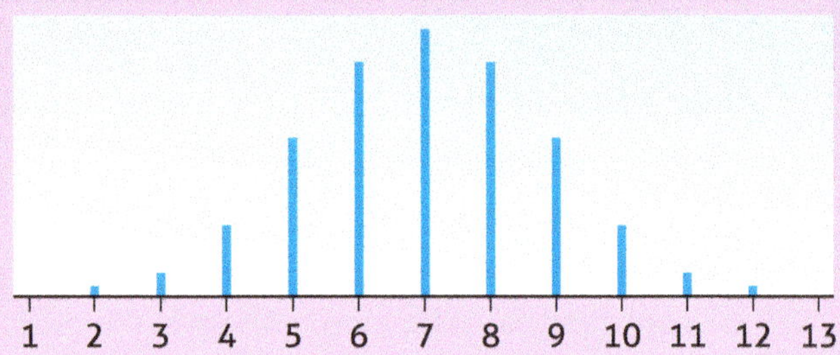

When you roll two dice, the distribution of their total is symmetrical, about 7, but it is not Normal. It has the shape of a triangle. You may, however, think that it looks a bit more like a Normal distribution than the uniform distribution from rolling one die.

If you increase the number of dice, the distribution of their total scores is neither uniform nor triangular but takes on a bell shape. The more dice you roll, the closer the shape is to the Normal distribution. This example illustrates two important general results:

- If you take large samples of a given size and find the total value of one of the variables, its distribution will be nearly Normal.
- Since the samples have a fixed size, it follows that the mean value of the variable also has a nearly Normal distribution. This can be a very useful result.

Worked example 17B

Kai uses a spreadsheet to simulate rolling 50 dice many times and recording the total each time.

He finds that the mean of the totals is 175 and their standard deviation is 12.

Draw and label a diagram to show the shape of the distribution of the totals.

Solution

With 50 dice, the distribution is very nearly Normal.

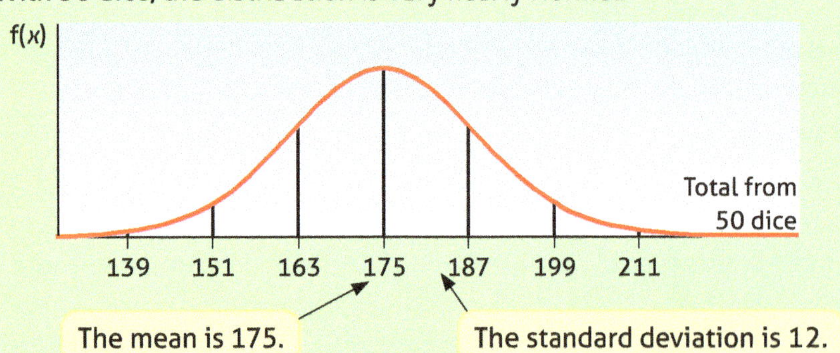

The total score from any number of dice must be a whole number so it is a discrete variable. By contrast, the Normal distribution is for a continuous variable, so in this case it can never be a perfect model. It can, however, be a good model in situations where the number of trials is large.

Exercise 17B

1 When you roll 50 dice, what is:
(a) the least possible total
(b) the greatest possible total?

2 Rosie looks at Kai's distribution from Example 17B and says that when you roll 50 dice the total is certain to be between 145 and 205. Is Rosie correct?

3 An ordinary die is rolled 120 times and the numbers it shows are recorded. Their total is 452.
(a) State the total that would be expected for a fair die.
(b) The standard deviation for the totals of 120 throws is 18.7.
Does the observed result suggest that the die is biased?

17C Nearly Normal distributions

In the previous section you met an example of a variable with an approximately Normal distribution. It is one of many situations where a variable does not have a perfect Normal distribution but one that is near enough to provide a useful working model.

Activity 17C The frequency of a particular outcome

You record the frequency of the number 4 coming up in repeated sets of 120 rolls of an ordinary 6-faced die.

What would you expect the distribution of these frequencies to be?

How could you check the frequency distribution?

You may have met a die with 10 faces, like the ones in the picture, perhaps playing *Dungeons and Dragons*.

What would the distribution of the number of 4s be when such a die is rolled 120 times?

Activity 17C points the way to a general result:

- The distribution of the frequency with which a particular outcome occurs in large samples of a given size is approximately Normal.

Rolling dice is only an example. There are many similar situations closer to real life, like that in Worked example 17C below, where it is helpful to use the Normal distribution to model the frequency of a particular outcome.

Worked example 17C

Large-scale studies around the world have concluded that 1 person in 10 is left-handed. An anthropologist suggests that the proportion of left-handers living in a remote part of the Amazon is higher than this. A study of 120 people living there found that 26 of them were left-handed.

(a) Write down the expected (or mean) number of left-handers in a sample of size 120.

(b) Given that the standard deviation of the mean of such samples is 3.3, find the z-score for the observed number of 26.

(c) Comment on whether this seems to support the anthropologist's suggestion.

Solution

(a) $\frac{1}{10} \times 120 = 12$

(b) The distribution is modelled by $N(12, 3.3^2)$.
So, $z = \frac{26 - 12}{3.3} = 4.24$.

(c) It is extremely unlikely that such a high z-value occurs by chance.

Exercise 17C

1 When you roll fair dice 60 times, the mean number of sixes is 10 and the standard deviation of the number of sixes is 2.89 (to 3 sf).

Holly suspects that one of her dice is biased towards sixes.

She rolls it 60 times and finds she has rolled 18 sixes.

(a) Find the z-score for this outcome.

(b) Is this evidence that the die is biased?

2 Marina is a marine biologist. One year she catches samples of 100 angel fish from a coral reef. She records how many of the angel fish in each sample are zebra angel fish.

Marina finds the mean number of zebra angel fish in her samples is 30 and the standard deviation is 7.

She returns briefly to the coral reef a year later, hoping that the number of zebra angel fish has increased. She only has time to take one sample of size 100 and finds there are 40 zebra angel fish in it.

Does Marina have enough evidence to say that the population of zebra angel fish has increased?

17D Using the toss of a fair coin as a model

Sporting events often begin with the toss of a coin. With probabilities of $\frac{1}{2}$ for both heads and tails it gives each side an equal chance of success. However, any situation involving a probability of $\frac{1}{2}$ can be modelled in the same way, whether or not a coin is actually involved. Where many repetitions are involved, the Normal distribution provides a good model for the number of successes.

Activity 17D Election time

At an election, the two leading candidates in one constituency have exactly the same number of votes.

So, the winner will be decided by spinning an old gold coin.

Both candidates ask for the coin to be tested to make sure it is fair.

The authorities consider two ways of testing the coin:

Method A: One person spins the coin 10 times. It will be declared biased if the number of heads is 0, 1, 2, 8, 9 or 10.

Method B: Ten people each spin the coin 10 times. It will be declared biased if the total number of heads is no more than 20 or at least 80.

(a) Which method is more likely to reject a fair coin?

(b) Suggest a possible improvement to Method B.

Note: You can use a spreadsheet to simulate both methods for a fair coin. Your method would be like that described in Activity 17B, with the command `=RANDBETWEEN(0,1)`, with 0 corresponding to a tail and 1 to a head.

When you toss only a small number of coins there will be more variation in the proportion of heads than when you toss a large number of coins (or a single coin a large number of times). For example, it is possible for a fair coin to come up with 8 heads in Method A, but incredibly unlikely that Method B will result in 80 heads (unless the coin is biased).

When you toss a fair coin n times you expect it to come up heads about $\frac{1}{2}n$ times and the same for tails. This won't always happen exactly, but it will usually be close, and the distribution of the number of heads is approximately Normal.

This graph shows the Normal model for the number of heads when 100 coins are tossed.

The mean number of heads when n coins are tossed is $\frac{n}{2}$.

This is also called the **expectation**.

The standard deviation of the number of heads is $\frac{\sqrt{n}}{2}$.

This formula is extension material for Statistical Problem Solving students.

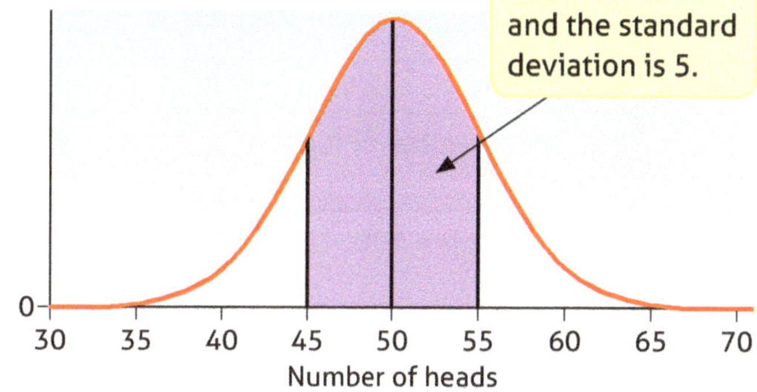

Some experiments can be modelled by the toss of a fair coin, provided that:

- there are two possible outcomes (sometimes called success or failure)
- the probability of each of the outcomes is $\frac{1}{2}$ the outcome of any trial is independent of those from previous trials.

Worked example 17D

A biologist finds a very large colony of a previously unknown species of bat living in a remote cave. She thinks that the ratio male : female is 1 : 1. She takes a sample of 1000 bats.

In answering parts (a), (b) and (c), assume that the biologist's model is correct.

(a) What is the expected number of females in a sample of 1000 bats?

(b) Describe the distribution of the numbers of females in samples of size 1000, stating the mean and standard deviation.

(c) There are 560 females in the biologist's sample of 1000 bats. What is the *z*-score?

(d) Comment on whether the evidence supports the biologist's model.

Solution

The biologist's model is assumed to be correct.

(a) The probability of a randomly selected bat being female is $\frac{1}{2}$.

So, the expected number of females is $\frac{1}{2} \times 1000 = 500$.

(b) The distribution is Normal.

Using the results mean is $\frac{n}{2}$ and standard deviation is $\frac{\sqrt{n}}{2}$, gives

Mean $= \frac{1000}{2} = 500$, standard deviation $= \frac{\sqrt{1000}}{2} = 15.8$.

So, the distribution is $N(500, 15.8^2)$.

(c) The *z*-score for the number of females in the sample is $z = \frac{560 - 500}{15.8} = 3.8$.

(d) 560 is more than 3 standard deviations from the mean, so the likelihood of the result happening by chance is very small, less than 1%. The evidence suggests that the biologist's model is incorrect.

Results that are more than 3 sd from the mean are very unusual.

Worked Example 17D follows the outline of a **hypothesis test**, so it provides an opportunity to become familiar with the steps and reasoning involved and some of the language you will need in the last two chapters in this book.

Note

Critical Maths students may treat this page as extension material.

A biologist would write down the steps below.

Step	Explanation and commentary
Let μ be the mean number of females in samples of size 1000.	Start by stating the variable you are dealing with.
Null hypothesis: $H_0 : \mu = 500$	The null hypothesis says the biologist expects half the bats to be female and half of 1000 is 500.
Alternative hypothesis: $H_1 : \mu \neq 500$	This says that the null hypothesis is wrong.
Significance level: 1%	The significance level is a way of saying how strong you require the evidence to be before you move away from the null hypothesis. In Worked example 17D the significance level isn't stated.
Assume the null hypothesis is true. Then the distribution of the number of females is $N(500, 15.8^2)$.	The distribution of the numbers of females in different samples of size 1000 would be Normal, with mean 500 and standard deviation 15.8.
The z-score for the observed number of bats is $\frac{560 - 500}{15.8} = 3.8$.	The *z*-score is the **test statistic**. You will use it to decide whether or not the evidence is strong enough to move away from the null hypothesis.
The probability of an outcome with a *z*-score greater than 3 is known to be less than 0.3% or 0.003.	This is using the well-known result that for a Normal distribution, 99.7% of values lie within 3 standard deviations of the mean, so have $-3 \leq z \leq 3$.
Since 0.3% < 1%, there is enough evidence to reject the null hypothesis. The biologist's model is unlikely to be correct.	The probability of a result as extreme as the observed result is compared to the significance level. If, as in this case, it is less than the significance level, the evidence suggests the alternative hypothesis is true.

The solution of Worked example 17D used the well-known results for the proportions of a Normal distribution that lie within 1, 2 and (in this case) 3 standard deviations of the mean. Another set of well-known results are the *z*-scores of the cut-off points for commonly used significance levels.

10% (5% at each tail) $z < -1.645$ or $z > 1.645$
5% (2.5% at each tail) $z < -1.96$ or $z > 1.96$
1% (0.5% at each tail) $z < -2.576$ or $z > 2.576$

These are called **critical values**.

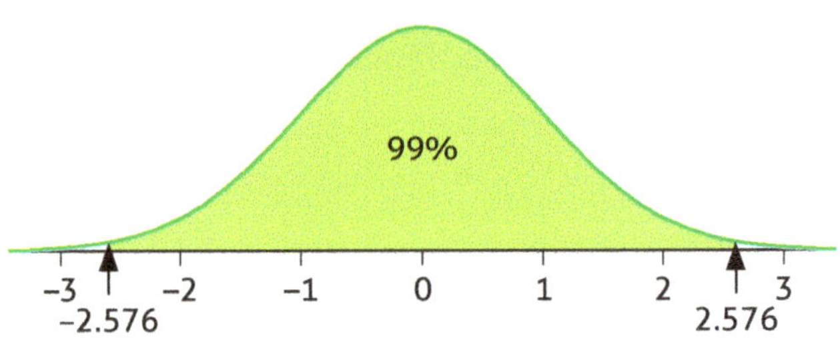

Using the critical value for 1%, the last two steps of this test would be:

The significance level for this test is 1%. So, the critical values are ± 2.576.	The critical values define the tails of the distribution where the observed result is more unlikely than you are prepared to accept. These tails are called the **rejection region**.
Since 3.8 > 2.576, the evidence does not support the null hypothesis. The biologist's model is unlikely to be correct.	So, you compare the test statistic with the nearer critical value.

Exercise 17D

1 David tosses a coin 400 times and finds it lands heads 225 times.

(a) How many heads would you expect if the coin is fair?

(b) Does David's result prove that the coin is biased?

2 A scientist tests a new drug on groups of 400 volunteers. They are allocated randomly to either a placebo or a treatment group, with equal probability of being in each group.

A large number of these randomised controlled trials are conducted in hospitals across the country.

(a) What is the mean number of people in the treatment groups?

(b) What is the standard deviation of the number of people in the treatment groups?

(c) The graph below models the likelihood of different numbers of patients being assigned to the treatment group. Number the horizontal axis appropriately.

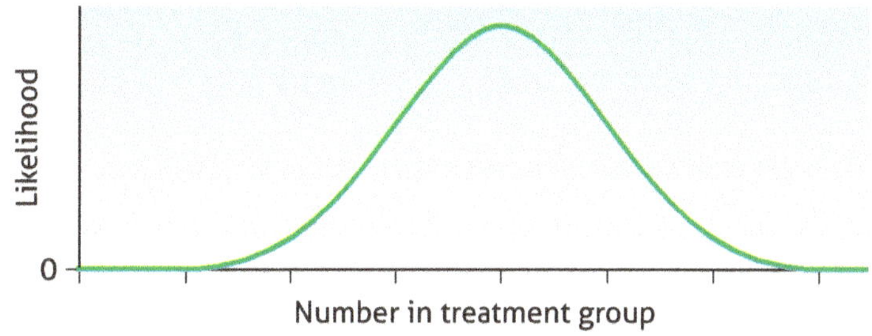

(d) In one hospital, 239 of the 400 participants were placed in the treatment group.

Is this evidence of bias in the process used for allocating participants into the two groups?

3 The entrance to a supermarket is in the middle of the front. The manager says that people entering the store are equally likely to turn left or right. He monitors the first 900 individual customers that enter his store one day.

Assume that the manager is correct.

(a) On average, how many of the 900 people will turn left?

(b) What is the standard deviation of the number of people who turn left?

(c) In fact, 397 people turned left. What can you say about the manager's statement? Suggest a possible explanation.

17E Using tables

So far, when you have needed to find probabilities from a Normal distribution the *z*-values have been whole numbers. This is now extended to other *z*-values. Critical Maths students may treat the use of Normal distribution tables as extension material.

Activity 17E When will my phone run out?

The battery life for a particular mobile phone is Normally distributed with mean 22 hours and standard deviation 1.6 hours.

Ava says her phone's battery never lasts longer than 24 hours.

How can you estimate the probability that a phone picked at random has a battery life of less than 24 hours?

In Activity 17E you were asked about the probability for $z < 1.25$. You could make an estimate from the graph of the Normal distribution but it would be so rough that it would leave you wanting something better. There are two methods: using a calculator or using a table of values.

On some calculators you can find Normal distribution probabilities, but not on all. Find out if your calculator has this facility. If the answer is yes, then find out how to use it to calculate the proportion of a population with a Normal distribution with a z-score that is less than 1.25.

If your calculator does not have the Normal distribution, you will need to use the table of values. It is based on the standardised Normal distribution.

The table gives the probability, p, of a random variable distributed as N(0, 1) being less than z. This is the shaded area in the image and it is given by the function $\Phi(z)$.

The total area under the curve is 1, as it is certain that a particular value will fall somewhere within the bell curve.

Worked example 17E shows you how to use a Normal distribution table.

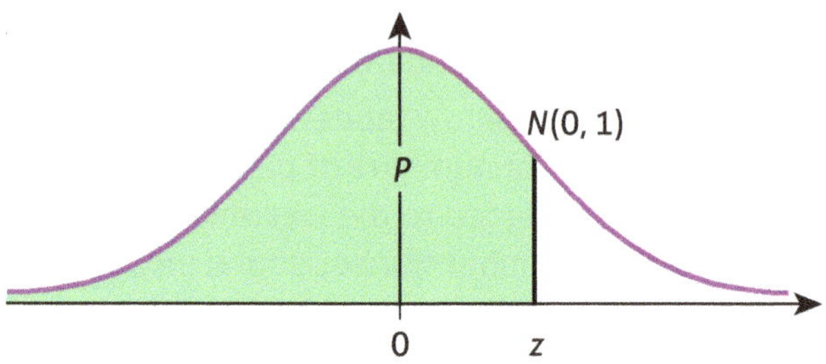

The symbol Φ is the upper-case Greek letter ph and is called 'phi'. It is pronounced 'fi', as in the nursery rhyme 'fee, fi, fo, fum'.

Worked example 17E

The battery life for a particular mobile phone is Normally distributed with mean 22 hours and standard deviation 1.6 hours.

(a) Find the probability that the battery in a randomly selected phone lasts less than one day.

(b) A shop sells 1000 of these phones. How many of the phones would you expect to have a battery life of more than 25.4 hours?

(c) What proportion of phones have a battery life of between 22 hours and 25.4 hours?

Solution

(a) First work out the value of z:

$$z = \frac{24 - 22}{1.6} = 1.25.$$

So you want to know what proportion of mobiles have a battery life that is 1.25 sd above the mean or less.

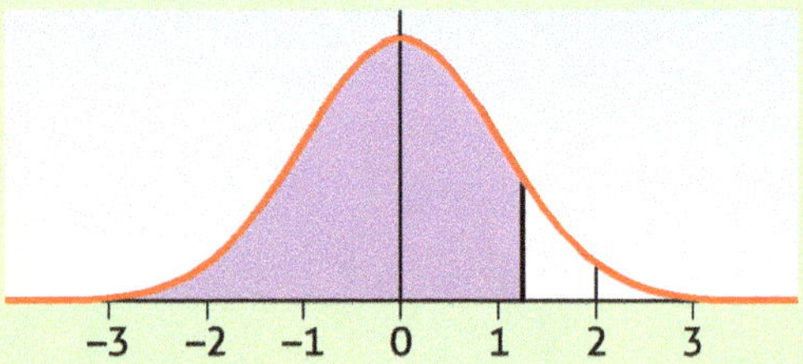

Use a sketch to help you work out what area you need.

Now read along the 1.2 row of the table (the red horizontal line) until you come to .05 (the blue vertical line). This gives you 0.8944.

z	.00	.01	.02	.03	.04	.05	.06	.07	.08	.09	1	2	3	4	5	6	7	8	9
0.0	.5000	5040	5080	5120	5160	5199	5239	5279	5319	5359	4	8	12	16	20	24	28	32	36
0.1	.5398	5438	5478	5517	5557	5596	5636	5675	5714	5753	4	8	12	16	20	24	28	32	35
0.2	.5793	5832	5471	5910	5948	5987	6026	6064	6103	6141	4	8	12	15	19	23	27	31	35
0.3	.6179	6217	6255	6293	6331	6368	6406	6443	6480	6517	4	8	11	15	19	23	26	30	34
0.4	.6554	6591	6628	6664	6700	6736	6772	6808	6844	6879	4	7	11	14	18	22	25	29	32
0.5	.6915	6950	6985	7019	7054	7088	7123	7157	7190	7224	3	7	10	14	17	21	24	27	31
0.6	.7257	7291	7324	7357	7389	7422	7454	7486	7517	7549	3	6	10	13	16	19	23	26	29
0.7	.7580	7611	7642	7673	7704	7784	7764	7794	7823	7852	3	6	9	12	15	18	21	24	27
0.8	.7881	7910	7939	7967	7995	8023	8051	8078	8106	8133	3	6	8	11	14	17	19	22	25
0.9	.8159	8186	8212	8238	8264	8289	8315	8340	8365	8389	3	5	8	10	13	15	18	20	23
1.0	.8413	8438	8461	8485	8508	8531	8554	8577	8599	8621	2	5	7	9	12	14	16	18	21
1.1	.8643	8665	8686	8708	8729	8749	8770	8790	8810	8830	2	4	6	8	10	12	14	16	19
1.2	.8849	8869	8888	8907	8925	8944	8962	8980	8997	9015	2	4	6	7	9	11	13	15	16
1.3	.9032	9049	9066	9082	9099	9115	9131	9147	9162	9177	2	3	5	6	8	10	11	13	14

So, P(battery lasts less than 24 hours) = $\Phi(z = 1.25) = 0.8944$.

89.44% of phones have a battery life of less than 24 hours.

(b) First work out the z-value:

$$z = \frac{25.4 - 22}{1.6} = 2.125.$$

So you want to know what proportion of mobiles have a battery life that is **more than** 2.125 sd above the mean.

Draw a sketch to help you see the area you need to find.

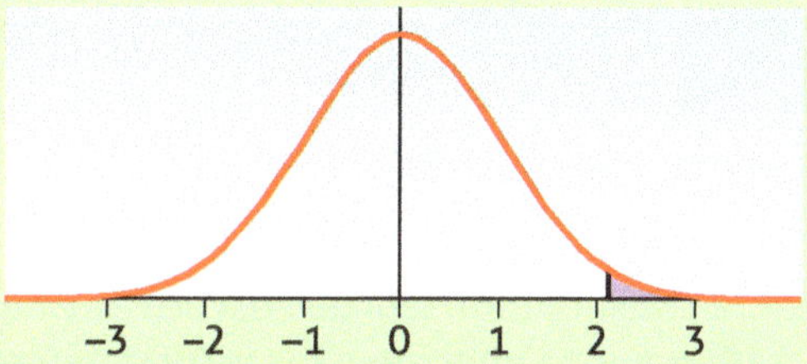

Now read off 2.125 from the table.

z	.00	.01	.02	.03	.04	.05	.06	.07	.08	.09	1	2	3	4	5	6	7	8	9
0.0	.5000	5040	5080	5120	5160	5199	5239	5279	5319	5359	4	8	12	16	20	24	28	32	36
0.1	.5398	5438	5478				536	5675	5714	5753	4	8	12	16	20	24	28	32	35
...																			
		9649	9656	9664	9671	9678	9686	9693	9699	9706	1	1	2	3	4	4	5	6	6
		9719	9726	9732	9738	9744	9750	9756	9761	9767	1	1	2	2	3	4	4	5	5
2.0	.9772	9778	9783	9788	9793	9798	9803	9808	9812	9817	0	1	1	2	2				
2.1	.9821	9826	9830	9834	9838	9842	9846	9850	9854	9857	0	1	1	2	2				
2.2	.9861	9864	9868	9871	9875	9878	9881	9884	9887	9890	0	1	1	1	2				

1 Read down to 2.1 ...

2 ... and then along to 0.02 ...

3 ... and then along to 5.

9830 + 2 = 9832

So, P(battery lasts less than 25.4 hours) = $\Phi(z = 2.125) = 0.9832$.

P(battery lasts more than 25.4 hours) = $1 - 0.9832$

$= 0.0168$.

Number of phones that last longer than 25.4 hours = 0.0168×1000

$= 16.8$.

(c) P(battery lasts less than 25.4 hours) = $\Phi(z = 2.125) = 0.9832$
P(battery lasts less than 24 hours) = $\Phi(z = 1.25) = 0.8944$

So, P(battery lasts between 24 hours and 25.4 hours)
$= \Phi(z = 2.125) - \Phi(z = 1.25)$
$= 0.9832 - 0.8944$
$= 0.0888.$

So, 8.88% of phones last between 24 hours and 25.4 hours.

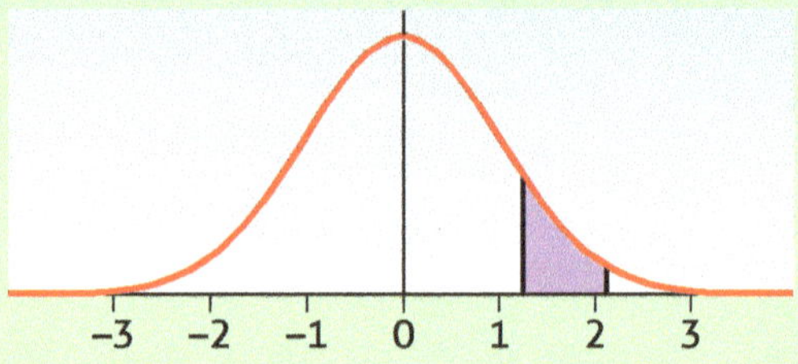

Exercise 17E

1 A variable X is Normally distributed with $X \sim N(0, 1)$.
Work out the percentage of values that have a z-value less than:
(a) 0.5 (b) 0.75 (c) 1.8 (d) 2.4 (e) 0.456.

2 The weights of bags of crisps are Normally distributed with mean 25 grams and standard deviation 0.2 grams.
A factory packs 10 000 bags of crisps a day.
How many of these packets of crisps would you expect to weigh between 25 grams and 25.5 grams?

17F Using symmetry

This section covers using the symmetry of the Normal distribution to solve problems when the z-score is negative. The use of tables in this work is extension material for Critical Maths students.

Activity 17F Finding probabilities

The weights of hedgehogs are Normally distributed with mean 600 grams and standard deviation 100 grams.

A wildlife conservation worker rescues any hedgehogs with a weight of less than 450 grams to give them extra food before the winter hibernation.

How can you estimate what percentage of hedgehogs need rescuing?

Worked example 17F

? Work out the percentage of hedgehogs that need rescuing in Activity 17F.

Solution

First work out the z-value:

$$z = \frac{450 - 600}{100} = -1.5$$

Use the Normal distribution table to read off the value when $z = 1.5$.

93.32% of hedgehogs have a mass less than 1.5 sd above mean.

So, 6.68% of hedgehogs need rescuing.

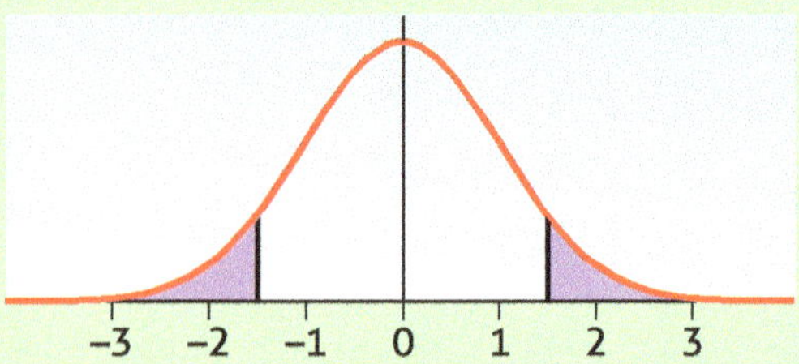

These two regions have the same area.
So, the proportion of hedgehogs that are underweight with z less than -1.5 is the same as those 'over' weight with z more than $+1.5$.

Exercise 17F

1 A variable X is Normally distributed with $X \sim N(0, 1)$.

Work out the percentage of values that have a z-score:

(a) less than −0.4
(b) more than −0.4
(c) less than −1.2
(d) more than −1.2
(e) between −0.4 and 1.2
(f) between −1.2 and −0.4.

2 The quantity of milk in bottles from a dairy farm is Normally distributed with mean 1.03 litres and standard deviation 0.05 litres.

Benji buys a litre of milk from the farm. What is the probability that his bottle contains less than 1 litre?

Test yourself 17G

1 A weather forecaster claims that the average number of correct forecasts he makes every 30 days is 24 and the standard deviation in the number of correct forecasts is 1.5.

Over one period of 30 days, the weather forecaster makes 20 correct forecasts.

(a) Calculate the z-score for this outcome.
(b) Does this figure provide evidence to doubt the validity of the weather forecaster's claim?

2 A quality control inspector at a factory finds that the weight of tins of baked beans is Normally distributed with mean 415 g and standard deviation 2 g. The inspector orders that any tins with a mass less than 414.5 g or greater than 418 g should be rejected.

What percentage of tins are rejected?

3 A sociology student collects data on how long students spend revising for an exam. She finds that the times are Normally distributed with a mean time of 4 hours and a standard deviation of 40 minutes.

Find the probability that a randomly selected student spent over 5 hours revising for the exam.

4 A driving instructor finds that the number of hours of driving practice her students need to be ready for the test is Normally distributed with mean 32 hours and the standard deviation is 3 hours.

What percentage of her students need between 30 hours and 35 hours tuition?

5 A company sells seeds and claims that 50% of its seeds will germinate.

To test the company's claim, random samples of 10 000 seeds were planted.

(a) Write down the average number of seeds that germinate in each sample.
(b) What is the standard deviation of the number of seeds that germinate.
(c) In one sample it was found that 362 seeds germinated. Is this evidence that the company's claim was false?

Practice question

A marine biologist surveys the lengths of orcas. She produces a histogram of her results.

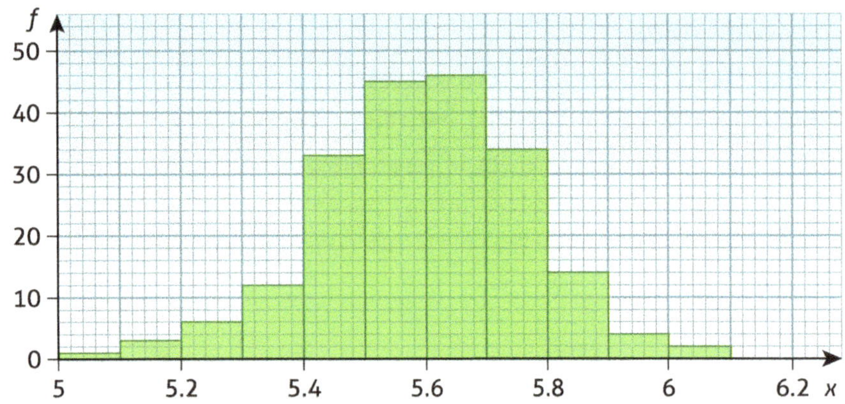

(a) What features of the histogram suggest that the Normal distribution might be an appropriate model?

(b) The marine biologist finds that the mean length of an orca is 5.6 m and the standard deviation is 0.16 m. Show that the total number of orcas with lengths between 5.4 m and 5.8 m is approximately what would be predicted by this model.

(c) Show that if the model is applied to all orcas then virtually none would be over 6.2 m.

Now that you have completed this chapter, you should know:

- When it is appropriate to use the Normal distribution as a model.
- How to recognise situations that can be modelled by the toss of a fair coin.
- That for a distribution that that can be modelled using n tosses of a fair coin the mean is $\frac{n}{2}$ and the standard deviation is $\frac{\sqrt{n}}{2}$.
- That proportions from small samples are more variable than those from large samples.
- How to standardise a value from a Normal distribution with a given mean and standard deviation, and so find its z-score.
- The notation $X \sim N(\mu, \sigma^2)$.
- How to use a z-score to decide whether a value from a Normal distribution is unusual.
- How to use Normal distribution tables to estimate the population proportions.

CHAPTER 18
Using statistics to solve problems

In Chapter 6 you met the statistical problem solving cycle. This chapter covers the activities involved at the various stages in greater depth.

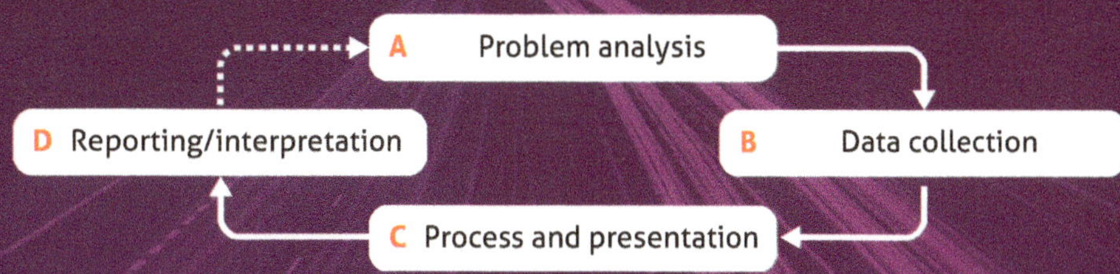

18A Problem analysis

Some problems can be solved without the use of statistics but many others require suitable data. In such a case the first step is to analyse the problem and decide what data will be helpful and how they can be used.

Activity 18A Problems need statistics

Here are three problems. Decide which of them would best be approached using the statistical problem solving cycle.

(a) A primary school is beside a busy road. Some parents campaign for speed humps to be introduced. The local council wants to decide whether to incur the expense involved.

(b) What is the largest prime number less than 100?

(c) Following a heavy rainstorm, the river through a small town bursts its banks and there is flooding. Should a population of beavers be introduced to build dams up-river?

In the first box of the cycle (Problem analysis) you plan out how you are going to tackle the problem. You may need to make some assumptions before you can get started. Then you decide on what data you are going to collect and how you expect to process the figures. When you present your findings, they must be in a form that addresses the original problem.

Worked example 18A

The speed limit on motorways in the UK is 70 mph.

However, many drivers go faster and do not get fined; speeds of 80 mph and greater are common. A government minister thinks this could be a source of income. He asks his officials how much extra money they could hope to receive if drivers caught exceeding 75 mph were fined.

Carry out a brief initial problem analysis.

Solution

Important issues

Assumptions	To get started you need to assume that the speeds at which people drive will continue to be the same as now. Otherwise, no data can be available.
Data collection: how many vehicles use the motorway network?	If you decide to collect your own data, a large number of measurements will be needed to estimate the number of vehicles using the motorway network. They will need to cover different stretches of the various motorways sampled, the times of day, the days of the week, etc. These would be primary data. However, before embarking on a complicated and expensive exercise in data collection, you would be well advised to check whether the data you need are already available from the motorway police. These would be secondary data.
Data collection: what proportion of vehicles exceed 75 mph?	The numbers exceeding the speed limit can be measured by cameras on the bridges. So, at any time the proportion can be estimated but it will vary with the traffic density. In congested conditions people are often unable to speed.
Processing: what proportion of those speeding will be caught?	The income will depend on the proportion of those speeding who are caught. The government's level of commitment to the scheme will be critical. To what extent are they prepared to divert police away from other duties?
Processing: what level of fine does the minister expect?	In order to estimate the income from fines it will be necessary to know how much the penalty will be and if it will be more for faster drivers.
Interpretation: is the answer provided good enough or is further work needed?	The figure calculated for the income can only be approximate, but it may be good enough to give the minister enough of an idea.

The estimate will be based on the initial assumption that people carry on driving just as they did before. This is unlikely to be true. Once fines start being issued, drivers can be expected to change their behaviour. Many people who currently get away with driving at 80 mph will reduce their standard speed to 75 mph or slightly under, so the government's income may well be less than expected. If the scheme is implemented, more data will be provided by the income from fines.

Exercise 18A

For each of the following situations carry out an initial problem analysis like that above.

1. The manuscript of a previously unknown book is found in a library. The author is uncertain but is likely to be one of three people. How can the writing style be used to decide which?
2. Is there any truth in the 13th-century saying. 'A moist and cool summer foretells a hard winter'?

18B Data collection: sampling

The next step, once the problem has been analysed, is to collect the relevant data. In some cases it is impossible to collect all the data, for example the weights of all the cod in the North Sea. A **sample** is needed. This is a set of data, items or people (called **subjects**) selected from the **population**.

The word population has a special meaning in statistics. It is the whole group about which information is required; it may be people, things or data. The sample is used to draw conclusions about the population, but this can only be done if the sample is genuinely representative of it.

Activity 18B How would you choose a sample?

Discuss how you might select a sample of each of the following groups. How representative do you think it would be?

(a) The cod in the North Sea.

(b) The homeless people in the UK.

(c) The mobile phones produced by a factory.

(d) The adults living in your town.

(e) University psychology students.

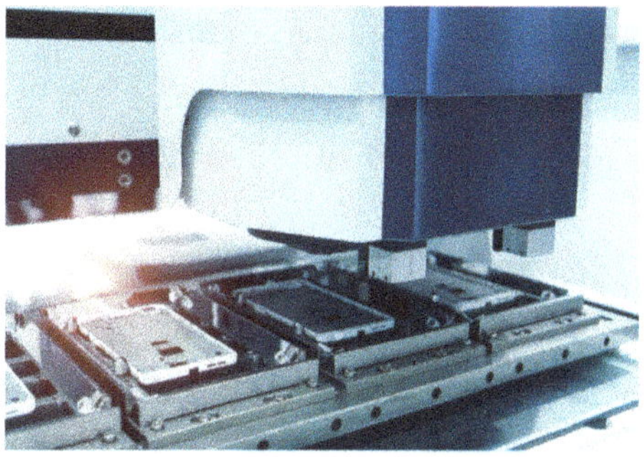

Here are some of the common methods of sampling:

Simple random sampling	In a simple random sample, subjects are selected at random from a list of the whole population. Every possible sample of the required size is equally likely to be selected.
Stratified sampling	The population is divided into subgroups or strata, for example the voters living in the different parliamentary constituencies in the UK. A random sample is taken from each of them. In a proportional stratified sample, the number of subjects chosen from each stratum is proportional to its size.
Cluster sampling	The sample is chosen from a number of subgroups, known as clusters. For example, a sample of size 100 of the cod in the North Sea might consist of 20 fish from each of 5 places where fishing boats were operating.
Systematic sampling	Systematic sampling can be used when there is an ordered list of the population, like an electoral roll. Each individual on the list is numbered, and items are selected at equal intervals, for example every 10th. The starting point is often chosen at random.
Opportunity sampling	The sample is selected from a source that happens to be easily available; for example, delegates at a conference or the bats in a particular roost.
Quota sampling	In a quota sample, each person collecting data is required to choose certain numbers of subjects from particular groups, for example 30 adult men, 30 adult women and 40 teenagers.
Self-selected sampling	A sample is self-selected when those involved choose to take part: examples are drug trials, online surveys and government consultations.

For each of these seven sampling methods, give one advantage and one disadvantage.

Sampling may raise ethical issues. There should be no possibility of harm – physical, psychological or financial – coming to any subject. Confidentiality and informed consent will often be important.

In a **random** sample each subject is chosen entirely by chance and each member of the population has a known, but not necessarily equal, chance of being included in the sample.

Exercise 18B

1 In each of the following situations a sample must be taken and used to obtain information. What sampling method would be most appropriate?

(a) A large airline wants to know what passengers think of the service on their flights.

(b) A newsworthy event has just occurred near a town centre. A TV news crew wants to include some representative local comments in its report.

(c) A town council wants to know how the various groups living there view the quality of its nursery provision.

(d) A scientist wants to investigate any damage that an invasive species may be causing to wildlife on our rivers.

(e) A clothing manufacturer is considering developing a new range and wants to assess how popular it would be with both females and males across a range of ages.

(f) An election is coming up and the local branch of a political party want to assess local opinion on a number of issues; they can only afford to interview 5% of the electorate.

(g) A pharmaceutical company is developing a new anti-wrinkle skin care cream and needs to carry out human trials.

(h) The government wishes to assess the level of air pollution in our cities.

(i) The authorities wish to check whether the conduct of a large company is consistent with the Employment Act.

(j) A GP wants to assess how many of her patients are suffering from stress.

(k) An environmental organisation wants to use its website to find out whether people would support a return to glass milk bottles in place of plastic ones.

18C Data collection: avoiding bias

The value of the data collected depends on how representative the sample is of the population being investigated. You need to be careful that:

- your sampling method does not introduce any bias
- any questions you ask are neutral
- your questions are crystal clear.

Activity 18C Opinion surveys

Online opinion surveys are popular. Which sampling method is involved?

Why might any results from these be biased, especially about controversial issues?

Face-to-face opinion polls in the street are also often used.

(a) Which sampling methods are likely to be involved?

(b) How can you avoid or reduce bias when carrying out such surveys?

Worked example 18C

An online shop sells individualised T-shirts. They have added a voluntary survey page which users can click onto, part of which is shown below.

Survey

Q.1 How old are you?

15 to 20 | 20 to 30 | 30 to 40 | 50 plus

Q.2 How easy is it to use and get around our website?

[ ]

Q.3 Did you make a purchase this visit?

YES | NO

[...]

Q.12 Did you not like our bargain basement T-shirts?

YES | NO

(a) Look at the four questions above. They are all unsatisfactory. In each case say why and suggest how it can be improved.

(b) The T-shirt company then decides to use a face-to-face survey.
They plan to use the improved questionnaire and administer it in a big local shopping mall.
Why is this not a not a suitable method for them to use?

Solution

(a)

Q.	Criticism	Improvement
1	The ages in the answer boxes overlap. A 30-year old could tick either of the middle two. There is no box for someone aged, say, 45.	Change 20 to 19 in the first box, 30 to 29 in the second box, etc. Include a new box 40 to 49.
2	There is no guidance in the answer space. It will be time-consuming to transcribe the answers and some will require interpretation.	Include boxes saying 'Very easy', 'Easy', 'Not very easy', 'Hard' and 'Very hard'.
3	This question is likely to introduce bias into the process, as those who made a purchase will be more likely to continue the survey and send it in than those who did not.	Omit this question.
12	It is unclear what this question is asking because it included the word 'not'. So some of the answers are likely to be for a different question.	Delete the word 'not' from the question.

(b) Many of the T-shirt company's sales are online so responses in a shopping mall may well be biased.
A different consideration is that people in a shopping mall are likely to be in a spending frame of mind.

Exercise 18C

1 Explain why each of the following questions is unsuitable and suggest how it could be improved.

(a) (In a face-to-face interview.) Many of us think that the railway service should be improved. Do you agree?

(b) Do you fill in your income tax return honestly?

(c) Do you disagree that local trams should not be made free?

(d) What is the fastest wild mammal in the UK: the fox or the fallow deer?

2 A town has three wards covering about the same land area. In Ward A there are large houses with gardens. In Ward B there are small houses close together. In Ward C there are mostly small houses and a few large ones have been divided into flats. The town council chooses 20 residents at random from each of the wards to take part in a consultation. Explain why this sample is biased and how it can be improved.

3 An MP wants to know how people in her constituency feel about internet scams. Her staff conduct a poll. They telephone those people whose names are at the top of the columns of the local telephone directory (mobile phones and unlisted numbers are not in these directories). They use the same scripted question until they get responses from 200 people. Of these, 60 say they are 'very worried'.

The staff report back to the MP that 30% of her constituency are 'very worried' about internet scams. What sources of bias are there in this situation and conclusion?

18D Processing and presentation: cleaning your data

Once you have collected your data, you need to check for any omissions or obvious errors before processing them. This is called cleaning the data.

Activity 18D How much rent?

Some students are investigating how much rent people pay each month in the different regions of England.

They find some data on the Office for National Statistics (ONS) website and copy it into a spreadsheet rather than download the spreadsheet on the website. This is their spreadsheet.

	A	B	C	D	E	F	G
1	Private rental market statistics						
2	Summary of 'Room' monthly rents recorded between 1 October 2019 to 30 September 2020 by region for England						
3							
4	Region	Number of rents	Mean	Lower quartile	Median	Upper quartile	
5	NORTH EAST	720	369	325	368	399	
6	NORTH WEST	3030	390	347	386	433	
7	YORKSHIRE AND THE HUMBER	1430	378	325	368	412	
8	EAST MIDLANDS	2470	380	342	376	412	
9	WEST MIDLANDS	1110	396	368	494	425	
10	EAST MIDLANDS	1410	436	350	430	500	
11	LONDON	1000	658	550	650	750	
12	SOUTH EAST	3100	457	520	425	390	
13	SOUTH WEST		428	377	420	460	
14	**ENGLAND**	**17780**	**625**	**360**	**403**	**464**	
15	Source: VOA's administrative database at 30 September 2020						

(a) There is an obvious omission from cell B13. How can you work out what it should be?

(b) How can you tell, just by looking at the figures and without doing any calculations, that there are errors in cells E9 and C14?

Estimate or calculate the numbers that should be in those cells.

Missing data

It is usually obvious when a data item is missing. In the case of cell B13 in Activity 18D you could work out what it should be, but that is often not the case.

In the case of primary data that you have collected yourself, the best solution is to repeat the missing measurement.

In the case of secondary data, you can return to the original source and check it but, for one reason or another, it may be incomplete. In that case you have to decide how important it is and whether to estimate it or to omit it completely.

If you really do need a value, for example as part of a time series, looking at values around the missing one can allow you to make a sensible estimate. You may, for example, take the mean of the values either side.

When two variables are involved, you can sometimes draw a scatter diagram and use a line of best fit to estimate the missing value.

Outliers and errors

An **outlier** is an item that differs significantly from the rest of the data.

It may be due to an error or to a chance result, or it may be telling you something really important. You should always look for outliers in the data you have collected and investigate them.

Two rule-of-thumb guides are used to identify a data point, x, as a possible outlier.

- For data with lower quartile Q1, upper quartile Q3 and inter-quartile range IQR, x is more than 1.5 times the IQR from the nearer quartile.

  $x < Q1 - 1.5 \times IQR$ or $x > Q3 + 1.5 \times IQR$

  Two possible outliers are shown on the box-and-whisker plot below.

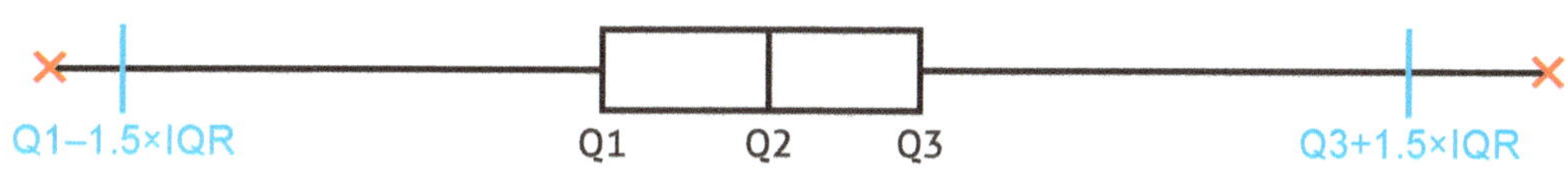

- For data with standard deviation s, x is more than $2s$ from the mean,

  $x < \bar{x} - 2s$ or $x > x + 2s$

Finding the standard deviation

You can only use the two standard deviation guide if you know the standard deviation of your sample data. There are several ways you can find it:

- You can enter the data into your calculator and use the appropriate facility. The details of this depend on the particular calculator you have.

Almost all calculators will give you two values. The larger one is the estimated value for the population from your sample; the smaller is the value if your sample is the whole population. The larger value will usually be the more appropriate for statistical problem solving.

- You can use a spreadsheet. Start by entering the data, say into cells C2 to C31. Then select a cell for the answer and enter the command =STDEV(C2:C31).
- There is also a formula but using it involves quite a lot of work, so is much more time consuming.

The spreadsheet command =STDEV(.... : ....) gives the larger value.

Worked example 18D

? Hamish is investigating long-term weather patterns as part of his research into climate change. He obtains these figures for the total rainfall, in mm, in January at Stornoway in the Hebrides. They are for every five years since 1878 in order, apart from 1883 where records are incomplete.

132	104	60	125	146	146	89	139	167	269
144	180	108	139	87	171	26	119	83	118
220	141	275	182	154	156	109	167		

(a) Find the median (or 2nd quartile) rainfall.

(b) Find the (i) 1st and (ii) 3rd quartiles for the rainfall.

(c) Use the IQR to identity any outliers in the data set. Should Hamish discard them?

Solution

The first step is to order the data and split it up into quartiles.

See Chapter 6 for methods of finding quartiles. Spreadsheets sometimes give slightly different results.

26 60 83 87 89 104 108 | **109 118 119 125 132 139 139** | 141 144 146 146 154 156 167 | 167 171 180 182 220 269 275

Q1 (between 108 and 109), Q2 (between 139 and 141), Q3 (between 167 and 167)

(a) Median (Q2) = (139 + 141) ÷ 2 = 140 mm.

(b) Q1 = (108 + 109) ÷ 2 = 108.5 mm and Q3 = (167 + 167) ÷ 2 = 167 mm.

(c) IQR = (Q3 – Q1) = 167 – 108.5 = 58.5.

Any outliers are
either below (108.5 – 1.5 × 58.5) = 20.75 mm, so none in this case,
or above (167 + 1.5 × 58.5) = 254.75 mm, giving 269 mm and 275 mm.

The two values are part of the weather pattern that Hamish is investigating, so he should not discard them. The exceptionally heavy rainfall in those years may be an important feature of climate change.

Exercise 18D

1 These are figures for the thickness of the ozone layer in suitable units for October for 20 years, as recorded by weather satellites and ground stations.

330	318	315	305	318	330	310	315	280	318
320	300	280	280	290	305	190	315	290	290

RealClimate

Any outliers for these experiments were discarded as probably due to instrument errors. Later work showed that they were caused by the thinning of the ozone layer; this has since been reversed by the worldwide banning of fluorocarbons.

(a) Plot the data on a chart.

(b) Enter the data in cells A2 to A21 of a spreadsheet.

(c) Use column B to give the ranks of the figures, with the smallest at 1. The command for cell B2 is =RANK(A2,A$2:A$21,1).

(d) Show that the IQR is 28. Hence, find any data items that are outliers.

(e) Now find the mean and standard deviation by placing the command =AVERAGE(A2:A21) in cell A23 and =STDEV(A2:A21) in cell A24.

Hence, show that the rule 'More than two standard deviations from the mean' identifies the same data items as part (c).

⚠ There are different ways of ranking. The =RANK(...,...,...) command treats the items as if they are competitors. So, if the figures here were scores in a quiz, the two people with 330 would both be ranked 1 (or 1=).

Another way of using Excel to find the IQR is to use the command =QUARTILE(A1:A21,1) to find the lower quartile and =QUARTILE(A1:A21,3) for the upper quartile.

2 When carbon steel is immersed in seawater, a layer of corrosion forms on its surface. This can be a serious problem for underwater pipes.

These are the results of some experiments on carbon steel immersed in seawater. The thickness of corrosion is measured in microns. A micron is 10^{-6} m.

Time immersed in sea (years)	0	1	3	7	8	10	12	14	15
Thickness of corrosion (microns)	0	100	215	550	660	1205	780	980	995

(a) Plot the data on a graph.

(b) Which data point may be an outlier? Justify your choice.

(c) Estimate the thickness of corrosion after two years of immersion.

(d) The thickness of corrosion is measured on a number of pipes that have been in use for 3 years. Here are the results.

195 250 270 255 215 220 310 165 300 200 290 285 290 285 270 235 245

Identify any possible outliers in these figures.

18E Processing and presentation: summarising your data

You are already familiar with many measures used to summarise a data set: mean, mode and median for central values and range, inter-quartile range and standard deviation for spread. The question is, which of these will best convey the information locked up in your data to your intended readership?

You are also familiar with the common ways of displaying data. Here, too, the question is which of them will be most helpful for people wanting to understand about your data.

Activity 18E Leaders' ages

These are the ages of the 20th-century US Presidents and UK Prime Ministers when they first took office.

20th-century UK Prime Ministers' age on first taking office

62 65 53 48 60 53 43 57 55 64 62 68
47 53 55 64 69 57 53

20th-century US Presidents' age on first taking office

64 52 46 51 62 61 55 54 55 43 56 56
69 51 42 51 60

Someone asks you what information is contained in these figures.

How would you summarise it as helpfully as you can?

Worked example 18E

The inequality ratio is a measure used by the United Nations to compare inequality in different countries. It is the number of times the average income of the richest 10% is greater than that of the poorest 10%. For the UK its value is 13.8.

These are the inequality ratios for countries in South America and the largest Asian countries.

South America

16.5	60.4	31.6	26.1	18.8	26.2	35.2	93.9	38.8

Largest Asian countries

4.5	8.6	8.3	12.3	10.6	8.6	12.6	15.5	8.5
17.2	8.2	12.7	6.5	6.9	6.4	7.8	6.6	

United Nations

Write a brief summary of the information for someone who does not have access to the data, illustrating what you say with a suitable display.

Solution

The Asian countries have far lower inequality indices with a mean value of 9.5 compared with 38.6 for the South American countries.

The figures for Asian countries are more tightly grouped with an IQR of 5.7 compared with 27.15 for those in South America. One South American country has an inequality index far greater than any of the other countries.

This is shown on the box-and whisker plots below.

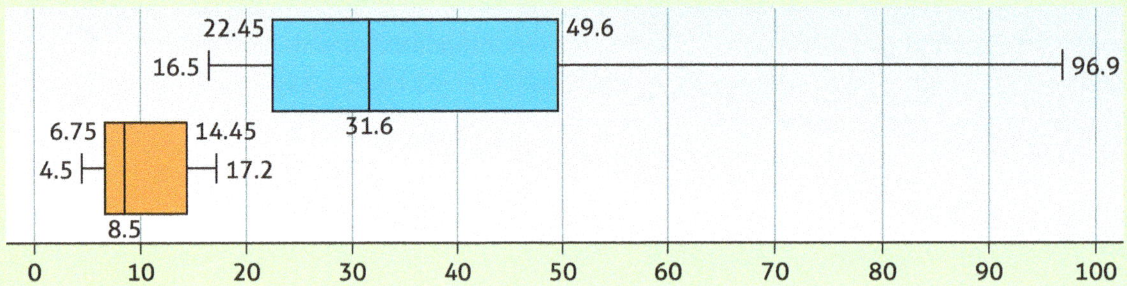

Exercise 18E

1 Here are the dates (CE) of various pieces of wood making up a Viking longship.

823	822	823	829	813	818
829	825	819	824	831	819
822	819	797	821	818	824

Write a short paragraph, supported by numbers, about the age of the longship.

Illustrate what you are saying with a suitable diagram.

2 These are the heartbeat rates for three groups of people. The groups are at rest, walking or jogging.

At rest (beats per minute)
98 84 99 102 81 83 110 83 65 97 93 86 86 79

Walking (beats per minute)
100 99 73 106 116 88 118 96 92 99 94 109 100 92 118 102 98

Jogging (beats per minute)
125 118 104 92 114 107 112 135 108 114 116 119 102

Write a short summary of the information provided by these data sets and display it in a suitable chart that shows their main features.

18F Process and presentation: aids to calculations

This section covers a number of techniques that you will find helpful for calculations to process your data.

Activity 18F The velocity of light

The velocity of light in a vacuum has been measured very accurately.

These measurements, in metres per second, were made in the 1970s.

299 792 457   299 792 455   299 792 459   299 792 456   299 792 453

What are the easiest ways of calculating (a) the mean and (b) the standard deviation of these numbers?

In Activity 18F, you probably notice that the first eight digits of all the numbers are the same. It is only the final digits that change: 7, 5, 9, 6, 3.

For these numbers:
mean = 6, standard deviation = 2.236.

So, the final number of the mean of the measurements of the velocity of light is 6. The mean of the measurements is 299 792 456 m s^{-1}.

The standard deviation is a measure of spread, so since the first eight digits are all the same they are irrelevant.

The standard deviation of the measurements is 2.236 m s^{-1}.

A similar situation arises with the height of water in a lake or river. It is measured above or below a fixed **datum level**. Calculations of the mean and standard deviation of the height are based on the datum level rather than the actual depth.

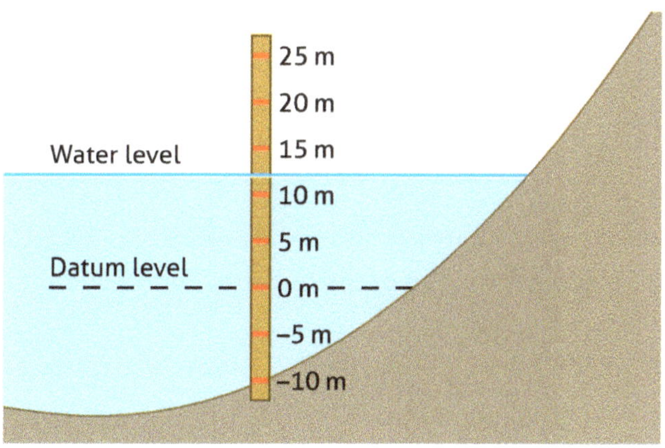

There are many other contexts where a datum level may be used, for example the value of a shop's daily sales.

Worked example 18F below shows the effect of a **change of units** on the mean and standard deviation.

Worked example 18F

? A group of patients are all suffering from the same condition in a hospital ward. Their temperatures in degrees Celsius are recorded as follows:

37.2 36.8 40.0 38.2 38.5 63.9 39.6 38.0 39.3 38.0.

(a) Identify the outlier in these figures and explain why it should be discarded.

The mean for the remaining nine figures is 38.4 and the standard deviation is 1.071.

(b) Convert the mean and standard deviation into degrees Fahrenheit as easily as possible.

Solution

(a) The outlier is 63.9°C. No one can live with such a high temperature so it must be a mistake.

It could be 36.9°C, written down wrong, but there is no evidence that this is the case.

(b) The usual formula for converting temperatures in Celsius, C, into Fahrenheit, F, is $F = 32 + \frac{9}{5} \times C$.

This is used to find the **mean** in degrees Fahrenheit.

Mean = $32 + \frac{9}{5} \times 38.4 = 101.12$

Since **standard deviation** is a measure of spread, only the factor $\frac{9}{5}$ is used.

Standard deviation = $\frac{9}{5} \times 1.071 = 1.928$

Worked example 18F illustrates a number of points.

It is an example of a **linear transformation**. It is of the type $y = a + bx$ where x is the original value and y is the new value. a and b are numbers; in the Worked example $a = 32$ and $b = \frac{9}{5}$.

- Means: $\bar{y} = a + b\bar{x}$, where $\bar{x}$ is the original mean and $\bar{y}$ is the new mean.
- Standard deviations: $s_y = bs_x$, where s_x is the original standard deviation and s_y is the new standard deviation.

Exercise 18F

1 In a game, everyone is given 5 points to start with. They then score 2 points for each time they obtain a hit. So, a player's total score is given by the formula $S = 5 + 2 \times H$.

(a) Explain the meaning of S and H in the formula.

Over the course of many games, it is found that at the end of a game the mean of the quantity H is 12 and the standard deviation is 3.6.

(b) Find the mean and standard deviation of S at the end of a game.

2 (a) Use a spreadsheet to find the mean and standard deviation of the numbers 6, 8, 9, 0, 6, 5, 0, 3, 4 and 10.

(b) Hence, write down the mean and standard deviation of the numbers 2560, 2580. 2590, 2500. 2560, 2550, 2500, 2530, 2540 and 2600.

(c) Use a spreadsheet to check your answers to part (b).

18G Process and presentation: modelling data

The data you collect as a sample will almost certainly give you an insight into the problem in hand. However, it will be particularly useful if the data come from a well-known distribution. In that case, standard results from that distribution can give you additional information. This section covers the cases where the sample data can be modelled by either the uniform distribution or the Normal distribution.

Activity 18G Finding a model

You have taken a sample.

(a) How would you use the data to decide whether the underlying population is likely to have a uniform distribution? What about a Normal distribution?

(b) What other way have you met for deciding whether a set of data fit Normal distribution?

Worked example 18G

These two grouped frequency charts show the weights and lengths of a sample of 42 new-born babies in a hospital one week.

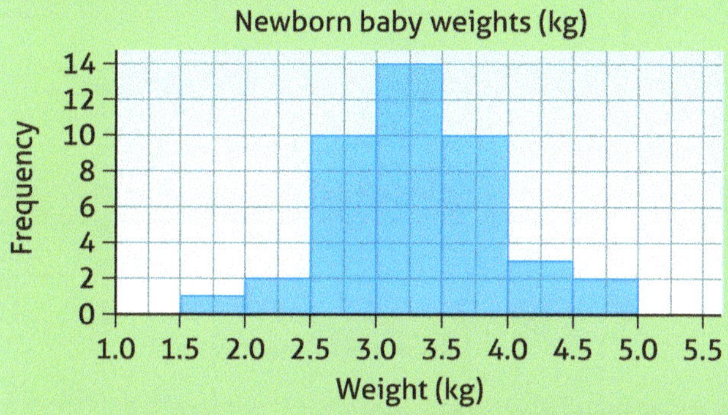

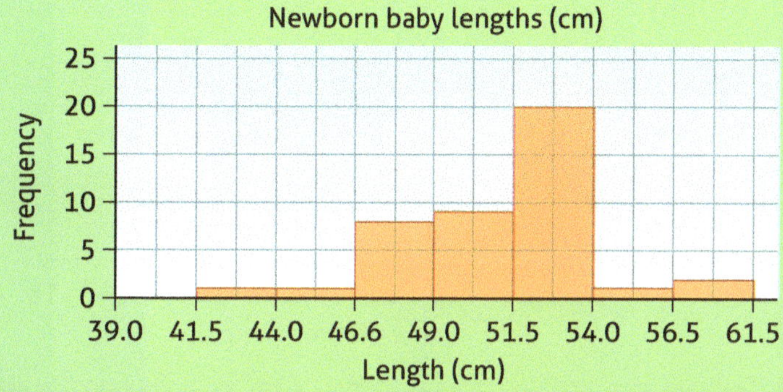

(a) Explain why it looks as though the Normal distribution is a good model for the weights but not for the lengths.

(b) The actual data show that the mean of the weights is 3.25 kg and the standard deviation is 0.64 kg.

The hospital has decided to give special attention to babies under 2 kg. They use the model to estimate how many babies this is in a year. How many do they expect?

Solution

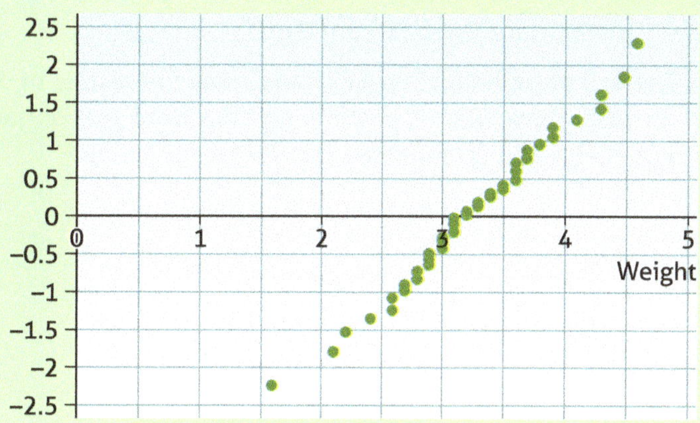

(a) The modal group for weights is in the middle but that for lengths is to the right.

So, the distribution for weights is symmetrical and bell-shaped, whereas that for lengths has negative skew.

The chart suggests that the Normal distribution would be a good model for the weights.

A further check is provided by the Normal probability plot, which shows the data lying close to a straight line.

(b) The distribution of babies' weights is modelled by $N(3.25, 0.64^2)$.

So, the z-score for a weight of 2 kg is $z = \frac{2 - 3.25}{0.64} = -1.95$ to 2 dp.

Using a calculator or tables gives the probability for $z < -1.95$ to be 0.026.

There were 42 babies in 1 week, so the number in a year is estimated to be $42 \times 52 = 2184$.

This gives the number under 2 kg as $0.026 \times 2184 = 56.8$.

So, the hospital estimates that there will be about 60 babies per year requiring special attention.

In this example, the model was probably not perfect but it was good enough to provide the hospital with a working figure. Given the uncertainty, they might well then decide on an interval of likely numbers, such as between 50 and 70.

Exercise 18G

1 For each of these distributions A–D, state whether it could be modelled by the uniform distribution, the Normal distribution or neither. Justify your decisions.

A

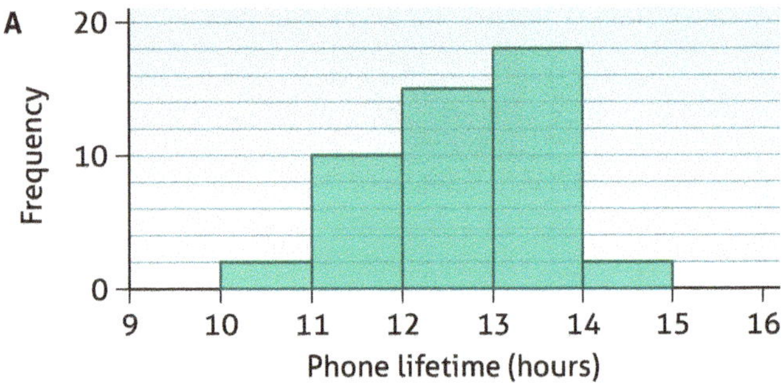

B

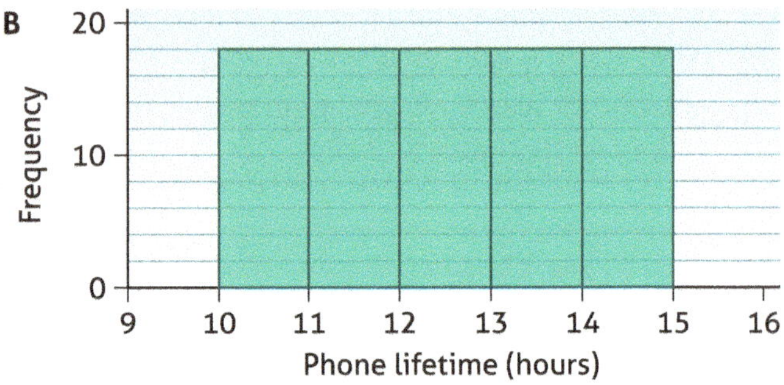

C

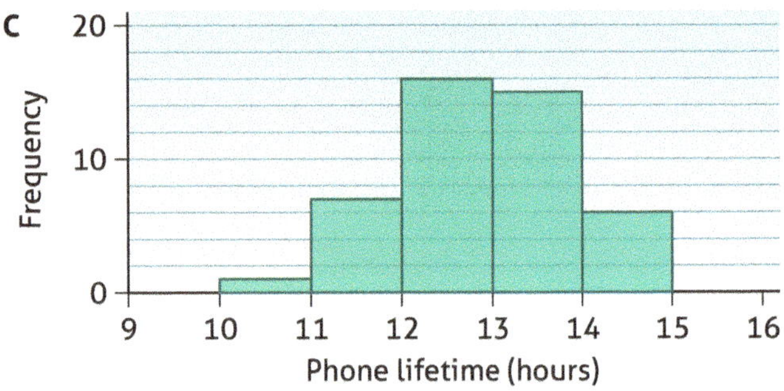

D

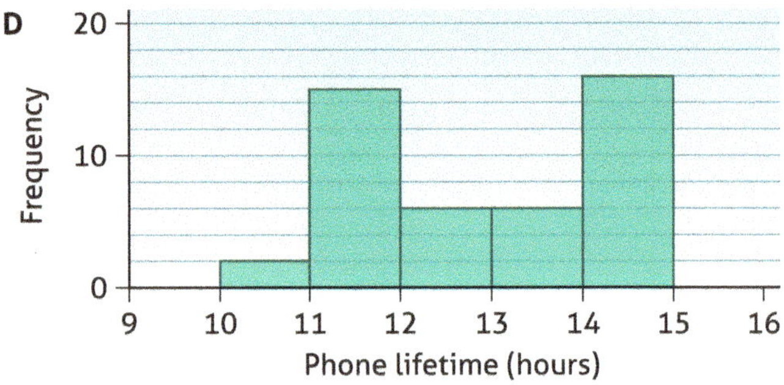

2 Angela is a marathon runner. She tries to maintain an even pace for every kilometre. She has a timer which keeps a record of her times for every kilometre.

(a) What distribution for her kilometre times would provide her ideal model?

Angela's times for her first 5 kilometres are:

4 m 16 s   4 m 14 s   4 m 18 s   4 m 17 s   4m 20s.

(b) Do you think Angela will complete the marathon in 3 hours? Justify your answer. (A marathon is 42.195 km long.)

3 The two grouped frequency charts show the results of 300 tests on the breaking strength, in kN (kilonewton), on each of two brands of climbing rope. Both sets of data have mean 17.0 kN; the standard deviation for rope A is 2.3 kN and that for rope B is 2.93 kN.

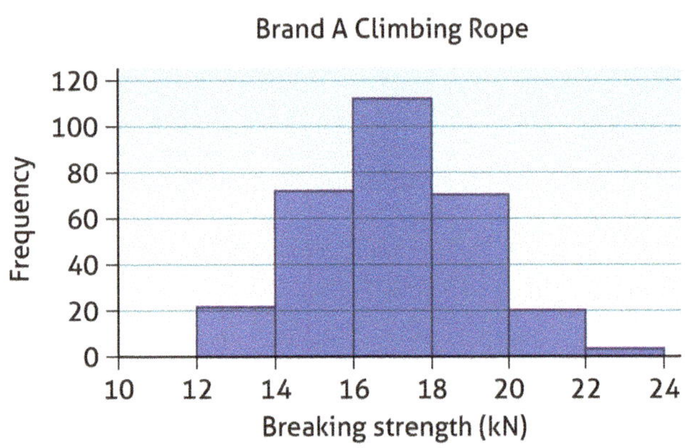

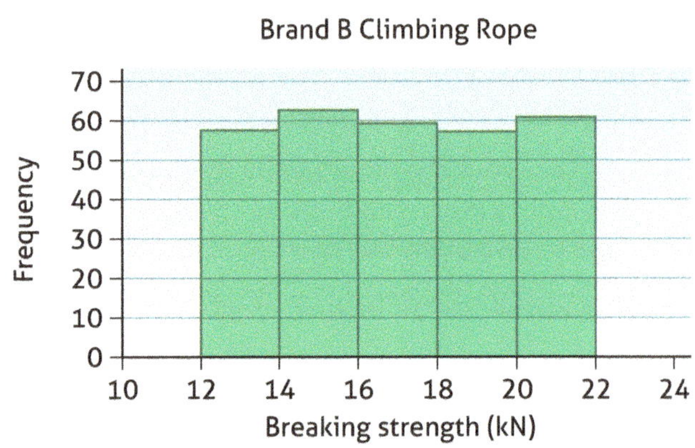

(a) Suggest a suitable model for the breaking strength of each brand of rope.

(b) Use your models to estimate the probability of each brand breaking if a force of 15 kN is applied to it. Which brand of rope is the more reliable? Justify your choice.

18H Reporting/interpretation

This section is about the fourth stage of the statistical problem solving cycle. It involves reporting your findings in terms of the original problem. The outcome may be satisfactory, but it is also possible that more work is required: this could be collecting more data but it may be that the original problem analysis needs to be revised.

Activity 18H True, false, maybe or just can't tell?

John investigated the old saying,
'As it rains in March, so it rains in June.'

This is his report.

Is 'As it rains in March, so it rains in June' true?

Total rainfall (to nearest mm) in June and March for 30 years.

June	20	70	41	55	35	20	28	43	4	96
March	83	43	21	21	78	22	53	12	88	58

June	71	74	30	29	25	91	53	12	88	58
March	5	64	92	28	25	66	32	80	46	21

June	104	102	18	65	30	36	53	38	17	103
March	54	52	13	80	43	12	29	36	33	56

The table shows that the statement is false.

What is good about this piece of work and what needs to be improved?

How well does it fit the statistical problem solving cycle?

Worked example 18H

? A city is about 3 hours away from London by train. The local Chamber of Commerce are concerned that the rail service is not reliable enough; too many trains run late. They decide to investigate the problem and then lobby the railway company.

In their investigation they concentrate on one key service. How often is it late and what are the causes? They monitor it for one calendar month and then prepare a report for the railway company. Here is the first draft of their Executive Summary.

Comment on whether this draft needs changing before it becomes part of the final report.

Executive summary Draft 1

File Edit View Insert Save

The service to and from London is not good enough.

In one month the 0712 train was late 10 times.
The following reasons were given.

Cause	Times
Leaves on the rails	3
Signal failure	2
Held up by slow-running train ahead	2
Medical emergency	1
Staff absence	1
Livestock on the track	1

It is deeply unsatisfactory that 90% of trains that run late are because of things that are your fault: not sweeping the rails, maintaining the signalling and fencing, etc.

Our businesses are suffering because of your incompetence.

Solution

The problem that the Chamber of Commerce is trying to solve is how to improve the rail service.

The conclusions stated in this summary were based on a sample of just one service over one month, so it is unlikely to be representative of all services through the year. However, the report reads as though the limited data really are representative.

The work so far should be treated as a **pilot study**, providing insights for a larger investigation. So, the Chamber of Commerce should go around the statistical problem solving cycle again, this time with a larger sample.

Stage	Comment
Problem analysis	The aim is to achieve a better train service for the city. The new investigation should focus on improvements that the company has the power to make; other problems can be acknowledged but not blamed on the company. This requires an analysis of responsibilities within the wider railway industry.
Data collection	To avoid possible bias, the data should cover other services and other times of the year. For example, leaves on the rails are only a problem for a few weeks in the late autumn. The lengths of the different delays should be recorded.
Process and presentation	The figure of 90% in the draft report is misleading, as many of the delays were caused by events outside the railway company's control; this mistake should be avoided with the new data. The report is likely to have more impact if it includes an attractive and interesting display of the data.
Reporting/interpretation	The Chamber of Commerce should consider the tone of their report. If the project is successful they may find themselves working with the railway company, so they do not want to alienate them.

In this example the first circuit of the cycle is a pilot study and that is used to inform a second and more comprehensive circuit. This is often done in investigations.

Exercise 18H

1 A town has two ambulance stations serving different areas. Both currently have 6 ambulances.

To save money the local Health Authority wants to reduce the number of ambulances by one.

They ask an accountant to recommend which station should lose one. He collects the following data from the two stations on the numbers of callouts during the next week.

	Mon	Tues	Wed	Thus	Fri	Sat	Sun
Station A	36	39	38	41	43	38	40
Station B	25	41	22	67	72	22	20

He reports back. 'Station B has fewer calls on an average day, so they should be the ones to lose an ambulance.'

The Health Authority considers the following three courses of action.

- They accept this decision.
- They accept the grounds for the decision but ask for more data.
- They should recommend a different way of coming to the decision.

(a) Give reasons for and against each course of action.

(b) State which one of the three you would recommend.

2 Jasper wants to buy a motorcycle but his parents are not happy with the idea. They are worried that he will drive too fast. Jasper finds the following data showing typical speeds of motorcyclists, in mph, on a motorway.

74.7	65.0	55.3	68.0	68.6	69.2	88.4	49.4	67.3	95.6
64.1	65.0	52.5	78.9	49.3	59.4	56.5	60.1	60.8	68.7
59.7	66.3	90.7	52.1	67.1	65.3	77.1	52.3	71.9	99.2
53.9	71.7	59.3	77.0	66.8	72.0	56.2	67.9	70.1	68.5

He says to his parents, 'See, most motorcyclists don't speed on the motorway. Only 30% go over 70 mph and the average speed is 67 mph. Motorcyclists are not maniacs.'

State which of the following courses of action would you recommend to Jasper's parents, and why.

- They should accept the data as evidence and raise no more objections.
- They should ask for more data about motorcyclists' speeds.
- They should seek data about the safety of motorcycles, not just the speeds at which they are driven.

Test yourself 18I

1 People say that the higher cats fall from high-rise flats, the fewer the injuries they suffer and the less of them die. This is sometimes called 'high-rise syndrome'.

How is it possible to investigate whether it is true?

2 The council of a town, with 100 000 voters, is going to build a sports centre. The council must survey the voters as to where they want it built. Four sampling methods are suggested.

A Select one of the town's 11 wards at random (these are more or less equal in size and population) then select every tenth voter in that ward and ask them.

B Ask the first 50 people who visit the town hall.

C Set up a stratified sample based on voter age then randomly select a sample from each group in proportion to the number of voters in that group. This will provide a group of about 200 voters.

Age range	18–24	25–44	45–64	65–84	85+
% of voters	38	28	18	14	2

D Use a local radio station on its lunchtime news programme to ask people their preferences and record them on air.

(a) How many voters will be sampled in A?

(b) Explain whether or not B is an appropriate sampling method.

(c) Calculate the number of voters in each age range in sample method C.

(d) State which of the sampling methods you consider to be the most suitable. Support your choice by explaining why the other three are less suitable.

3 A local council is planning to carry out a telephone survey about council services. They do an initial investigation to find the number of rings before either people answer their phone or the phone switches to voicemail.

Here are the results of the investigation.

2	8	3	10	4	4	10	3	2	3	9	3
2	4	4	10	4	4	3	3	4	4	3	3
5	2	3	2	2	8	5	4	3	3	4	5
6	3	8	3	5	2	4	3	1	1	3	7
9	5	4	4	1	3	2	4	9	3	4	5

(a) Display the data in a suitable form.

(b) Find the median and inter-quartile range of the number of rings.

(c) Jan estimates that over half the calls are answered in 3 or fewer rings.

Is she is correct?

The council does not want to make use of voicemail so they want to terminate calls before the receiver's voicemail cuts in.

(d) Estimate after how many rings most voicemails cut in. Explain how you decided.

4 The figures below are a sample of wave heights measured from a satellite over the South Atlantic.

The mean is 6.46 m and the standard deviation is 3.23 m.

1.7	1.9	2.0	2.1	2.6	2.6	2.8	2.9	3.0
3.0	3.2	3.6	3.6	4.7	4.8	4.8	5.0	5.1
5.3	5.6	5.7	6.1	6.3	6.5	6.7	7.0	7.2
7.2	7.6	7.7	7.9	8.2	8.5	8.6	8.7	9.1
9.4	10.2	10.2	10.3	11.5	11.8	11.9	12.8	13.4

(a) An oceanographer says, 'This shows that the distribution of wave heights is $N(6.46, 3.23^2)$.' Comment on whether he is correct.

(b) It was found that a fault in the satellite's measuring system under-measured the wave heights by 1 m.

What effect would this have on the sample's (i) mean, (ii) median and (iii) standard deviation?

5 The planet HD 10546b orbits a star 320 light years from Earth. It is possible to estimate the planet's radius. Here are some typical measurements; the unit used is the radius of Jupiter.

Radius of HD 100546 b (times radius of Jupiter)

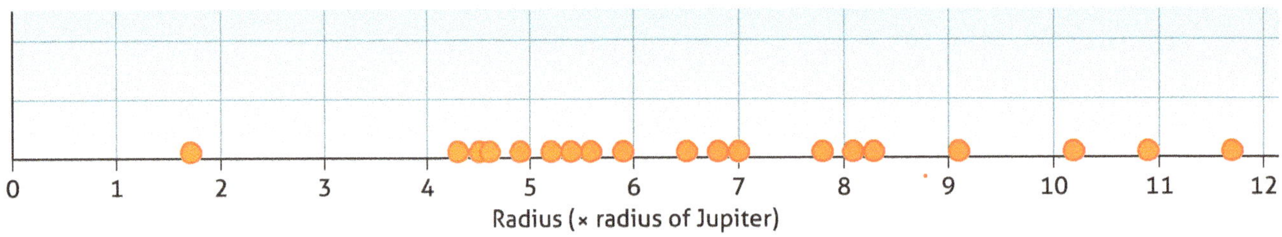

(a) State, with reasons, whether the Normal distribution is a good model for the data.

(b) Enter the data into a spreadsheet and use it to estimate their mean and standard deviation.

(Take the values to be 1.7, 4.2, 4.3, 4.5, 4.9, 5.2, 5.4, 5.6, 5.9, 6.5, 6.8, 7.0, 7.8, 8.1, 8.3, 9.1, 10.2, 10.9, 11.8).

(c) Use the criterion 'Mean ± 2 × Standard deviations' to identify which values may be considered outliers.

(d) Find the median and quartiles of the data.

(e) Use the criterion 'Nearer quartile ± 1.5 × IQR' to identify which values may be considered outliers. Comment.

(f) State whether it would be helpful to have more data to estimate the planet's radius.

Practice question

A national park includes 20 km of river. An invasive species of plant has become established on both banks. The park wardens plan to invite local volunteers to weed out the plants.

Before doing so they carry out a trial in which they take 1 km of one bank of the river and divide it into 40-metre sections. They time how long it takes someone to weed out the plants in each section. Their times in minutes are as follows:

60	69	71	76	80	84	90	92	94	95
96	98	100	100	103	106	107	108	114	116
123	125	126	129	138					

The mean time is 100 minutes and the standard deviation 20 minutes.

(a) Copy and complete this grouped frequency table.

Time (m)	60–69	70–79	80–89	90–99	100–109	110–119	120–129	130–139
Frequency	2	2						

(b) Draw a frequency chart to illustrate the data.

The wardens suggest that the data can be modelled by the distribution $N(100, 20^2)$.

(c) Give two reasons why this model is plausible.

In the rest of the question assume the wardens' model is true.

(d) (i) What percentage of people will complete a 40-metre section within 2 hours?
(ii) What percentage will take between 2 hours and 3 hours?

120 people volunteer to spend up to 6 hours a day weeding out the remaining plants.

(e) Show that there are 39 km of riverbank still to be weeded.

(f) Show that if each volunteer weeds 2 sections a day, the task will require a 5th day, but that if each does 3 sections a day then 3 days will be sufficient.

(g) Comment on whether it is reasonable to expect each volunteer to complete 3 sections per day.

Now that you have finished this chapter, you should know:

- How to plan a statistical investigation while recognising any inherent problems.
- How to use the statistical problem solving cycle as a framework.
- How to view approaches to sampling critically.
- How to evaluate a statistical investigation and any subsequent conclusions for bias.
- How to simplify, if appropriate, calculations for mean, range and standard deviation.
- How to recognise distributions that are Normal or very near to Normal.
- How to examine a statistical investigation report critically.

CHAPTER 19

Hypothesis testing and the chi-squared test

This chapter follows on from the Note on page xxx in Chapter 17. You should read through it again to familiarise yourself with how a hypothesis test is conducted and the vocabulary used.

There are many different types of hypothesis tests. In the first two sections, the Normal distribution is used as a basis for the steps involved. The chapter then moves on to put the ideas into practice with the chi-squared test on a contingency table.

19A Testing an idea

Sometimes an idea comes to you and you would like to know whether it is true or false. Perhaps it is a theory or **hypothesis** that you have thought of, or that someone else has proposed; either way you would like to put it to the test. Always ask yourself if data will help you to decide whether it is true or false.

Activity 19A Peppered moths

The picture shows the two variations of peppered moths: dark and light.

In the past, almost all peppered moths were light but at the time of the Industrial Revolution dark moths started to appear. They were better camouflaged in the polluted environment. Now there are places where one or the other colour predominates.

What does a high proportion of dark peppered moths tell you about the area where you live?

Worked example 19A

? Oscar lives in a town where a local environmentalist says there are equal numbers of light and dark peppered moths. Oscar wants to test whether this is the case, so he sets a number of moth traps one night. He catches 35 light peppered moths and 15 dark ones. [If the environmentalist is correct, then for samples of size 50 the mean number of dark moths will be 25 and the standard deviation 3.5.]

Test at the 5% significance level whether the results show that the proportions of dark and light moths are different from each other in Oscar's town.

Solution

Steps:

The variable is the numbers of dark moths in samples of size 50. Its mean is μ.

State the variable you are going to deal with.

Null hypothesis: $H_0 : \mu = 25$

Half the moths are expected to be dark, so μ is 25.

Alternative hypothesis: $H_1 : \mu \neq 25$

Notice that this alternative hypothesis covers both the cases when the proportion of dark moths is greater than $\frac{1}{2}$ and when it is less than $\frac{1}{2}$.

Significance level: 5%

2-tail test

As illustrated below, both tails need to be considered.

Assume the null hypothesis is true. Then the distribution of the number of dark moths is $N(25, 3.5^2)$

The standard deviation of 3.5 was given in the question.

The *z*-score for the observed number of dark moths is $z = \frac{35 - 25}{3.5} = 2.86$.

The *z*-score is the **test statistic**.

For a significance level of 5%, the **critical values** are −1.96 and +1.96.

An alternative approach, using a *p*-value, is shown in the next section.

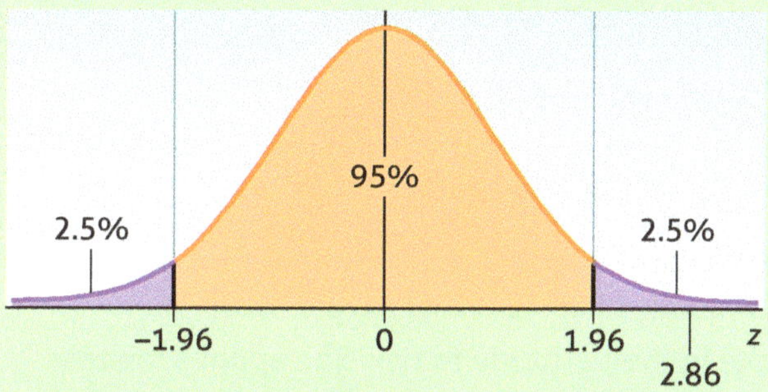

The two tails on the graph are the **critical region**. This is a **2-tail test** so both tails need to be considered. Since the significance level is 5%, each tail is 2½%.

If the test statistic lies in the critical region, the null hypothesis is rejected. The critical region is also called the **rejection region**. When the null hypothesis is rejected the test is often said to be **significant**.

The **acceptance region** lies between the critical values.

The test statistic of 2.86 is compared to the nearer critical value, so to 1.96 rather than −1.96.

Since 2.86 > 1.96, so is in the right hand tail, there is enough evidence to reject null hypothesis in favour of the alternative hypothesis.

The conclusion of the test is that the proportions of light and dark moths are almost certainly not the same in Oscar's town.

Always end by stating the conclusion in terms of the original problem in simple English.

The table below gives some critical values for a 2-tail test using the Normal distribution.

Significance level	10%	5%	2%	1%
Each tail	5%	2.5%	1%	0.5%
Critical values	−1.645, +1.645	−1.960, +1.960	−2.326, +2.326	−2.576, +2.576

Hypothesis testing and statistical problem solving

Hypothesis testing is an example of statistical problem solving and follows the same steps.

You should start with an idea and express it as a hypothesis, a theory that in this case is based on a variable with data drawn from a **population**.

Then you collect **sample** data.

Once you have the sample data, you are ready to go on and process them by carrying out the test.

The test provides **evidence** that relates the original problem to the relevant population. Your conclusion should be stated in those terms.

Using a sample to draw conclusions about the population from which it is drawn is an example of **inference**.

⚠ It is a common mistake to use the data to formulate the hypothesis and then use the same data to test it.

Exercise 19A

1 You wish to investigate each of the following suggestions.

(a) The life expectancy of women in a particular town has changed over the last 100 years.

(b) There has been a change in the number of earthquakes with magnitude 3+ in the UK.

(c) Students answering the 100 questions in a multiple-choice examination with 4 options have guessed all the answers.

For each one you should:

(i) State a suitable variable to use in a hypothesis test.

(ii) Suggest a possible source of suitable data.

(iii) Write down your null and alternative hypotheses.

2 Alice is a psychology student. She is investigating people's approach to taking risks, using a set of 10 Yes/No questions. A score of 0 means the person is extremely risk averse; someone who scores 10 is a risk-taker. Each question has been designed so that overall people are equally likely to answer Yes or No.

Alice has a theory that singers are different from other people in their attitude to risk. She applies the test to 20 members of her choir and they score a total of 89. Here is an outline of the hypothesis test she uses to test her theory; not all the information is given. [For samples of size 20 people with an average level of risk aversion, the mean total would be 100 and the standard deviation 7.07.]

> The variable is the total score for 20 people. Its mean is μ.
>
> $H_0: \mu = 100$, H_1 ______
>
> Significance level: 10%
>
> 2-tail test
>
> The distribution of the scores by groups of this size is $N(100, 7.07^2)$.
>
> The z-score for the observed total score is $z =$ ___ $= -1.56$.
>
> The critical values for the test are ______.
>
> Since $-1.645 < -1.56$ ______________
>
> ____________________

The symbol ___ means that something is missing and you are expected to provide it.

(a) Write down the alternative hypothesis.

(b) Explain the meaning of $N(100, 7.07^2)$.

(c) Show how the z-score is calculated.

(d) Write down the critical values for the test.

(e) State the conclusion from the test both in terms of the hypothesis being tested and also Alice's theory.

19B 1-tail tests

There are situations where you are only interested in one of the two tails of a distribution. This section is about hypothesis tests that deal with this.

Activity 19B Champion hens

A poultry keeper wants to keep hens for breeding that are exceptionally good egg layers.

She looks on the internet and finds a statement that for a certain breed, 'The vast majority of the hens lay between 280 and 320 eggs a year.'

She uses this information to decide that the number of eggs a hen of this breed lays in a year can be modelled by the distribution $N(300, 10^2)$. Explain the meaning of this and say how the poultry keeper decided on it.

The poultry keeper decides to carry out a hypothesis test on whether a typical hen from her flock lays more eggs than the model would predict. However, a hen that lays unusually few eggs is of no interest at all; she wants to know whether it lays more. So, the test would need to be based on the right-hand tail of the distribution. It would be a 1-tail test, as shown below in Worked example 19B.

Worked example 19B

A poultry keeper has a flock of a particular breed of hens and wants to know if they lay an exceptional number of eggs. Four hens from the flock are taken at random as a sample and their eggs are counted for one year; they lay 1254 eggs. Information about the breed indicates that the distribution of the total number of eggs per year from 4 hens is $N(1200, 20^2)$. Carry out a 1-tail hypothesis test at the 1% significance level to check the claim that the hens in her flock lay exceptionally many eggs.

Solution

The variable is the number of eggs laid by samples of size 4 from the poultry keeper's hens in a year.
Its mean is μ.

Null hypothesis: $H_0 : \mu = 1200$
Alternative Hypothesis: $H_1 : \mu > 1200$
Significance level: 1%
1-tail test

Notice that for a one-tail test, the alternative hypothesis is an inequality sign rather than $\neq$.

Assume the null hypothesis is true.
Then the distribution is $N(1200, 20^2)$
The z-score for the observed number of eggs is $z = \frac{1254 - 1200}{20} = 2.7$

The **critical value** for a one-tail test with 1% significance level is 2.326.
Since 2.7 > 2.326 there is enough evidence to reject the null hypothesis.
The evidence supports the claim that hens in the poultry keeper's flock are laying exceptionally well.

In this example the z-score was found to be 2.7. At that point there is another way of proceeding. Using your calculator, or Normal distribution tables, you can find that the **probability** of a result at least as extreme as this is 1 − 0.9965 = 0.0035. This is 0.35%, so is less than the significance level of 1%. This tells you the same result: that there is enough evidence to reject the null hypothesis in favour of the alternative hypothesis.

The probability of a result at least as extreme as that observed is called the ***p*-value**.

It is common for statistics software packages to report the results of hypothesis tests as p-values.

There are a lot of steps in carrying out a hypothesis test. Look at those below and discuss a suitable order for them. Some definitely come before others but others could be done a bit earlier or a bit later. So, there is no single right answer. Remember you should always know what you are going to do with data before collecting them.

A Collect data.

B Decide on the variable you are going to use.

C Work out the test statistic.

D State the null hypothesis.

E State what you conclude about your hypothesis.

F State the idea, theory or hypothesis you are investigating.

G Assume the null hypothesis is true.

H State the alternative hypothesis.

I Find the critical value(s).

J State what you have learnt about the underlying population.

K Decide whether the test is 1-tail or 2-tail.

L State the significance level.

M Decide what hypothesis test you are going to use (or what distribution you will use).

N Compare the test statistic to the critical value.

Exercise 19B

1 For a certain hypothesis test the test statistic lies between −1 and 1. Possible values are shown on the diagram below.

Copy the diagram and colour the acceptance region red and the rejection region blue in each of the cases when the test is:

(a) 1-tail with critical value 0.8.

(b) 2-tail with critical values ±0.8.

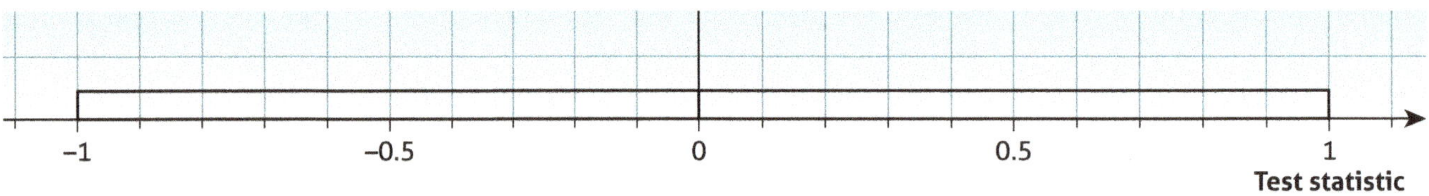

2 For each of situations (a)–(c) a hypothesis test is to be carried out.

(a) A survey reports that the average household size in the UK is 2.4. A town's planning officer thinks that households in their town tend to have more people than the average. The town council want to investigate whether this is true.

(b) A housing charity has records of the number of rough sleepers in different areas, from a survey 5 years ago. A politician says, 'There are now more people on the streets.' The charity wants to know whether this is true.

(c) A company interviews people on the street to find out their opinions on various issues. They are concerned about possible gender bias among their interviewers. This could be preferring to interview people who are the same sex as themselves, or preferring the opposite sex.

For each of situation (a)–(c):

(i) Say whether a 1-tail or 2-tail test is appropriate.

(ii) State suitable null and alternative hypotheses.

(iii) Suggest what data they might collect.

19C The χ^2 (chi squared) test

Now that you are familiar with the ideas and language used in hypothesis testing, it is time to move on to the χ^2 (chi-squared) test. This is commonly used in several subjects.

Chi, written χ, is the Greek letter Ch. It is pronounced 'ky'.

Activity 19C Park birds

Look at this **contingency table**. It refers to the birds in two parks.

How many blackbirds are there in Park A?

Complete the missing numbers.

Observed frequency, f_o	Robins	Blackbirds	Sparrows	Total
Park A	8		10	32
Park B	12			48
Total	20	32	28	80

Notice the word **frequency**. It means the number of items (in this case birds) in a particular group. In this table, the frequencies are observed and so written as f_o with the o for observed.

Now look at the figures for the two parks.

Do you think that the distribution of the birds is essentially the same between them with just chance fluctuations?

Or do you think there are major differences?

The questions in the Activity are what you will use the χ^2 (chi-squared) test to address.

The null hypothesis in this case is that the underlying distributions of the birds in the two parks are the same. This is contrasted with the alternative hypothesis that the distributions are different.

It is a 1-tail test.

The calculations in the test are based on the proportions in the different groups, so the two hypotheses can be written:

Null hypothesis, H_0: The proportions of the types of birds are independent of the park.

Alternative hypothesis, H_1: The proportions are not independent of the park.

In this book you meet the χ^2 **(chi-squared) test on a contingency table**. In more advanced study you may come across other χ^2 tests.

Exercise 19C

1 Different students state the null hypothesis for the χ^2 (chi-squared) test on the birds in Activity 19C as follows. State whether each of the following statements is correct or incorrect for H_0.

(a) All the birds are equally likely.

(b) The proportions of the different birds are the same in the two parks.

(c) Blackbirds and sparrows are equally abundant in Park B.

(d) Robins are the least common of the three birds.

(e) There are no predators about.

(f) The proportions of the different birds are independent of the park.

2 The proportions of foxes, badgers and deer are the same in two woods, P and Q. Copy and complete this table showing how many of them are in each wood.

Frequency	Foxes	Badgers	Deer	Total
Wood P				80
Wood Q				120
Total	40		100	

19D Observed and expected frequencies, the test statistic

An important first step in the calculations for any hypothesis test is to assume the null hypothesis is true and work out the consequences. Then you can compare the observations with this base line and so can work out the test statistic.

Activity 19D The order of a contingency table

This contingency table has **order** 2 × 3.

Observed frequency, f_o	Robins	Blackbirds	Sparrows	Total
Park A	8	14	10	32
Park B	12	18	18	48
Total	20	32	28	80

(a) What is meant by an order of 2 × 3?

(b) Where do the numbers 2 and 3 come from?

(c) If the null hypothesis is true, the proportions of the types of bird in each park will be independent of the park. If you assume this is so and work out the expected frequencies, you can present them in a contingency table. What order will it be?

Here is an example of how the expected frequencies are calculated.

It is for the sparrows in Park B.

The total number of sparrows is 28. → $28 \times \frac{48}{80} = 16.8$

There are 48 birds in Park B out of a total of 80.

So $\frac{48}{80}$ is the proportion of birds that are in Park B.

In Worked example 19D, this method is used for all three types of birds in both the parks.

Worked example 19D (1)

? (a) Calculate the numbers of robins, blackbirds and sparrows you would expect in the two parks if the 80 birds are in the same proportions in both parks.

(b) Draw up the contingency table for the expected frequencies.

Solution

(a) Assuming the null hypothesis is true, the expected numbers are as follow:

Park A Robins $20 \times \frac{32}{80} = 8$ Blackbirds $32 \times \frac{32}{80} = 12.8$ Sparrows $28 \times \frac{32}{80} = 11.2$

Park B Robins $20 \times \frac{48}{80} = 12$ Blackbirds $32 \times \frac{48}{80} = 19.2$ Sparrows $28 \times \frac{48}{80} = 16.8$

(b) So, the contingency table for the expected frequencies, f_e, is:

Expected frequency, f_e	Robins	Blackbirds	Sparrows	Total
Park A	8	12.8	11.2	**32**
Park B	12	19.2	16.8	**48**
Total	**20**	**32**	**28**	**80**

Notice that the numbers in the Total column in the expected frequency table are the same as those in the Observed frequency table. And that is the case for the two Total rows as well.

This too is a 2 × 3 contingency table.

Notice that some of the answers come out with parts of birds, like the 16.8 sparrows in Park B. Although it seems natural to round such numbers to the nearest whole bird, this would be a mistake as the calculation of the test statistic that follows is based on unrounded numbers.

Once you know the Observed and Expected frequencies you can work out the **test statistic.**

It is denoted by X^2 and is calculated using the formula $X^2 = \sum \frac{(f_o - f_e)^2}{f_e}$.

To see how this is used, look carefully at Worked example 19D (2).

In this formula, Σ means the sum of (i.e. added together) and f_o and f_e are the six pairs of values in the two contingency tables.

Worked example 19D (2)

(a) Use the observed and expected frequencies of the 3 types of birds in the 2 parks to calculate the value of X^2.

(b) State which cells contribute most to the value of X^2 and which contribute least.

(c) The null hypothesis is that the proportions of the types of birds are independent of the park. If this is true, would you expect a high value or a low value of X^2?

Solution

(a)

	Observed frequency, f_o			Expected frequency, f_e			
	Robins	Blackbirds	Sparrows	Robins	Blackbirds	Sparrows	Total
Park A	8	14	10	8	12.8	11.2	**32**
Park B	12	18	18	12	19.2	16.8	**48**
Total	**20**	**32**	**28**	**20**	**32**	**28**	**80**

To find the value of X^2 in this case, there are six different elements to add together:

$\frac{8 - 8^2}{8}$ (= 0) + $\frac{14 - 12.8^2}{12.8}$ (= 0.1125) + $\frac{10 - 11.2^2}{11.2}$ (= 0.12857...)

$\frac{12 - 12^2}{12}$ (= 0) + $\frac{18 - 19.2^2}{19.2}$ (= 0.075) + $\frac{18 - 16.8^2}{16.8}$ (= 0.08571...)

So, $X^2 = 0.402$ (to 3 sf).

(b) The cell that contributes most to the value of X^2 is Park A Sparrows at 0.12857.

The cells that contribute least are Park A Robins and Park B Robins, both with 0.

(c) If the null hypothesis is true you would expect much the same proportions of the different birds in the two parks, so the value of X^2 would be low.

When you work out the value of X^2, the expected frequency of each group should be at least 5. If it is less than 5 for one of your groups you should, if possible, combine two (or more) groups together to get one with an expected frequency of at least 5.

Notice the use of … . This means that the number hasn't been rounded and should be stored in your calculator's memory. Always work with unrounded numbers and wait till the very end of a calculation before you do any rounding. However, you should give your final answer to a sensible level of accuracy.

Exercise 19D

1 A sociologist is interested in people's alcohol drinking habits in three areas of a town. She carries out a survey of 120 people living in Area X, 80 in Area Y and 100 in Area Z.

She uses the classifications: Abstainer (A), Social drinker (S), Binger (B) or Dependent (D) to describe them.

This is one out of several ways that are used to describe people's drinking habits.

Her results are given in the contingency table below. Some of the entries are missing.

	Observed frequency, f_o				
Habit / Area	A	S	B	D	Total
Area X	16	78	18	8	120
Area Y	8			25	80
Area Z	35		20	5	100
Total	59		53	38	300

(a) Copy the contingency table and fill in the missing entries.

(b) Write down the order of this contingency table.

(c) Construct the contingency table for the expected frequencies.

(d) Write down null and alternative hypotheses the sociologist would use for a chi-squared test.

(e) Calculate the test statistic, X^2.

(f) State which cells contributed most to the value of X^2.

19E Critical values for the χ^2 (chi-squared) test

Before you can carry out the χ^2 (chi-squared) test you need to know the **critical value** for the significance level you are going to use, but to find this you first need to know the **degrees of freedom**.

Activity 19E Degrees of freedom

Look at this newspaper report.

It contains three numbers: 10, 6 and 4.

(a) Were all three needed?

(b) How many were strictly necessary?

GOALS GALORE

There were 10 goals in the match. Bolton Wanderers bagged 6 and Preston North End scored 4.

The number of free variables in a situation is called the **degrees of freedom**.

In a contingency table with r rows and c columns, there are $(r - 1) \times (s - 1)$ degrees of freedom.

The number is often denoted by ν. ← ν is the Greek letter n. It is written as nu and pronounced 'new'.

The table of the birds in the parks has 2 rows and 3 columns, so the degrees of freedom are given by $\nu = (2 - 1) \times (3 - 1) = 2$.

You need to know the degrees of freedom to find the critical value. An extract from the relevant table is given below. Assume that in this case the significance level is 5%.

Percentage points of the χ^2 (chi-squared) distribution

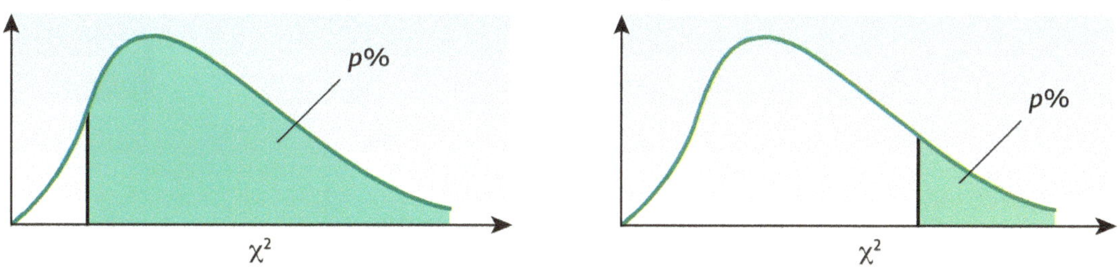

Degrees of freedom ↓ (rows, ν); Significance level → (columns, $p\%$)

$p\%$ / ν	99	97.5	95	90	10	5.0	2.5	1.0	0.5
1	.0001	.0010	.0039	.0158	2.706	3.841	5.024	6.635	7.879
2	.0201	.0506	0.103	0.211	4.605	5.991	7.378	9.210	10.60
3	0.115	0.216	0.352	0.584	5.251	7.815	9.348	11.34	12.84
4	0.279	0.484	0.711	1.064	7.779	9.488	11.14	13.28	14.86
5	0.554	0.831	1.145	1.610	9.236	11.07	12.83	15.09	16.75
6	0.872	1.237	1.635	2.204	10.64	12.59	14.45	16.81	18.55
7	1.239	1.690	2.167	2.833	12.02	14.07	16.01	18.48	20.28
8	1.646	2.180	2.733	3.490	13.36	15.51	17.53	20.09	21.95
9	2.088	2.700	3.325	4.168	14.68	16.92	19.02	21.67	23.59

You will see that for $\nu = 2$ and a significance level of 5%, the critical value is 5.991. The right-hand diagram shows the critical region shaded blue. If the test statistic turns out to be greater than 5.991, and so lie in the blue region, the null hypothesis will be rejected in favour of the alternative hypothesis. However, if the test statistic is less than 5.991, the null hypothesis will be accepted.

You will notice that the left-hand side of the table covers very high probabilities. These are only relevant for more advanced work, so you can ignore them.

Exercise 19E

1 In a χ^2 test on a contingency table there are 7 degrees of freedom. Write down the order of the contingency table. There are two possible answers.

2 In Exercise 19D, you were given the observed frequencies for the drinking habits of people in three parts of a town.

You worked out the expected frequencies and the value of X^2.

(a) Write down the order of the contingency table.

(b) Calculate the degrees of freedom for this contingency table.

(c) Find the critical value for a test at the 1% significance level.

19F The complete hypothesis test

By now you have met all the elements that you need for the test on the birds' data. In this section you can see how they are all brought together and set out as a hypothesis test.

Activity 19F What is the outcome of the hypothesis test?

What are the possible conclusions from the hypothesis test on the birds and how will you express them?

Worked example 19F

? Birdwatchers count the robins, blackbirds and sparrows in two nearby parks. In one park their numbers are 8, 14 and 10 respectively and in the other park 12, 18 and 18.

Use this information to carry out a χ^2 test, at the 5% significance level, on the proportions of these birds in the two parks.

Solution

The variable used in this test is the number of birds.

Null hypothesis, H_0: The proportions of the types of birds are independent of the park.
Alternative hypothesis, H_1: The proportions are not independent of the park.
Significance level: 5%
Test: χ^2 test on a contingency table, 1-tail.

Always start by setting out the hypotheses, the significance level, the type of test and whether it is 1-tail or 2-tail.

The observations are given in this contingency table.

Observed frequency, f_o		Robins	Blackbirds	Sparrows	Total
	Park A	8	14	10	**32**
	Park B	12	18	18	**48**
	Total	**20**	**32**	**28**	**80**

Now go on to calculate the test statistic.

Assuming that H_0 is true, the expected frequencies are given by:

Expected frequency, f_e	Robins	Blackbirds	Sparrows	Total
Park A	8	12.8	11.2	**32**
Park B	12	19.2	16.8	**48**
Total	**20**	**32**	**28**	**80**

The calculation of the expected values was carried out on page xx. You would normally include it in what you write down.

Using $X^2 = \sum \frac{(f_o - f_e)^2}{f_e}$, the test statistic is $X^2 = 0.402$.

The calculation of X^2 was carried out on page xxx. You would normally include it too.

The degrees of freedom are given by $\nu = (2 - 1) \times (3 - 1) = 2$.

For $\nu = 2$ and a 5% significance level, the critical value is 5.991.

Now find the critical value from tables (or maybe your calculator). Before you can do this, you must work out the degrees of freedom.

Since $0.402 < 5.991$, the alternative hypothesis is rejected in favour of the null hypothesis.

It is very important to show the comparison between the test statistic and the critical value. You will usually use one of the inequality signs, < or >.

Notice that the larger the value of X^2 the more likely it is that the null hypothesis is false and the alternative hypothesis true.

The evidence suggests that the proportions of the types of birds are independent of the park.

End by stating the conclusion in English, in terms of the original problem.

Exercise 19F

1 In Exercise 19D on page 238 you were given a contingency table about people's drinking habits. Write out the complete χ^2 hypothesis test based on this table.

Test yourself 19G

1 Simon is considering setting up an online dating agency.
He has the theory that most girls prefer tall men to short men.

To test this, he shows pictures of two young men, one tall and the other short, to a sample of 100 girls. He asks each girl to choose between them. (Girls with no opinion were not included in the sample.) 58 chose the tall man and 42 the short man. Simon uses these figures in the hypothesis test on the right, at the 5% significance level.

Note that If girls have no overall preference then, for samples of size 100, the mean number preferring tall men will be 50 and the standard deviation 5.

(a) Explain why it would be wrong to state the alternative hypothesis as $H_1 : \mu \neq 50$.

(b) Complete the distribution, N(__ , __).

(c) Show the missing part of the calculation of the z-score.

(d) Simon's friend writes '1.6 < 1.96' instead of '1.6 < 1.645' near the end.

Explain why this is wrong.

(e) The wording at the end is incomplete. Suggest a suitable concluding statement about Simon's claim.

> The variable is the number of girls in a sample of 100 who prefer tall men. Its mean is μ.
>
> Null hypothesis: $H_0 : \mu = 50$
>
> Alternative hypothesis: $H_1 : \mu > 50$
>
> Significance level: 5%
>
> 1-tail test
>
> Assume the null hypothesis is true.
>
> Then the distribution is N(__ , __).
>
> The z-score for the observed number is
>
> $z =$ ______ = 1.6.
>
> The critical value is 1.645.
>
> Since 1.6 < 1.645 there is not enough evidence to reject the null hypothesis.
>
> ______________________ Simon's theory that girls prefer tall men.

2 A research student has collected a large quantity of data. She knows what hypothesis test she wants to use. She enters all her figures into the appropriate function of a software package. The answer comes back 'p-value = 0.031 295'.

(a) State whether the result is significant for a 1-tail test 5% significance level.

What about a 2-tail test?

(b) The software package does not report the test statistic. Explain why this is not important.

3 Dylan is carrying out a survey of the ages at which people finished their formal education. He asks samples of people in three towns: A, B and C.

He places people in four categories: aged before 16, 16/17, 18/19, after 19.

He presents his findings in the contingency table below.

Observed frequency, f_o	Before 16	16/17	18/19	After 19	Total
Town A	11	14	10	5	40
Town B	4	6	30	20	60
Town C	1	9	10	5	25
Total	16	29	50	30	125

Dylan plans to carry out a χ^2 test on this contingency table.

(a) Give the expected frequencies as a contingency table.

(b) The expected frequencies reveal a problem for Dylan.

Explain the problem and give two possible courses of action that Dylan can take.

(c) Choose one of the courses of action and then carry out the complete hypothesis test with a 5% significance level.

4 This diagram illustrates the acceptance region and the critical region for a χ^2 test with 7 degrees of freedom.

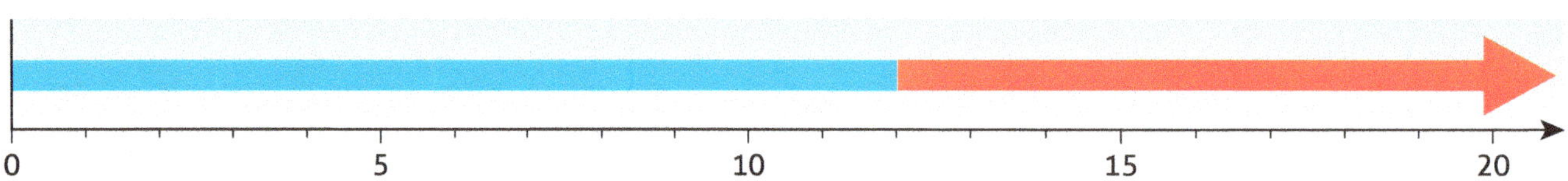

(a) State which region is blue and which is red.

(b) State the significance level of the test.

(c) Explain the meaning of the red arrow on the right of the diagram.

5 The data are missing from this contingency table.

Category	A	B	C	D	Total
P					90
Q					110
R					100
Total	70	75	80	75	300

(a) Find a set of numbers that can go in the empty cells and give the correct totals. (There are many possible answers.)

(b) What is the smallest number of empty cells that you can fill while leaving no choice for the numbers in the other cells?

(c) Explain the connection between your answer to part (b) and the degrees of freedom.

Practice question

For many years, Jane's family has owned a small wood and a surrounding field. One day Jane finds the following note written long ago by her late grandfather.

Results of a survey of animals living in our wood and field:
12 hedgehogs, 17 rabbits, 9 badgers, 6 hares, 15 squirrels (13 red, 2 grey), 5 foxes.

Jane wants to know whether the proportions of the different animals have changed, so carries out a survey of her own. Her findings are in the observed frequencies in Table 1 below.

Jane decides to carry out a χ^2 test on the figures, with a 5% significance level.

(a) Write down the null and alternative hypotheses Jane should use.
State whether it is a 1-tail or 2-tail test.

(b) Find the degrees of freedom and the critical value.

(c) Table 1 gives Jane's observed frequencies and some of the expected frequencies.
Complete Table 1.

	Observed frequency, f_o			Expected frequency, f_e		
	Past	Now	Total	Past	Now	Total
Hedgehogs	12	2	**14**			
Rabbits	17	21	**38**	20.583 ...	17.416 ...	**38**
Badgers	9	6	**15**	8.125	6.875	**15**
Hares	6	5	**11**	5.958 ...	5.041 ...	**11**
Red squirrels	13	0	**13**	7.041 ...	9.583 ...	**13**
Grey squirrels	2	15	**17**	9.208 ...	7.791 ...	**17**
Foxes	6	6	**12**			**12**
Total	**65**	**55**	**120**	**65**	**55**	**120**

Table 1

The contributions to the test statistic X^2 are shown in Table 2. They are given to 4 decimal places.

	Contributions to X^2	
	Past	Now
Hedgehogs	2.5723	3.0400
Rabbits	0.6238	0.7372
Badgers	0.0942	0.1113
Hares	0.0003	0.0003
Red squirrels	5.0417	9.5833
Grey squirrels	5.6427	6.6687
Foxes	0.0385	0.0455

Table 2

(d) Calculate the value of X^2.

(e) State the conclusion from the test.

(f) A friend of Jane's suggests that, since they occupy the same environmental space, red squirrels and grey squirrels should be counted together.
How would this affect the outcome of the test? Give detailed working to support your answer.

Now that you have completed this chapter, you should know:

- How you conduct a hypothesis test.
- The vocabulary associated with hypothesis testing.
- How to carry out a χ^2 test on a contingency table.

CHAPTER 20

Bivariate data

There are many numbers that you could use to describe each of your friends. Two obvious ones are their age in years and height in metres, while others include how far away they live and how long you have known them. Another important piece of information is the person's gender but male and female are categorical variables rather than numerical. Put this information together for all your friends and you have a set of multivariate data.

This chapter covers the simple case where there are just two variables and the data are described as bivariate and can be displayed on a scatter diagram.

20A Correlation and association

This section is about the vocabulary you will use in this chapter.

Activity 20A Do richer people live longer?

Each point on this scatter diagram represents a mainland American country (North, South or Central). The variable on the horizontal axis is a country's GDP per capita in US$, so is a measure of people's wealth. The variable on the vertical axis is life expectancy in years.

The diagram was drawn by a spreadsheet, which also added the line of best fit.

How close are the points to the line of best fit?

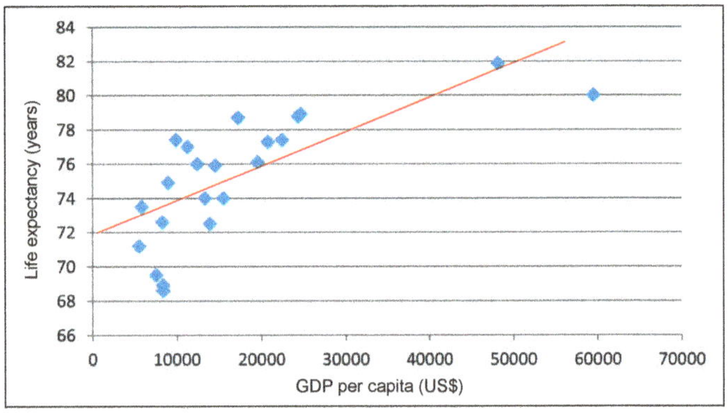

Just looking at a scatter diagram usually tells you if there is a relationship between the two variables.

In the scatter diagram in Activity 20A, many of the points are not particularly close to the line of best fit. You can draw a curve through the points that is a much better fit than the line, as in the image on the right

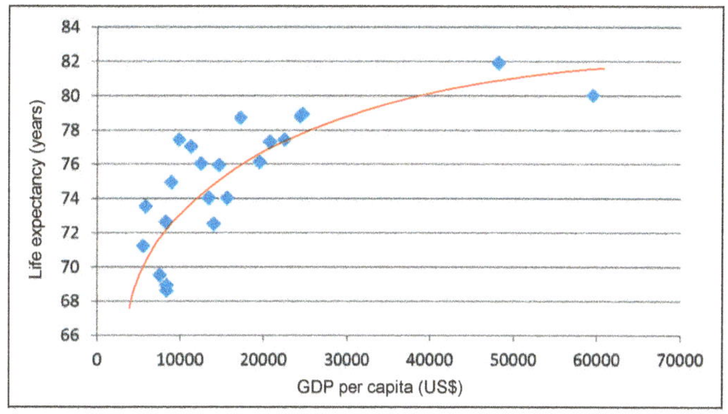

Clearly, subject to the variability you expect with real data, there is a relationship between the variables, but it is not a straight line. A relationship like this is called an **association**.

If, as in this case, both variables increase together, the association is **positive**. If one decreases while the other increases, it is **negative**.

A special case of association occurs when the points on the scatter diagram lie close to a straight line and this is called **correlation**.

If the points on the scatter line are close to the curve or line, the relationship is said to be **strong**. If there is no pattern on the scatter diagram, there is neither association nor correlation between the variables.

> You will often hear the word correlation used loosely to describe situations where the relationship is not a straight line.

Worked example 20A

? Describe and comment on the relationship between the variables in each of the scatter diagrams given below.

(a)

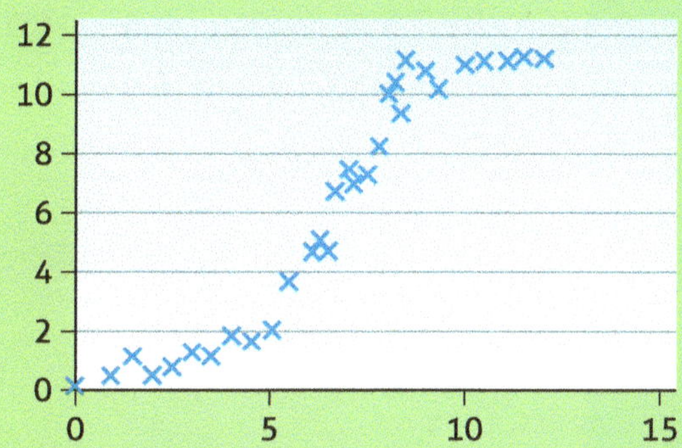

(b)

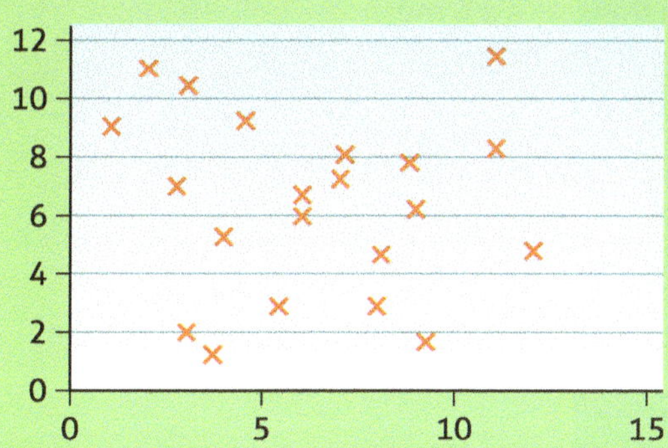

(c)

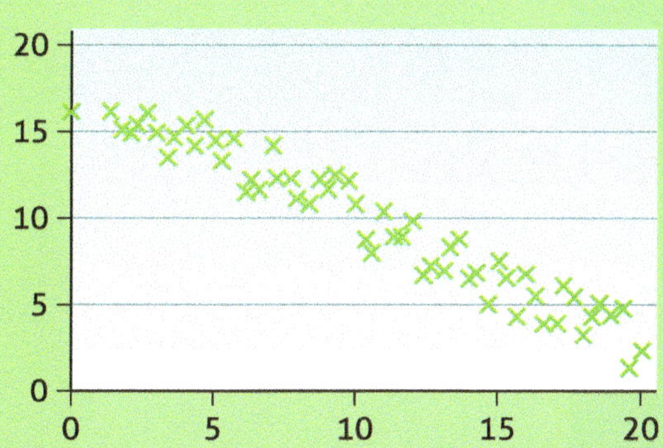

Description (solution)

(a) Strong positive association.

There is clearly a relationship between the variables. The pattern is a curve almost flat at the ends and steep in the middle. It is not a straight line.

So it is best described as an association.

Both variables increase together so it is positive. The points lie close to a curve so it is strong.

(b) Neither association nor correlation.

There is no pattern on this scatter diagram.

(c) Strong negative correlation.

The points on this scatter diagram lie either side of a straight line, so there is correlation. The points lie close enough to the line for the correlation to be described
as strong.

As the variable on the horizontal axis increases, that on the vertical axis decreases, and vice-versa, so the correlation is negative.

A common mistake is to think that because there is a relationship (association or correlation) between two quantities, one must cause the other. In many cases there is something else that is causing both.

For example, there is a strong association between the number of people who own laptops and the number using satellite navigation. Neither is the cause of the other; both result from the development of modern technology and people being wealthy enough to afford them.

This idea is captured in the phrase:

Correlation does not imply causation.

Exercise 20A

1 There is an association between the quantities **A** and **B** where **A** is the clearness of the water in a stretch of a river and **B** is the number of fish living in that stretch of the river.

(a) Explain why neither of the following is a correct description of the situation:

(i) **A** implies **B**, **A** $\Rightarrow$ **B** (ii) **B** implies **A**, **B** $\Rightarrow$ **A**

(b) A different description is shown by 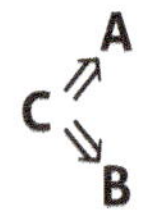

Suggest what **C** could be.

2 A pharmaceutical company is conducting preliminary human tests of a possible new sleeping pill. The subjects are volunteers who describe themselves as weak sleepers. They are given measured doses of the pill every day for a week, and their mean number of hours sleep are recorded in the data below. Some receive placebos and their doses are recorded as zero.

Dose (mg)	0	0	0.5	1.0	1.5	2.0	2.5	3.0	3.5	4.0	4.5	5.0
Sleep (hours)	8.1	8.5	3.9	5.5	5.7	4.9	5.2	6.5	6.5	7.3	6.9	7.4

(a) Enter these data into a spreadsheet and use it to draw a scatter diagram.

(b) Identify any outliers in the data. State, with reasons, whether they should be discarded.

(c) Describe the relationship shown by the scatter diagram.

The company carries out another preliminary test, this time with 10 different self-selected subjects. They too describe themselves as weak sleepers. They record their sleep times for a week without medication; then they take a 4 mg pill every night for another week and record their sleep times. Their mean sleep times for each of the two weeks are given in the table below.

Subject / **Sleep (hours)**	**A**	**B**	**C**	**D**	**E**	**F**	**G**	**H**	**I**	**J**
No medication	4.2	4.7	6.1	6.0	3.2	5.3	2.4	6.5	7.0	5.5
Medication	3.8	5.0	6.9	6.9	2.4	6.0	2.2	7.2	8.1	5.9

(d) Comment on the effectiveness of the medication, supporting your answer with suitable calculations and diagrams.

3 Copy this Venn diagram. Place the letters C, A and N in the appropriate regions to show the relationships between Correlation, Association and No relationship.

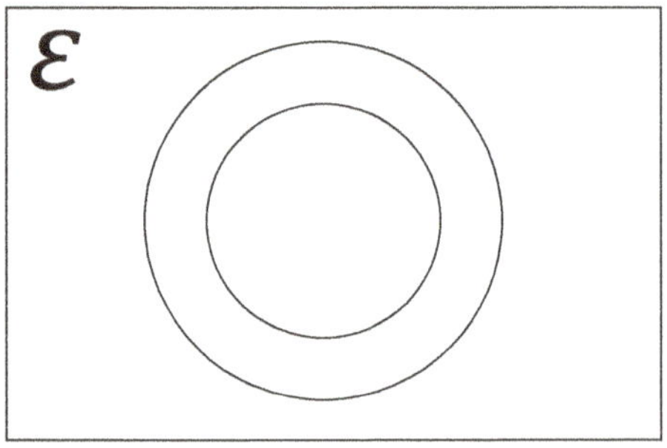

20B Measuring association: Spearman's rank correlation coefficient

You have now met the idea of association and seen that it can be strong or weak and positive or negative. This section shows you how to express this as a number.

Activity 20B Book club

Jo and Tara have started a book reading club with their friends. The members read the same book each month and give it a percentage score. Jo thinks that they are getting more generous in their score. She plots the scores for the last eight books on the scatter diagram below. She rings her own scores in yellow and Tara's in red.

(a) Does the scatter diagram support Jo's theory?

Tara says the diagram shows that her scores are very similar to Jo's.

(b) Suggest a way to quantify Tara's idea.

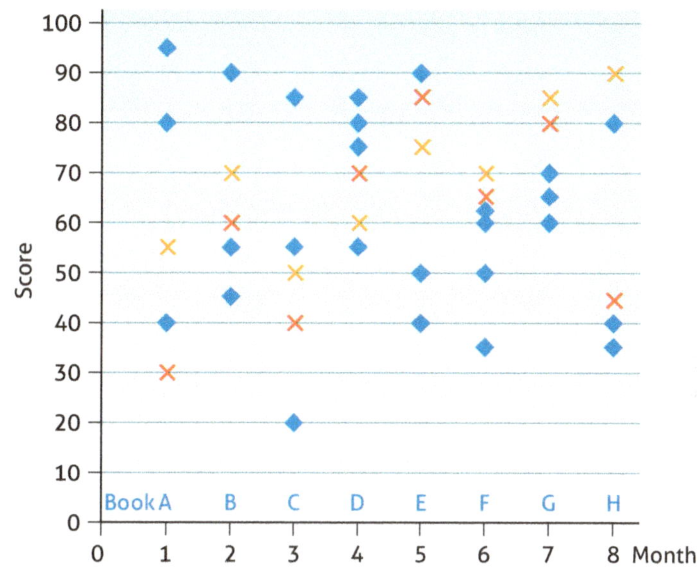

One way of comparing the scores that Tara and Jo gave the books is to look at their ranks. These are given in the table below. Their highest scores are numbered 1 and their lowest 8.

Ranks								
Book	**A**	**B**	**C**	**D**	**E**	**F**	**G**	**H**
Tara	8	5	7	3	1	4	2	6
Jo	7	4.5	8	6	3	4.5	2	1

Notice that Jo's ranks for books B and F are tied.
If they had been slightly different they would have been 4 and 5, so they are both given the rank of 4.5.

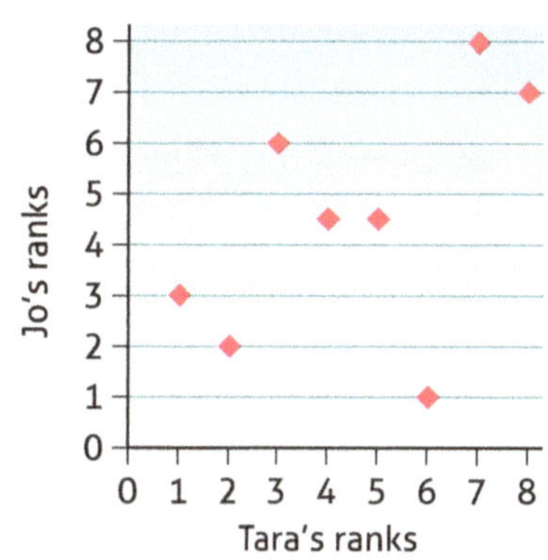

The scatter diagram on the right gives you a visual comparison of these ranks. Tara's are on the horizontal axis and Jo's on the vertical.

You can see that they are in broad agreement apart from on one book. The scatter diagram in Activity 20B shows that this is book H.

A measure for comparing ranks is given by Spearman's rank correlation coefficient.

It is denoted by r_s and it tells you how close the points on the ranks scatter diagram are to a straight line.

- If the ranks lie perfectly on a straight line with positive gradient, its value is + 1.
- If there is a perfect fit for a line of negative gradient its value is −1.
- Otherwise, the value is between −1 and +1.

Worked example 20B

? Calculate Spearman's rank correlation coefficient for the ranks of Tara's and Jo's book scores.

Solution

The calculation is best set out in a table.

The two sets of ranks are denoted by x and y and the difference between them by d. So, $d = x - y$.

Book	T's score	T's rank, x	J's score	J's rank, y	$d = x - y$	d^2
A	30	8	55	7	1	1
B	60	5	70	4.5	0.5	0.25
C	40	7	50	8	−1	1
D	70	3	60	6	−3	9
E	85	1	75	3	−2	4
F	65	4	70	4.5	−0.5	0.25
G	80	2	85	2	0	0
H	45	6	90	1	5	25
				Σ	**0**	**40.5**

The final line gives the values of Σd and Σd^2.

The value of Σd should work out to be 0, as it does here. This calculation is just a check that you haven't made a mistake working out the values of d.

The formula for Spearman's rank correlation coefficient is

$r_s = 1 - \dfrac{6\Sigma d^2}{n(n^2 - 1)}$,

To carry out the calculation you need the value of Σd^2.

where n is the sample size. in this case $n = 8$.

In this case,

$$1 - \frac{6\Sigma d^2}{n(n^2 - 1)} = 1 - \frac{6 \times 40.5}{8 \times (8^2 - 1)} = 0.517\,857...$$

So, $r_s = 0.518$ (to 3 dp).

Since the possible values for r_s lie between −1 and +1, the result that $r_s = 0.518$ shows that such association as there is between Tara's and Jo's ratings is positive. The question of how strong it is involves interpreting the value of r_s and this is covered in the next section.

Exercise 20B

1 Tara says she prefers longer books to shorter ones. The table shows the lengths of the reading club's first eight books and Tara's scores for them.

Book	A	B	C	D	E	F	G	H
Length (pages)	252	450	689	590	934	365	650	425
Score (%)	30	60	40	70	85	65	80	45

(a) Show this information in a table giving the ranks of the lengths of the eight books and the ranks of the scores Tara gave them.

(b) Draw a scatter diagram to show the contents of your table in part (a).

(c) Comment on what the scatter diagram shows you.

(d) Calculate Spearman's rank correlation coefficient for the data.

(e) Is the value of r_s consistent with Tara's statement about preferring longer books?

2 Two selectors, Amalie and Berin, are trying to decide who should take up the final 3 places in a cricket touring team. They must choose from 8 players and so, independently, they rank the players, as shown in the table below.

	Amir	Bertie	David	Leroy	Mac	Mohan	Oscar	Wilfred
Amalie	1	5	8	4	6	3	7	2
Berin	7	5	3	4	2	8	1	6

(a) Calculate Spearman's rank correlation coefficient for these ranks.

Explain how this shows that further consideration is needed for the selection.

A third selector, Robin, is called on to rank the 8 players.

	Amir	Bertie	David	Leroy	Mac	Mohan	Oscar	Wilfred
Robin	2	6	5	3	7	1	8	4

(b) Calculate the values of r_s between (i) Robin and Amalie, and (ii) Robin and Berin.

(c) It is decided to accept the opinions of the two selectors who are in greatest agreement.

State, with reasons, which selectors these are and which 3 players should be members of the touring team.

20C Using Spearman's rank correlation coefficient as a test statistic

This section shows how you can use r_s in a hypothesis test.

Activity 20C A testing time

Ticki is a psychology student. She thinks that people who are good at crosswords should also be good at sudokus. 'It is all using the same part of the brain', she says.

To test her hypothesis, she gives the same crossword and sudoku to 8 different people and times them on each.

Here are their times in minutes, to the nearest whole number.

	Sally	Bella	Kangwa	Louis	Sam	Mat	Ollie	Hector
Crossword	22	45	38	46	35	23	27	62
Sudoku	31	30	38	36	22	24	38	38

(a) What steps do you have to take to work out the value of r_s?

(b) Show that, rounded to 3 decimal places, $r_s = 0.381$.

(c) Explain why it does not matter whether x is used for crosswords and y for sudokus or vice versa.

A value of r_s close to 1 shows there is positive association. Is the value of 0.381 that you found in the Activity close enough to support such a claim? In the Worked example below this question is posed and answered as a hypothesis test.

Worked example 20C

Use the data in Activity 20C to carry out a hypothesis test at the 5% significance level regarding whether there is an association between being good at crosswords and being good at sudokus. Explain the steps involved.

Solution

Start by setting up the test.

Test using Spearman's rank correlation coefficient.

Null hypothesis, H_0: There is no association between being good at crosswords and sudokus.
Alternative hypothesis, H_1: There is a positive association.
Significance level: 5%
1-tail test

If the alternative hypothesis had been, 'There is an association', without saying whether it is positive or negative, then the test would have been 2-tail.

The test statistic has already been found.
It is $r_s = 0.381$.

The critical value is found from the table of critical values for Spearman's rank correlation coefficent.

The part of this table covering values of n from 1 to 20 is shown here.

1-tailed	5%	2.5%	1%	0.5%
2-tailed	10%	5%	2%	1%
n				
1				
2				
3				
4	1.0000			
5	0.9000	1.0000	1.0000	
6	0.8286	0.8857	0.9429	1.0000
7	0.7143	0.7857	0.8929	0.9286
8	0.6429	0.7381	0.8333	0.8810
9	0.6000	0.7000	0.7833	0.8333
10	0.5636	0.6485	0.7455	0.7939
11	0.5364	0.6182	0.7091	0.7545
12	0.5035	0.5874	0.6783	0.7273
13	0.4835	0.5604	0.6484	0.7033
14	0.4637	0.5385	0.6264	0.6791
15	0.4464	0.5214	0.6036	0.6536
16	0.4294	0.5029	0.5824	0.6353
17	0.4142	0.4877	0.5662	0.6176
18	0.4014	0.4716	0.5501	0.5996
19	0.3912	0.4596	0.5351	0.5842
20	0.3805	0.4466	0.5218	0.5699

Using $n = 8$, and a 5% 1-tail test gives a critical value of 0.6429.

Since $0.381 < 0.6429$, there is not enough evidence to reject the null hypothesis in favour of the alternative hypothesis.

Ticki's theory that there is an association between being good at crosswords and sudokus is not supported.

Notice that the closer the value of Spearman's coefficient is to +1, the more likely it is that there is a positive association. Similarly, for −1 and a negative association.

Exercise 20C

1 Laura is a keen gardener. One year she applies different measured amounts of fertiliser to each of 10 gooseberry plants and in the autumn she weighs their fruit yield. The results are shown in this table.

Plant	A	B	C	D	E	F	G	H	I	J
Fertiliser (g)	10	20	30	40	50	60	70	80	90	100
Yield (kg)	1.5	2.7	4.1	3.1	4.6	5.5	6.2	6.2	3.8	5.4

Laura wants to know whether there is a relationship between the amount of fertiliser and the yield. She uses Spearman's rank correlation coefficient to carry out a hypothesis test with a 5% significance level.

(a) State whether the test should be 1-tail or 2-tail.

(b) Write down the null hypothesis and the alternative hypothesis.

(c) Calculate the test statistic for Laura's data.

(d) Carry out the test and state the result.

2 Here is a letter to a local newspaper.

How can our local cricket team score more runs? The answer is very surprising.

They must eat more ice cream. I noticed that the ice cream van at the ground was busy on several recent Saturday afternoons when the team were batting well. So, I asked them if they had records of recent sales and luckily they had for the last 6 matches. When they are compared with the runs our team scored on those days there is an almost perfect correlation.

Eating ice cream makes you bat better.

For those who don't believe me, the figures speak for themselves

Date	June 2	June 9	June 16	June 23	June 30	July 7
Ice cream sold (£)	230	78	315	355	123	247
Runs	143	65	275	266	98	156

The correlation coefficient is 0.94 and that means I am 99% certain to be right.

Write a commentary on this letter, pointing out any major statistical errors in it.

20D Correlation

This section moves on from association to correlation, and how to use a spreadsheet to find the measure that is commonly used.

Activity 20D Recognising correlation

The screenshot on the right shows the ranks of 8 items of bivariate data.

(a) Say how you can tell if there is association between the original data items:

(i) without doing any calculations

(ii) with a further calculation (which you should do).

(b) Copy the ranks into your spreadsheet and in cell C1 place the command =CORREL(A1:A8,B1:B8).

What do you notice?

	A	B
1	8	2
2	5	3
3	7	4
4	6	1
5	2	6
6	4	7
7	1	8
8	3	5

So far you have met three methods for judging whether two variables are associated.

- Plot a sample of points on a scatter diagram and look for a pattern.
- Rank the data for the two variables, plot the points for the two ranks on a scatter diagram and look for a straight line. In that case there is a correlation between the ranks.
- Calculate Spearman's rank correlation coefficient and look for a value close to +1 or –1.

In the Activity you met a fourth method: to use the spreadsheet command = CORREL(... : ... , ... : ...) on the ranks of the two variables.

You will have noticed that this gives the same answer as the Spearman calculation. That is because they are actually carrying out the same calculation, albeit by apparently different methods.

The command =CORREL(... : ... , ... : ...) gives the **product moment correlation coefficient** of a set of bivariate data. It is usually used as a measure of the correlation between the actual variables, rather than their ranks (as in the Spearman case).

The outcome is denoted by the letter *r* and is a number between −1 and +1.

The term product moment correlation coefficient is often abbreviated to **pmcc**. It is also often called the **Pearson product moment correlation coefficient.**

- −1 corresponds to perfect negative correlation; the points lie on a straight line with negative gradient means no correlation.
- +1 indicates perfect positive correlation; the points lie on a straight line with positive gradient.

Worked example 20D Planet density

? The screenshot shows the radii in thousands of kilometres and the density in grams per cubic centimetre of the planets in the solar system.

Many thousands of planets of other solar systems (exoplanets) have now been discovered.

The scatter diagram below shows a sample of those for which it has been possible to estimate the radii and densities. The pmcc for this sample is −0.39.

The scatter diagram also shows the eight planets in the solar system, marked in red.

	A	B	C
1	Planet	Radius	Density
2		(× 1000 km)	(g cm^{-3})
3	Mercury	2.44	5.5
4	Venus	6.05	5.25
5	Earth	6.38	5.52
6	Mars	3.39	3.94
7	Jupiter	71.1	1.33
8	Saturn	59.65	0.71
9	Uranus	25.9	1.7
10	Neptune	24.75	1.77

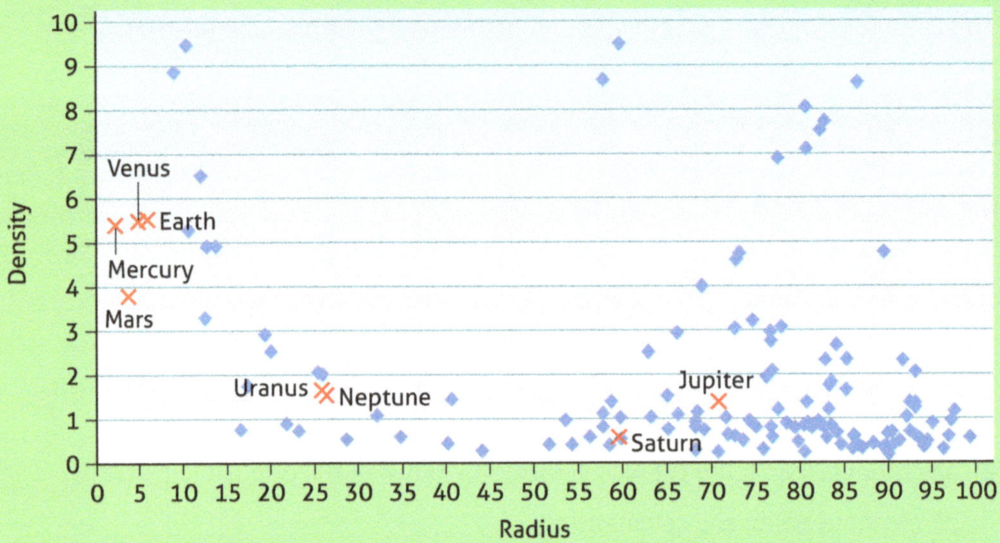

(a) Find the product moment correlation coefficient for the planets in the solar system. State what you conclude.

(b) Comment on the difference in this value of the pmcc from that for the planets in our solar system.

Solution

(a) Using =CORREL(B3:B10,C3:C10). gives $r = -0.8426$.

This is a strong negative correlation.

(b) The pmcc for the exoplanets is considerably lower.

The scatter diagram shows that there are many outliers and their effect is to lower the value of the pmcc.

The scatter diagram also shows an absence of small exoplanets, like the inner planets in the solar system. When discovered they may affect the pmcc.

The pmcc can be used as the test statistic in a hypothesis test for correlation in just the same way as the Spearman's rank correlation coefficient is used in a test for association. Notice, however, that the tables used for the two tests are not the same.

Some calculators have the facility to return the value of the pmcc when you input a set of bivariate data. Check whether this is true for your calculator and, if so, make sure that you know how to use it.

Exercise 20D

1 Enter the following data into a spreadsheet.

x	23	35	51	51	62	33	78	75	82	65	89	95
y	16	22	53	50	14	42	67	70	64	60	72	79

(a) Use your spreadsheet to draw a scatter diagram of the data, and say what it shows.

(b) Identify any outliers.

(c) Use your spreadsheet to work out Pearson's product moment correlation coefficient.
Comment on how it relates to the data.

2 The variables in this scatter diagram are as follows.

Horizontal axis: Life expectancy in years.

Vertical axis: Birth rate, babies per 1000 population per year.

All the countries in the world are represented.

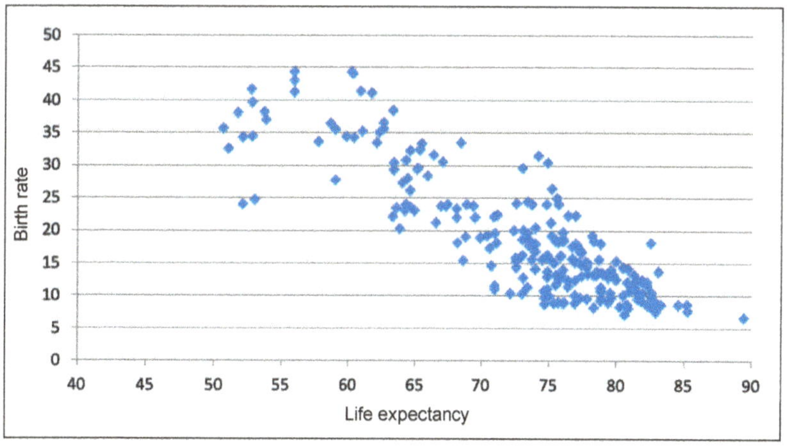

(a) Describe the relationship between the variables.

(b) The product moment correlation coefficient is −0.8592.

Comment on **(i)** the value of the pmcc and **(ii)** its negative sign.

(c) Alex says, 'This shows that having lots of children tires you out and makes you die younger.'
Sandra says, 'This shows that some people value their own lives more than their families.'
Comment on why neither of these statements explains the pattern in the scatter diagram.
Give your own possible explanations.

Test yourself 20E

1 Three friends are discussing rugby union scores.

Owen says, 'In general, the more points the winning side scores, the fewer the losing side gets'.

Fiona says, 'That is not true. There are high-scoring games and low-scoring games, so the more points the winning side gets, the more the losing side gets'.

Hari says, 'You are both wrong. The only connection is that the winning side scores more.'

As a pilot investigation they look at the scores for the matches in the 2019 Six Nations.

	France	Ireland	Italy	v Scotland	Wales
England v	44–8	32–20	57–14	38–38	13–21
	France v	14–26	25–14	27–10	19–24
		Ireland v	26-16	22–13	7–25
			Italy v	20–33	15–26
				Scotland v	11–18

(a) Show the scores on a scatter diagram with the winning scores on the horizontal axis and the losing scores on the vertical axis.

(b) Comment on what the scatter diagram suggests about each of the three friends' statements.

(c) Suggest steps they could take for a more thorough investigation.

2 Amir wants to buy a house in a particular part of the country. He obtains details of 10 houses from a local estate agent and lists them in the table below.

House	A	B	C	D	E	F	G	H	I	J
Price (£1000)	275	125	1239	399	599	449	177	455	445	625
Bedrooms	3	1	7	4	6	5	2	5	3	5
Land area (ha)	0.1	1.2	16.5	0.3	2.2	0.2	5.6	1.4	2.1	5.5

(a) Rank the values for each of the three variables.

(b) Calculate the value of r_s for:

(i) price and number of bedrooms

(ii) price and land area

(iii) number of bedrooms and land area.

(c) Comment on whether the three values of r_s you found in part (b) are about what you would expect.

3 Izaak is an angler. He spends 4 hours every Saturday fishing the same stretch of water. A friend has a theory that the temperature influences how likely fish are to bite (and so be caught). One autumn Izaak puts his friend's theory to the test; he records the temperature while he is fishing and the weight of fish that he catches. These are his readings.

Temperature (°C)	25	18	12	16	21	5	7	–1	10
Weight of fish (kg)	8.1	5.3	7.7	5.5	5.2	2.9	4.1	1.6	6.6

Isaak uses his data for a hypothesis test using Spearman's rank correlation coefficient. He uses a 5% significance level.

(a) State the null and alternative hypotheses.

(b) Calculate the value of r_s.

(c) Find the critical value.

(d) Complete the test. What would Izaak tell his friend?

4 It has been suggested that there is an association between people drinking a particular fizzy soft drink and an increase in their weight.

To test this theory, Aziz, a medical student, asked his fellow students to be subjects in an experiment. There were 12 volunteers, all of them healthy. They were given as much of the fizzy drink as they wanted.

Over a fortnight they counted the number of cans of the fizzy drink that they drank. They were weighed before and after this period and their weight gain or loss was recorded.

Number of cans	14	28	20	69	16	21	33	80	45	42	56	70
Weight gain (kg)	−0.2	0.6	0	0.9	1.3	−0.1	0	1.5	1.2	0.7	1.6	2.2

(a) Name the sampling method that Aziz used.

Give two criticisms of its use in this case.

Aziz used the data in a hypothesis test based on Spearman's rank correlation coefficient. The significance level was 5%.

(b) State the null and alternative hypotheses.

(c) Carry out the test and state the result.

Aziz talks to a journalist about the result and his newspaper carries the story under the headline:

Proved beyond doubt. Fizzy drinks are bad for you

(d) Give three criticisms of the newspaper headline.

5 (a) Calculate the value of Spearman's rank correlation coefficient for the pairs of numbers in the two columns of this spreadsheet.

Set up a spreadsheet of your own and copy the screenshot into it.

Place the command =RANK(A1,A$1:A$8,0) in cell C1. Then copy it down to C8.

(b) (i) State what the numbers in column C represent.

(ii) Explain the significance of each part of the command =RANK(A1,A$1:A$8,0).

Copy cells C1 to C8 across to D1 to D8.

Place the command =CORREL(C1:C8,D1:D8) in cell D10.

(c) Write down the number that appears in cell D10 and say what it means.

Place the command =CORREL(A1:A8,B1:B8) in cell B10.

(d) Write down the number that appears in cell B10 and say what it means.

	A	B
1	2	10
2	6	14
3	8	23
4	3	9
5	9	25
6	11	21
7	4	2
8	7	11

Practice question

A college runs a course in word processing. At the end of the course students are given a scientific report to type as quickly and as accurately as they can.

The college has a theory that the faster people type, the more errors they are likely to make. They decide to use the data from a sample class of 8 people to carry out a hypothesis test, based on Spearman's rank correlation coefficient. They will use a 5% significance level.

This spreadsheet shows the outcomes from the group.

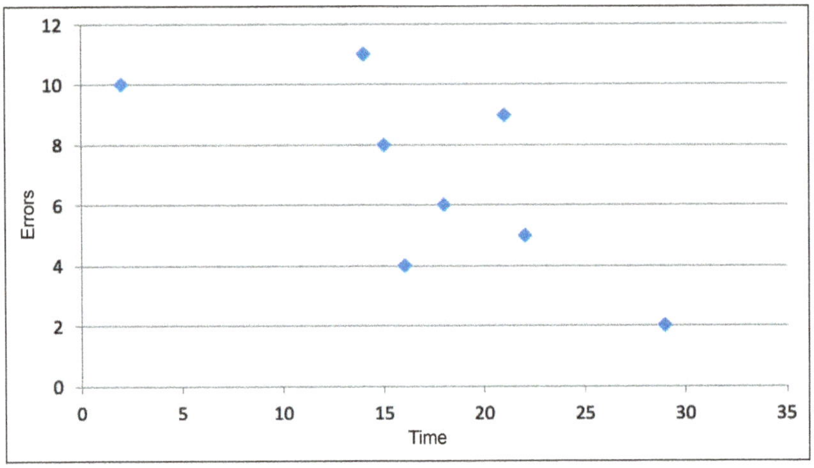

(a) Copy and complete this table giving the times and errors of the 8 people.

Person	A	B	C	D	E	F	G	H
Time	21	15	18	14	22	2	16	29
Errors	9	8	6	11	5		5	

(b) Identify the outlier and give a reason why it should be discarded.

(c) State the null and alternative hypotheses for the test.

(d) Copy and complete this table.

	Time	Time rank, x	Errors	Errors rank, y	$d = x - y$	d^2
A	21	3	9	2	1	
B	15		8			
C	18	4	6	4		
D	14	7	11	1		
E	22	2	5	5.5	−3.5	
G	16	5	5	5.5		
H	29	1		7	−6	
				Σ		

Then calculate the value of r_s.

Complete the test and state the conclusion.

The tutor finds past records for 126 people who have taken the test. He enters them into a spreadsheet and uses it to work out the product moment correlation coefficient. To 3 significant figures, it is −0.791.

(e) What can the tutor conclude from this value of r?

Now that you have completed this chapter, you should know:

- The meanings of the words 'association' and 'correlation'.
- How to calculate Spearman's rank correlation coefficient, r_s.
- How to interpret r_s.
- How to use r_s as the test statistic in a hypothesis test.
- How to find the product moment correlation coefficient (pmcc) and to interpret it.

Answers

CHAPTER 1 Probability and risk

Activity 1A

(a) Likelihood refers to how likely or probable a risk is to happen. Impact is how big an effect the risk happening would have on the business.
(b) As both likelihood and impact increase, the risk rating increases from green to red.
(c) Risk assessments help the business identify where to concentrate their attention.

Exercise 1A

1 (a) B; (b) D; (c) D (or C) – depends where the office is.
2 Risks in order (from least to most likely) are as follows:

Suffering from depression

Going into a care home in old age

Getting cancer

Developing high blood pressure

3 Answers will vary. Genetic factors and lifestyle will affect risks.

Activity 1B

(a) Just over 14%.
(b) 1 in 12 is just over 8%.

Exercise 1B

(a) 1 in 1000	1 in 100	1 in 5	1 in 2
(b) 10%	25%	55%	77%
(c) 1 in 30	1 in 6	23%	76%

Activity 1C

Time is an important factor when expressing risk because most risks will become more likely for longer time periods.

Exercise 1C

1 (a) True; (b) False; (c) False; (d) False

Activity 1D

Data on how many people are diagnosed with cancer out of the total population.

Exercise 1D

(a) Equally likely; (b) Not equally likely; (c) Not usually equally likely; (d) We usually assume that it is equally likely; (e) Not equally likely

Activity 1E

Although the test is wrong for the first two people tested, a larger sample size shows that it is somewhere between 80% and 90% accurate.

Exercise 1E

1 (a) 0.483; (b) 0.517; (c) 0.151; (d) 0.392; (e) Answers (a) and (b) add to 1. For this data table, the person chosen must either be male or female – these categories are mutually exclusive and exhaustive.
2 (a) 0.146; (b) 0.767
3 (a) 0.688; (b) 0.897; (c) 0.104

Activity 1F

For (a) and (b) in Exercise 1E Q 3, knowing that a 17-year old is at a selective state school increases the chance of this being a student that goes on to HE.

Exercise 1F

1 (a) Independent; (b) Dependent; (c) Independent
2 Not independent

Activity 1G

Your own work.

Exercise 1G

1 (a) $\frac{23}{200} = 0.115$ (b) $\frac{89}{200} = 0.445$ (c) $\frac{68}{200} = 0.34$
(d) $\frac{20}{200} = 0.1$
2 Answers will vary.
3 The larger simulation is likely to give more reliable results.
4 (a) 0.125 (b) 0.375 (c) 0.375 (d) 0.125

Activity 1H

See Worked example 1H for one way to work out the answer.

Exercise 1H

1 0.874
2 0.066
3 0.1513

Test yourself 1I

1 Road safety has clearly improved as deaths of car occupants and injuries have both gone down when all the population, the billions of km driven and the number of motor vehicles have all gone up. Deaths and injuries per billion km driven is probably the best measure of risk to use.

Year	Deaths per billion km	Injuries per billion km
2000	3.55	680
2013	1.6	372

2 (a) 36%
(b) There are several possible things the company might do:
- decrease the likelihood (e.g. by taking on more workers)
- decrease the impact (e.g. by seeking to make profit in other ways at the same time or sharing the work and risk with another company)
- not take on the new opportunity.

3 0.001

4 $\frac{8}{27}$

Practice question

(a) 12%; (b) 16%; (c) 2%; (d) 45 to 64 is 20 years; 65 to 74 is 10 years, so more people aged 45 to 64; (e) We don't know how many males and females there are in the population aged 25–44 but it's reasonable to assume roughly equal numbers. Roughly twice the number of men live alone compared to women, so John is right.

Chapter 2

Activity 2A

(a) £10.10 (b) 10 km

Exercise 2A

1 (a) £13 (b) $C = 2.20 + 1.35d$

2 (a) £0.15 or 15p (b) £68.20 (c) 196 minutes

3 (a) 8.4 m (b) 2.86 seconds

Activity 2B

By summing the final column using =SUM(G4:G13), or the AutoSum.

Exercise 2B

1 (a) (the year) 2016 (b) B5 (c) =C6-D6, or 62.7 - 36.3

Activity 2C

x^2 and A2^2, $2x - 2$ and A2*2 -2, $\frac{x}{2}$ and A2/2, $\sqrt{x}$ and SQRT(A2)

(a) × (b) The number in cell A2 (c) ÷

Exercise 2C

1 =A1*100

2 =3.14*A1^2 or =PI()*A1^2

3 =SQRT(A1^2+B1^2)

4 The mean of the numbers in A1, A2, A3, A4 and A5.

5 The mean of the numbers stored in B1 through to B10.

Activity 2D

(a) Changing cell A2 to =A1+5 gives the sequence 1, 6, 11, 16 ...

(b) Changing the number in cell A1 to 2 gives the sequence 2, 6, 10, 14 ...

Exercise 2D

1 (a) A1 contains 3 and A2 contains =A1+2
(b) A1 contains 1 and A2 contains =A1*2
(c) A1 contains 2187 and A2 contains =A1/3
(d) A1 contains 65536 and A2 contains =SQRT(A1)

2 (a) Multiply by 2 and add 1 so 3 is in A1 and B1 contains =2*A1+1
(b) Square and add 1, so 2 is in A1 and B1 contains =A1^2+1

Activity 2E

(a) Row 1: Each number is the sum of the previous two numbers.
Row 2: Each number is 4 more than the previous number.

(b) =D1+E1

(c) Each term increases by the 4 in A2.

(d) The A doesn't change when the command is copied.

(e) =$A2+C2

Exercise 2E

1 (a) The given spreadsheet.
(b) 67, the total sales of tuna sandwiches for the 6 days.
(c) 65, the total sales of tomato sandwiches for the 6 days.
(d) £624.75, the total income from the sales of the sandwiches for the week.

2 (a) =D8+A$1 (b) 0 (c) D2 contains 9 and G2 contains 7.

Activity 2F

7

Exercise 2F

1 (a) $t = 3.5$ (b) $t = 3.6$ (c) $x = 1.5$

2 (a) $x = 2.5$ (b) $x = 1.4$ (c) $x = -1.5$

3 (a) $x = -4$ or $x = 6$ (b) $x = 1$ or $x = 4$ (c) $t = 3$ or $t = 4$

4 (a) $x = 8.2$ or $x = -1.2$ (b) $t = 4$ or $t = 3.2$ (c) $x = 4.6$

Activity 2G

(a) 21.5

(b) Your own answer.

(c) Your own answer.

Exercise 2G

1 (a) 9.09 (b) 1.53

2 (a) $x = -1.94$ or $x = 2.28$ (b) $x = 12.67$

Activity 2H

(a) $A = BC + D$ and $C = \left(\frac{A - D}{B}\right)$ $BC = A$ and $C = \frac{A}{B}$

$A = B(C + D)$ and $C = \frac{A}{B} - D$, $A = \frac{B}{C}$ and $C = \frac{B}{A}$

(b) Making C the subject of the first four formulae gives the second four.

Exercise 2H

1 $n = \frac{C - 2.35}{0.25}$

2 (a) $u = v - at$ (b) $t = \frac{v - u}{a}$

3 $t = (\pm)\sqrt{\frac{d}{5}}$

4 (a) $c = y - mx$ (b) $x = \frac{y - c}{m}$

Test yourself 2I

1 (a) 110 minutes or 1 hour 50 minutes
(b) =A2*45+20

2 (a) 8.83 m s^{-1} (b) 4.91 mm

3 (a) 310 mm (b) $t = \frac{d^2}{49} + 12$ (c) =7*SQRT(A2-12)

4 (a) 32.3 cm (b) B1 (c) =B$1*0.754^A3 (d) 9

5 (a) An adult dose of 20 ml (b) 9.6 ml (c) 9 years old
(d) $20 \times \frac{1}{(12+1)} = 1.538...$

(e) When 3 or 4 is substituted into the formula the fractions of 20 are $\frac{1}{4}$ or $\frac{1}{5}$ and both 4 and 5 divide 20 exactly.
(f) =C$1*A3/(A3+12)
(g) The cell reference, C1, for the adult dose does not change.
(h) The formula gives an answer of 60% of the adult dose. 18-year olds are considered adults and have reached close to their adult weight, so the dose is likely to be insufficient.

Practice question
(a) 54°F (b) $C = \frac{1}{2}(F - 30)$ (c) −30° (d) −40° (e) 2.9 %

Chapter 3

Activity 3A

(a) 55%
(b) 2017: Total = 99.9%
2018: Total = 100.1%
The totals aren't quite 100% due to rounding.
(c) E.g. 1.71% + 17.11% + 26.44% + 54.74% = 100%
E.g. 1.85% + 16.15% + 26.70% + 55.30% = 100%

Exercise 3A

1

Waste type	2016	2017	2018	% change from 2017 to 2018
Total recycling *of which:*	10 217	10 139	9 840	−2.9%
Dry recycling	6 042	5 917	5 866	−0.86%
Separately collected food waste	355	386	414	7.3%
Other organics	3 820	3 836	3 560	−7.2%
Total residual	12 535	12 266	12 151	−0.94%
Total waste from households	22 752	22 405	21 991	−1.8%
Waste from households recycling rate	44.9%	54.7%	55.3%	

2 (a) (i) 4.4% (ii) 32.1% (iii) 20.3%
(b) Mia's total recycling rate is 64.3%, which is 9% higher than the 2018 national recycling rate of 55.3%.

Activity 3B

Your own answers. Examples may include: Estate agents often charge a percentage of the sale value of a house. You pay interest on the loan taken out (mortgage) to buy a house. You pay stamp duty, which is a percentage of the total value of your house.

Exercise 3B

1 (a) £54,000 (b) £636.50 (c) £5000
2 (a) £1500, £0, £27,500
(b) £3000, £1800, £11,250

Activity 3C

You can have percentages over 100% in certain circumstances, e.g. price increases, although not in this case.

Exercise 3C

1 (a) 0.62
(b) 1.38 is an increase of 38%; 0.038 is a decrease of 96.2%; 0.38 is a decrease of 62%; 1.62 is an increase of 62%.
2 (a) £63.36 (b) £24.75
3 (a) 176 beats per minute (b) 79 beats per minute
(c) 43.6%

Activity 3D

(a) High-status cars will lose their value fastest.
(b) Cars are unlikely to depreciate at a constant rate and will lose value fastest when newest.

Exercise 3D

1 (a) (i) £36,750 (ii) £38,600 (to nearest £100)
(b) No
2 (a) Sometimes cells are set up as text or money, so the cells need to be formatted to match the data that are being entered.
(b) The formula =A2+1 adds 1 to the value in the cell A2. The formula =B2*0.7 finds 70% of the value in cell B2.
A3: 1; B3: 11550
(c) (i) =A5+1 (ii) =B5*0.7
(d) Year 8; the car is unlikely to keep depreciating at the same rate.

Activity 3E

(a) It would be too time consuming to include everything on the list.
(b) Your list will depend on your lifestyle.

Exercise 3E

1 1 litre of milk: 73 p (to nearest penny)
Pizza: £4.60 (to nearest 10p)
Average fuel bill: £1285
2 (a) 2019: Index number = 110
2020: Index number = 92
(b) Because index numbers are based on percentages.
3 2019: £892.50
2020: £748

Activity 3F

(a) 400 rabbits (to 2 sf)
(b) N_0 = initial number of rabbits; $1 + \frac{r}{100}$ = multiplier; n = number of years

Exercise 3F

1 (a) $x = 40$ (b) $x = 3$ (c) $x = 20$ (d) $x = 10$
2 (a) 1.18% (b) 9.327 billion
3 Reduction rate in number of clients is 22%; growth rate in profit is 12.6%. Company is focusing on a smaller number of more lucrative clients.

Activity 3G

Saved: £255 ÷ 0.85 − £255 = £45

Exercise 3G

1 (a)

Item	Normal selling price	Sale price
Jacket	£80	£68
Jeans	£40	£34
Shirt	£20	£17
Trainers	£50	£42.50
Hoodie	£35	£29.75

(b) Check that applying the discount to the normal selling price gives the sale price.

2 £8.50

Test yourself 3H

1 Seals: −3.66%; Penguins: 3.73%. Your own comment about seals and penguins.
2 (a) Paper 1: Instructing others; Paper 2: Accompanied diving
(b) Accompanied diving
(c) 122
3 £256,250
4 (a) £2856.10 (b) £20,000
5 2.8%

Practice question

(a) Due to rounding
(b) 129.1 mm (to 1 dp)
(c) 84.0 mm (to 1 dp)
(d) 9.03%; July is a bit drier.
(e) 2016: 100; 2017: 102.0; 2018: 96.1; 2019: 111.8
The index numbers are all close to 100, so the total annual rainfall from 2016 to 2017 is fairly consistent.

Chapter 4

Activity 4A

(a) 20,000 (but possibly 18,000)
(b) Advantage easy to work out in your head; disadvantage not accurate.

Exercise 4A

Assuming that 18,000 or 20,000 are used as approximations for Rp to £s.
1 (a) £3.33 or £3 (b) £5 or £4.50 (c) £1.39 or £1.25
2 Either 140,000 is below 8 × 18,090.91 or 150,000 is above 8 × 18,090.91
3 Your answers in Q1 will be underestimates if you used 20,000 Rp to the £1.
If you used 18,000 Rp to the £ your answers will be overestimates.

Activity 4B

(a) Car mileage (including fuel), food and drink.
(b) To the nearest £1.
(c) Value of likely sales.

Exercise 4B

Estimated answers may vary but should be about:
1 £12 2 £30 3 £650

Activity 4C

(a) Nearest kg. His weight will vary through the day.
(b) For a rough idea, e.g. when flying.
(c) If undergoing medical treatment.

Exercise 4C

1 (a) 4200 (b) 0.03 (c) 200
2 (a) 6.1 (b) 0.08 (c) 10.1
3 (a) 20 000 (b) 400 (c) 0.7 (d) 0.007
4 (a) 73 000 (b) 4000 (c) 0.70 (d) 0.071

Activity 4D

(a) $\frac{50}{250} \times 100 = 20$
(b) Probably, but it depends on why she wants to know.

Exercise 4D

1 Estimated answers may vary but should be close to:
(a) 3700 (b) 2.9 (c) 0.2
2 100 ÷ 300 × 100 ≈ 30%

Activity 4E

(a) £6
(b) £5
(c) The first is an overestimate; the second an underestimate.
(d) The first is closer, £6.
(e) Your own answer and reasoning.

Exercise 4E

1 (a) 50,000 ÷ 20,000 = £2.50; 50,000 ÷ 18,000 = £2.78; the cost lies between £2.50 and £2.78.
(b) £9.12 to £10.95
(c) £9.12 to £10.13
(d) Probably £10 but that may differ for individuals.

Activity 4F

(a) 16 m
(b) Just over 14 m, say 14.05 m.

Exercise 4F

1 $3.5 \times 4.5 = 15.75$; $2.5 \times 3.5 = 8.75$; $8.75 \leq A \leq 15.75$
2 Lower bound is 6.29; upper bound is 6.41; $6.29 \leq S \leq 6.41$

Activity 4G

(a) Between 27 and 33 nuts.
(b) Yes, it is reasonable.
(c) Your own answer.

Exercise 4G

1 (a) $11.5 \leq x \leq 12.5$ (b) $10.8 \leq x \leq 13.2$ (c) $114 \leq x \leq 126$
(d) $0.09 \leq x \leq 0.11$
2 (a) 50 ± 10% (b) 60 ± 10% (c) 250 ± 2%
(d) 20 ± 2.5%
3 (a) 55 ± 10 (b) 150 ± 3 (c) 4 ± 0.3 (d) 40 ± 2

Activity 4H

(a) Both round to 12 m^2.
(b) The length and width (the input data) are given to 2 sf.

Exercise 4H

1 947 (3 sf)
2 53 km/h (using 7.2 hours and 382 km) or 53.1 km/h (using 432 minutes and 382 km).

Activity 4I

(a) About 20 million
(b) About 20 million have a birthday on the same day of the year as you. The accuracy depends on that of the figure you took for the world population. The number with the same day and year as you depends on how old you are.

Exercise 4I

1 The number of 2 kg boxes he needs to buy.
2 (a) Determine the volume of water in a lake.
(b) The surface of the lake is circular, and the cross-section is the same all the way down. The circle has radius 15 m and the depth of the water is 6 m.
(c) Calculate $V = \pi r^2 h$, where $r = 15$, $h = 6$.
(d) $V = 4241.15$
(e) The lake contains 4200 m^3 to 2 sf of water.
(f) 'Select information from the experiment, experience or observation and compare with theoretical results.'
(g) The shape is more like an inverted cone.
(h) 1400 m^3 (to 2 sf)

Test yourself 4J

1 (a) 64 mm
(b) Upper bound 79.5 mm, lower bound 59.5 mm
2 313.5 ml
3 Upper bound = 2200 m^2, lower bound =1350 mm^2
4 0.049 km/h (2 sf)
5 (a) Assumptions may include: number of meals; cost of meals; cost of drinks; number of excursions; cost of each excursion; cost of souvenirs; tips.
(b) Evidence such as: prices in the area she actually stays in; size of meals; protocols with tips.

Practice question
There are many ways of approaching these questions, so answers will vary considerably. The model answer that follows is just one possibility.

(a) *Annual earnings = £2*
= 2 x 20 × 12d
= 480d
This has to support him for 365 days.
He wouldn't have worked Sundays, so about 300 working days.
Daily pay $= \frac{500d}{300}$ *which is approximately* $1\frac{1}{2}d$.

(b) *Assume no meals are supplied and 50 weeks in a year.*

2 pigs @ 2s	*4s*
50 chickens @ $\frac{1}{2}d$ *each*	*2s 1d*
100 dozen eggs @ $\frac{1}{2}d$ *per dozen*	*4s 2d*
80lbs cheese @ $\frac{1}{2}d$ *per lb*	*3s 4d*
80 gallons of ale @ 1d per gallon	*6s 8d*
Bread (estimated from prices at the inn)	*13s 4d*
Vegetables (unknown but very little)	
Total	*£1 13s 7d*

Comments: He may have had a family to support and they may have grown their own vegetables, kept some chickens and trapped rabbits to eat. They may have kept a cow for milk and drunk that, rather than ale.

(c)

Food	*£1 10s to £2*
Cottage	*5s*
Tunic(s)	*2d to 1s*
Shoes	*3d to 1s*
Clothes	*2d to 1s*

Lower bound = £1 15s 7d
Upper bound = £2 8s

(d) *The lower bound suggests yes, he could.*
The upper bound makes it look difficult!
(e) *The assumptions on the costs are very rough and could be revised up or down. The yearly income depends on the labourer being fully employed for the year. There may be less work in the winter months. He could become ill and unable to work.*

Chapter 5

Activity 5A

Your own discussions.

Exercise 5A

1 (a) (i) £1070 (to nearest £10) (ii) £1230 (to nearest £10)
(b) (i) £354.44 (ii)£334.44
(c) Your own answer.
2 (a) and (b) Your own answers.

Activity 5B

£3

Exercise 5B

1 £2428.98
2 £2706
3 2.50% (3 sf)

Activity 5C

(a) Probably the one with the highest AER.
(b) What happens if you need to withdraw money before the year-end may be important.

Exercise 5C

1 (a) £11,614.72 (b) 5.12% (to 3 sf)
2 (a) 4% (b) 4.04% (to 3 sf) (c) 4.06% (to 3 sf)
(d) 4.07% (to 3 sf) (e) 4.08% (to 3 sf)
(f) 4.085% (to 4 sf)
Advantage: Higher AER

Activity 5D

(a) (i) With the loan he pays a total of £1200, so interest of £200, and then the laptop is his.
(ii) With the credit card he pays £15 per month but the laptop is never his.
(b) This assumes there is no repayment element in the credit card method of payment.
(c) Your own concerns, such as whether the laptop will go out of date.

Exercise 5D

1 (a) (i) 29.6% (ii) £648.05
(b) (i) 10.4% (ii) £551.91
(c) (i) 28.3% (ii) £641.63
2 (a) 1732.7% (b) £200 + £223.60 = £423.60

3 (a) (i) =A2+1 (ii) =D2*0.015 (iii) =D2-C3+B3
(b) (i) 24 months (ii) £1200
(c) 11 months
(d) (i) 94 months (ii) £862.20

Activity 5E

1 **Option A:** £18,460.75
Option B: £17,840

Exercise 5E

1 (a) =B4/B5 (b) =B8/B9 (c) =E4+E5+E6
(d) = 52*B15*B16 (e) =H4−E8
2 (a) −£9800 (b) £3200 (c) Write-off the van over a longer period of time.

Activity 5F

(a) A high selling price means profit per bracelet will be high but demand will be low; a low selling price will increase demand but decrease profit per bracelet.
(b) Market research to find out what type of bracelets will sell well and the best selling price to optimise profit.

Exercise 5F

1 (a) (i) 70 bracelets (ii) 22 bracelets (iii) 10 bracelets
(b) (i) £31.50 (ii) £19.80 (iii) £12.50
2 (a) (i) =A2+5 (ii) =A2*B2 (iii) =7*B2 (iv) =C2-D2
(b) (i) £25 optimises profit
(ii) Isobel can make 30 bracelets and a profit of £540.
(iii) Out of her selling price of £25, £7 is the cost of materials, leaving £18 for 2 hours of work. £9 an hour is quite a low rate of pay.

Test yourself 5G

1 Option 1 you pay £942 over 24 months and no more.
Option 2 you pay £921 over 24 months but continue monthly payments after that.
Your own answer.
2 (a) £4244.83 (b) (i) £4246.71 (ii) 2.02%
3 (a) 2.02% (b) (i) =0.02*D2 (ii) =D2+B3-C3
(c) No answer required (d) £21.23
4 (a) (i) £620 per year or £11.93 per week
(ii) £3500 per year or £291.67 per month
(iii) £15,500 per year.
(b) (i) £14.04 per week
(ii) £47.27 per week
(iii) £101.16 per week
5 (a) Monthly tax: £128.33 Monthly NI: £106.84
(b) £438.16

Practice question
(a) 15 cakes
(b) £40
(c) £10
(d) (i) =A2 + 10 (ii) =A3*B3 (iii) =10*B2 (iv) =C2–D2
(e) £10
(f) (i) £756 (ii) 14 (iii) She makes £13.50 an hour, so OK.

Chapter 6

Activity 6A

A: 7, 8, 10, 13, 15, 17, 18, 23
B: 1, 2, 11
C 5, 6, 9, 19, 21, 24
D: 3, 4, 12, 14, 16, 20, 22

Exercise 6A

1 Your own answers.
2 A: Set up the hypothesis 'Teams that score more concede fewer goals'.
B: The data were collected from the results table for the season.
D: It looks as though the hypothesis may be true, but with only 20 points it would be better to collect more data.

Activity 6B

Eruptions occur frequently, at intervals between three-quarters of an hour and an hour and a half.

Exercise 6B

1

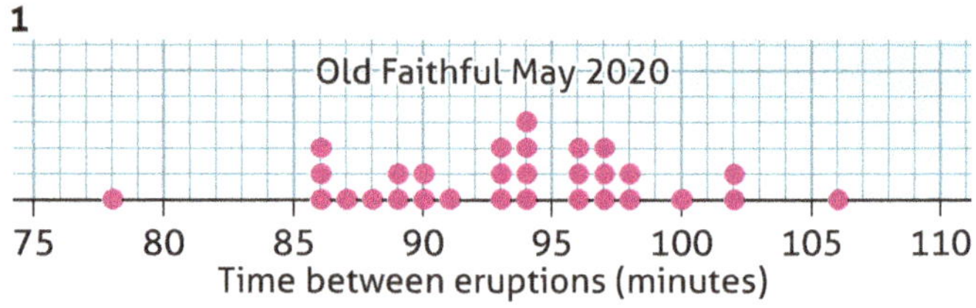

2 Similarity: A sequence of fairly regular eruptions.
Differences: Times between eruptions were longer in 2020; Times between eruptions were less spread out in 2020.
3 Mean, median, IQR and range would be useful to support the impressions from the dot plots.

Activity 6C

(a) At the very least: range from 145 hours to 325 hours, but most Junes had over 200 hours of sunshine.
(b) Mean, median calculations would be useful – but mode of which there are 3 would not be helpful.

Exercise 6C

1 (a) Mean = 2, Median = 3, Mode = 3 (b) Median or mode
(c) Mean
2 (a) New mean = 5.4
(b) (i) =C2*C3 (ii) =B2*B3 (iii) =SUM(B4:E4)
(iv) =B6/SUM(B3:E3)

Activity 6D

(a) Star ratings are often combined (as in section 6C) so are treated as numerical data. But in many star rating situations, 5 is regarded as excellent, 4 as good, down to 1 for poor – in these cases categorical as you cannot add a good and a poor!
(b) The age boxes are effectively categorising people's age – although by taking the boxes as giving age intervals you could estimate the mean age of those answering such a question.

Exercise 6D

1 Categorical

2

ID	*Last name*	*First name*	*Sentence (years)*	*Date transported*	*Ship*	*Number of convicts on ship*
Categorical	Categorical	Categorical	Discrete numerical	Categorical	Categorical	Discrete numerical

3 (a) Categorical
(b) More transported, fewer executed, imprisoned went up and then back down.

Activity 6E

Age (years)	Frequency	Age (years)	Frequency	Age (years)	Frequency
10 to 14	1	25 to 29	13	40 to 44	0
15 to 19	2	30 to 34	3	45 to 49	3
20 to 24	5	35 to 39	6	50 to 54	0

Exercise 6E

1 (a) (i) 3 (ii) 5 (iii) None
(b) Ages of female convicts transported from the Old Bailey 1800

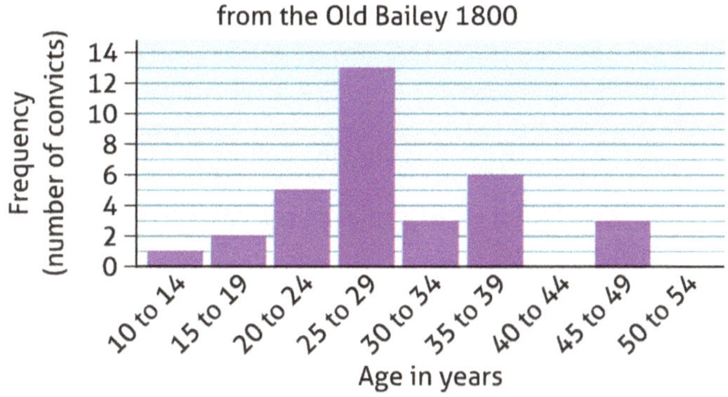

2 Your own answer.

Activity 6F

(a)

Force *P* N	Frequency
$1400 \le P < 1500$	2
$1500 \le P < 1600$	6
$1600 \le P < 1700$	18
$1700 \le P < 1800$	20
$1800 \le P < 1900$	4
Total =	50

(b) The suggested intervals would not be nearly as easy to handle. So, there is danger of items being placed in a wrong group or the groups being misinterpreted.

Exercise 6F

1 (a) 15 (b) $65 \le d < 68$

2 (a) Estimated mean = 66.7 m
(b) Actual mean = 66.592 m, so estimate of 66.7 is close.
(c) 8 m

Activity 6G

The heights of the bars do not take account of the widths of the intervals of values they represent, so they do not provide useful information.

Exercise 6G

1 (a) (i) 15 years (ii) 27 years (iii) 21 years
(b) True, 60 > 57.

Activity 6H

(a) 150 m^3s^{-1}
(b) Typically, 4 roomfuls.

Exercise 6H

1 (a)

Level (*h* cm)	Frequency	Less than	Cumulative frequency
$100 \le h < 150$	43	150 cm	**43**
$150 \le h < 200$	54	200 cm	**97**
$200 \le h < 250$	179	250 cm	**276**
$250 \le h < 300$	56	300 cm	**332**
$300 \le h < 350$	17	350 cm	**349**
$350 \le h < 400$	10	400 cm	**359**
$400 \le h < 450$	4	450 cm	**363**
$450 \le h < 500$	1	500 cm	**364**
500 £ *h* < 550	1	550 cm	365

(b) 97 days (c) 268 days
(d)

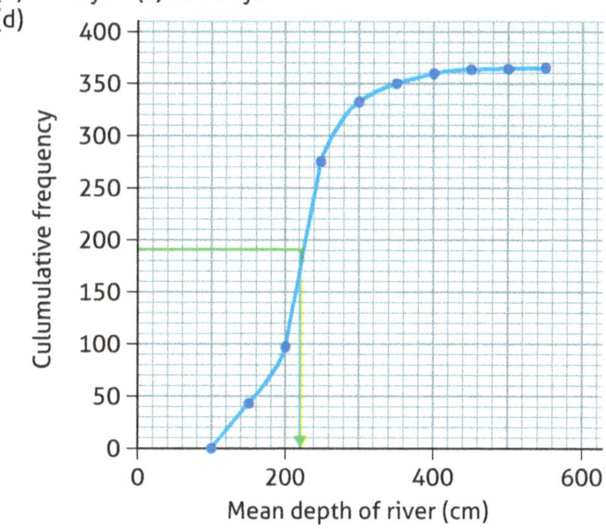

(e) 220 cm

2 (a) The numbers in the frequency column should be 0, 4, 12, 40, 14, 11, 3 and 1. Those in the cumulative frequency column should be 0, 4, 16, 56, 70, 81, 84 and 85.

(b)

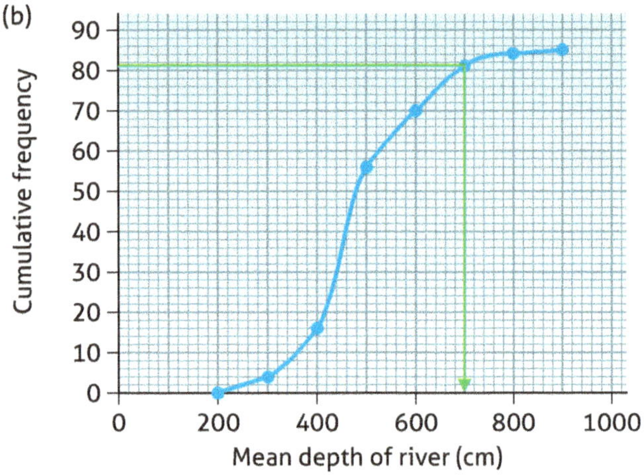

(c) About 700 cm

Activity 6I

Earthquake strength, *M*	Cumulative frequency	Interval for *M*	Frequency
< 1	0	$0 \le M < 1$	0
< 2	38	$1 \le M < 2$	38
< 3	120	$2 \le M < 3$	82
< 4	185	$3 \le M < 4$	65
< 5	200	$4 \le M < 5$	15

Exercise 6I

1 (a) 1.5 (b) 6.1 (c) 1.25 (d) 6
2 (a) 75% (b) 50% (c) 597

Activity 6J

(a) 25%
(b) Both rates increased. The gap between males and females remained but reduced in the 1800s.
(c) Secondary. It is based on records not live observations.

Exercise 6J

1

	(a) Type of chart	(b) Message	(c) Alternative
(i)	Percentage bar chart	• People who are proficient at English have better health than those who are not. • *People whose main language is English* are also healthier	The data support the first bullet point but not the second. Another bar chart is needed. The bar charts need to be presented in two pairs.
(ii)	Time series line graph	• Proportionally more disabled people feel lonely. • The figures are quite stable.	Compound bar charts for each year.
(iii)	Time series line graph	• The price of silver increased overall but there were dips.	The information is shown clearly by this graph.
(iv)	Scatter diagram	• Infant mortality is higher in countries with lower average wealth.	It is quite hard to read the scales.
(v)	Pie chart	• Most people are in good health.	It is hard to read the 3D pie chart and to get any sense of the actual size of the angles. An ordinary 2D pie chart would be better, or a bar chart.

Test yourself 6K

1 (a) 12 (b) 17 (c) (i) 26 (ii) 23 (iii) 48 (d) 40
2 (a) The main points are:
- The children of upper-middle and high-income groups have the higher rates of vaccination.
- Vaccination rates have dropped very slightly over time for the children of upper-middle and high-income groups.
- Vaccination rates have risen slightly over time for the children of lower-middle-income and low-income groups.

(b) The percentage scale starts at 70%, this tends to over-emphasise the differences.
With the years starting at 2018 and going down to 2009 tends to suggest that things are getting worse (we read from left to right).
(c) It does not make a clear, fair point – it suggests a large difference in vaccination rates between either end of the income bands and that this situation is getting worse.

3 The present conviction rate of 80% is higher for both genders compared with 1741–1759, when it was about 60%.

4 (a) The pie chart is hard both to read and interpret. It shows the proportions of the total brightness of the top 10 stars but not their individual values. A bar chart for each individual star would have been a better way to display the data.
(b) She has:
(A) posed the problem and indicted what is needed
(B) collected the relevant data
(C) put the data into a spreadsheet and produced a scatter diagram.
(c) (A) Increase the size of the data set so that any trends are more visible.
(B) Collect more data.

(C) Add the data into the original spreadsheet and produce another scatter diagram. This process may involve investigating the effect of removing some data points which might be outliers masking a trend.
(D) Consider the results in (C) in the light of the original task. Depending on the significance or clarity of these conclusions it might to useful to a further round of the cycle.

5 (a) 48%

(b) (i) The mean of the 4 rates is 38 (2 sf).
This does not give an overall picture as the groups are of different sizes.

(ii) One way is to weight the raw literacy weights by sample size.
This gives a mean of 55 (2 sf), which is greater.
A better method would involve a sample in which the numbers for the four occupations are proportional to how many of them there were in the wider population at that time.

(c) Based on the given literacy rates 58 > 56 > 27 > 10.
However, the figures for weavers and tailors are almost the same, so the ordering is not reliable.

Practice question

(a) Using mid-interval values, mean = $\frac{752}{173}$ = 4.35 m s^{-1}.
Reasons include that all the same measurements were at the same time of day.

(b) Cumulative frequency table is shown in the two right-hand columns.

Windspeed, w, ms^{-1}	Frequency	Mid-interval	Frequency × Mid-interval	Less than	Cumulative frequency
$0 \le w < 2.0$	12	1	12	0	0
$2.0 \le w < 3.0$	16	2.5	40	2.0	12
$3.0 \le w < 4.0$	20	3.5	70	3.0	28
$4.0 \le w < 4.5$	30	4.25	127.5	4.0	48
$4.5 \le w < 5.0$	60	4.75	285	4.5	78
$5.0 \le w < 6.0$	25	5.5	137.5	5.0	138
$6.0 \le w < 10.0$	10	8	80	6.0	163
	173		752	10.0	173

(c)

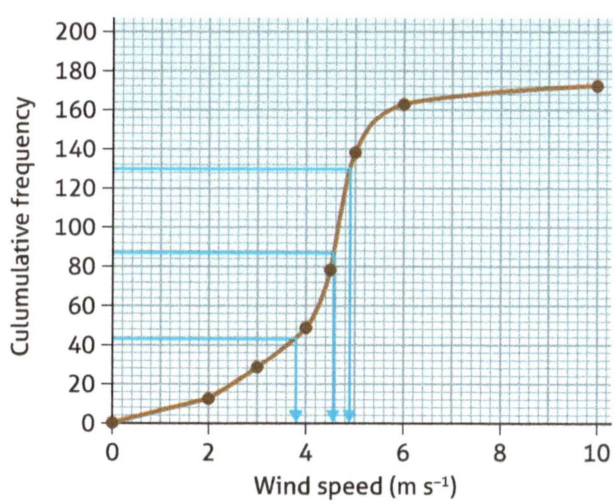

(d) Median = 4.6 m s^{-1} Q.1 = 3.7 m s^{-1} Q.3 = 4.9 m s^{-1}
IQR = (4.9 – 3.7) = 1.2 m s^{-1}

(e)

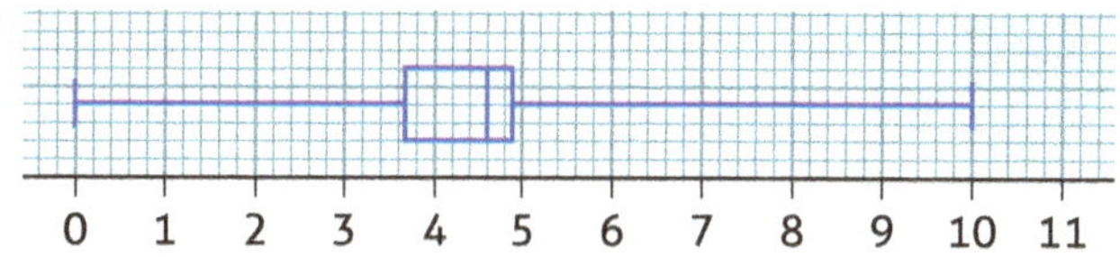

(f) Mean and median are, in m s^{-1}, 4.35 and 4.6 respectively.
Median is about 0.25 m s^{-1} larger than the mean.
The distribution is certainly skewed as can be seen from the box and whisker plot. The median appears more typical.

Chapter 7

Activity 7A

(a) 16:00 (b) 21:25 (c) Clocks change from GMT to BST
(d) 20:15 (e) Your own answer.

Exercise 7A

1 (a)

Year	Male	Female
2001	76	80.6
2002	76.1	80.8
2003	76.3	81.1
2004	76.5	81.4
2005	76.9	81.8
2006	77.1	82.1
2007	77.2	82.4
2008	77.5	82.7
2009	77.7	83
2010	78.1	83.4
2011	78.2	83.7
2012	78.3	83.9
2013	78.4	84
2014	78.4	84

(b) Life expectancy is rising for both females and males.
Life expectancy for females is consistently over 4 years greater than for males.

2 (a)

Month	Jan	Feb	Mar	Apr	May	Jun	Jul	Aug	Sep	Oct	Nov	Dec
Length of day	7 h 55 m	9 h 10 m	10 h 50 m	12 h 50 m	14 h 45 m	16 h 20 m	16 h 40 m	15 h 25 m	13 h 40 m	11 hr 40 m	9 h 40 m	8 h 15 m

(b) Sunrise almost nothing, sunset about 2 hours.

Activity 7B

(a) A (b) B C D A (c) You can no longer compare their gradients.

Exercise 7B

1 Gradient (b)

2 (a) $\frac{1}{3}$ (b) $\frac{8}{3}$ (c) $-\frac{3}{2}$

3 (a) Speed in kilometres per hour.
(b) Rate of water loss in litres per minute.

Activity 7C

1 (a) A few people arrive early, most people come in reasonable time and a few scrape in.
(b) The applause starts quietly but builds up to a sustained volume.
(c) Immunity rises slowly initially but then builds up quickly as more people get the disease until it levels out.
(d) Your own answer.

Exercise 7C

1 (a) 7 m s^{-2} (b) At the start.

2 The vertical scale makes the graph misleading. The rate of increase is less than 1%.

Activity 7D

(a) Points (i), (iii) and (iv)
(b) Below
(c) Substitute the x-coordinate of the point in the equation of the line and compare the values of y.

Exercise 7D

1 (a)

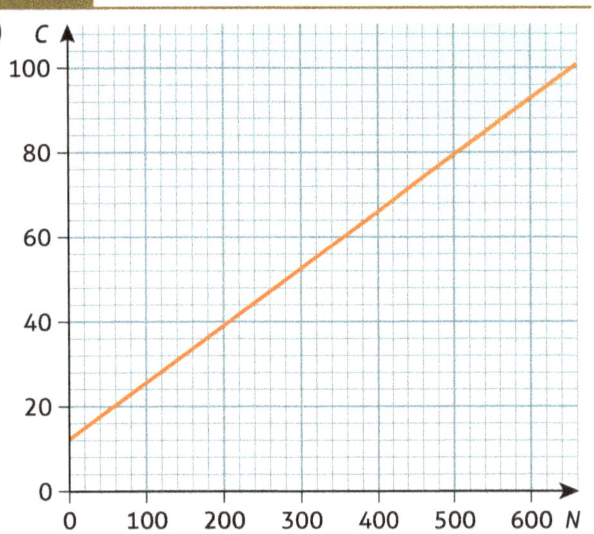

(b) 206

2 (a) Your spreadsheet will be self checking. The points on the scatter diagram will be those on the graph in the question.
(b) Highlight the columns you have used and then do Insert and Scatter.

Activity 7E

Starting from a velocity of 1 m s^{-1} it accelerates at a constant rate.

Exercise 7E

1 (a) iv (b) 40 m (c) $\frac{4}{3}$ m s^{-1} (d) $-\frac{4}{3}$
(e) The mouse returns to its nest.

2 (a) 15 m s^{-1} (b) It has a constant velocity of 10 m s^{-1}
(c) acceleration of 2.5 m s^{-2} (d) 1 m s^{-2} (e) 30 m

Activity 7F

Number of people sharing a pizza and the size of each portion. The relationship is indirect proportion; the other two are direct proportion.

Exercise 7F

Direct proportion: (a), (c)
Indirect proportion: (f)
Neither: (b), (d), (e), (g), (h)

Activity 7G

(a) and (b) The graph is described in answer (d).
(c) The vertical scale is the same.
(d) Graph (a) is a curve and graph (b) is a straight line.
(e) Both have equation $y = \frac{1}{2}x^2$.

Exercise 7G

1 (a) $k \approx 3.8$

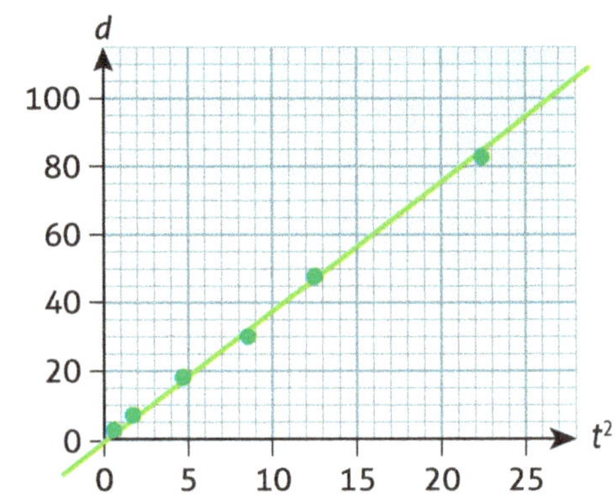

(b) 49 m (c) 4.1 seconds

2 (a)

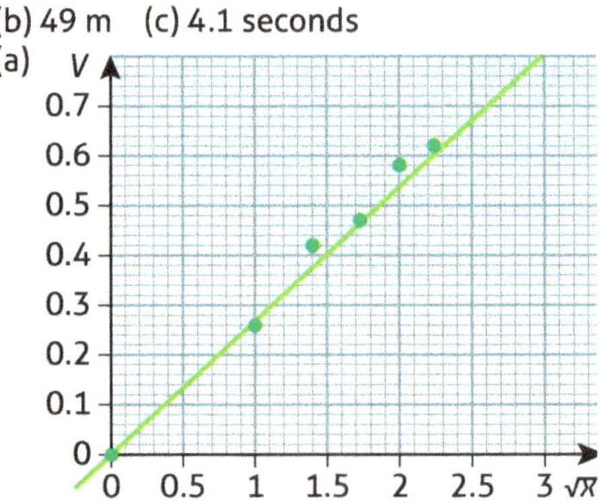

$c \approx 0.28$

(b) 3.19 m

Test yourself 7H

1 (a) Direct proportion (b) Inverse proportion (c) Neither (d) Neither (e) Neither (f) Neither
2 (a) 7000 lek (b) £70
(c) The number of lek is directly proportional to the number of £.
(d) 140 (e) The exchange rate from £ to lek.
3 (a) =SQRT(A3)
(b)

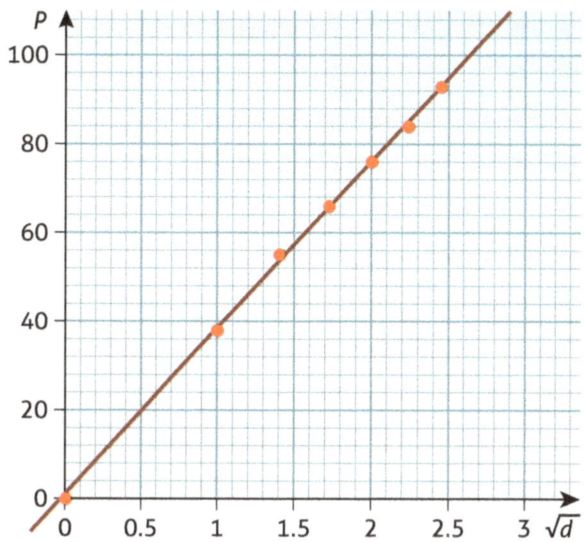

(c) $k \approx 38$
(d) $P = 38\sqrt{(d)}$
(e) $P = 100.5$
(f) This suggests they have forgotten slightly more than everything after 7 days, so the model no longer applies.
4 (a)

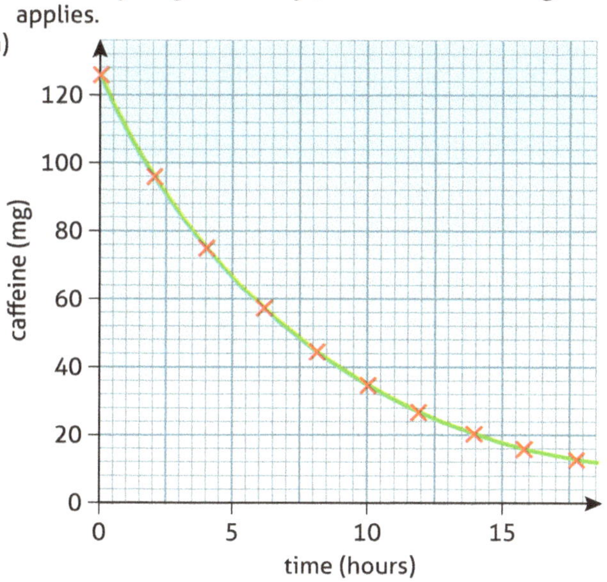

(b) −6 mg per hour
5 (a) 35 mph (b) 230 miles (c) Between 10 and 11, and then between 12 and 1 o'clock (d) 38.3 mph

Practice question
(a) (1, 2), (11, 8) (b) 0.6 (c) $y = 0.6 \times 6 + 1.4 = 5$ (d) (6, 5)
(e) (−2.33... , 0); a negative number of daisies is impossible.
(f) Your own answer.

Chapter 8

Activity 8A

(a) Both the histograms should be roughly bell shaped with a peak in the middle and decreasing tails at either end.
(b) The histogram for boys should be further to the right than that for girls.
(c) Your own answer.

Exercise 8A

1 (a) 0.45 seconds (b) 47 grams
2 (a) 75 apples (b) (i) 90 grams (ii) 90 grams

Activity 8B

(a) An average is a typical middle value, whereas the spread tells you how far away from the middle you can expect to find values.
(b) They are saying different things about the development of the ducklings.

Exercise 8B

1 (a) Crop 1: mean 10 cm; standard deviation about 2 cm
Crop 2: mean 30 cm; standard deviation about 4 cm
(b) Crop 2, as cucumbers are generally longer than carrots.
2 (a) 50%
(b) 5 minutes or less
(c) (i) 1 sd above the mean
(ii) 16%

Activity 8C

His IQ is more than 3 standard deviations above the mean, so, yes, he is very talented.

Exercise 8C

1 (a) (i) $z = -2$ (ii) $z = -2$ (iii) $z = 1.6$
(b) (i) $x = 11$ (ii) $x = 19$ (iii) $x = 184$
2 (a) −1.2 0.4 2.0 −0.4
−4.0 0.6 −3.2 1.2
(b) Yes, some bags are more than 3 sd below the mean.

Activity 8D

The mean wingspan has a z-score of 0, so the mean is about 1.97 m. Reading off from $z = 1$ gives 2.05 m, so the standard deviation is around 2.05 – 1.97 = 0.08 m.

Exercise 8D

1 (a) The plot lies along a straight line.
(b) 60 beats/minute
(c) 10 beats/minute

Test yourself 8E

1 (a) Charts A, B and E
(b) C not bell-shaped; D not a straight line; F not symmetrical
2 (a) The vertical scale is frequency and not frequency density.
(b) It is symmetrical. The mode is in the middle. It is bell-shaped.
(c) 200 cm
(d) 68% of 60 = 40.8; consistent with 20 + 21 players between 195 cm and 205 cm.
95% of 60 = 57; consistent with 8 +20 + 21 + 8 players between 190 and 210 cm.

3 (a) (i) z = −2 (ii) z = 1.333...
(b) Yes. Sam is unusually quick as his *z*-score is -2.67.
4 (a) Normal probability plot is a straight line.
(b) Yes, as he is more than 2 sd below the mean.
5 (a) 114 mm (b) 10 mm (c) 139 mm
(d) 2.5% of patients

Practice question
(a) 0.68 (b) 5%
(c)

Weight of egg	*z*-score	Size class
55 g	0	Medium
62 g	1.167	Large
68 g	2.167	Reject
46.6 g	−1.4	Small

(d) (i) 65.8 g (ii) 52.6 g

Chapter 9

Activity 9A

(a) E.g., the number of bacteria in your body.
(b) E.g., the size of each of the bacteria.

Exercise 9A

1 (a) 3.5×10^3 (b) 1×10^6 (c) 4.5×10^{-4}
2 A1: $5.67 \times 10^8 = 567\,000\,000$
B1: $1.23 \times 10^{-6} = 0.000\,001\,23$

Activity 9B

(a) Do $6.4 \times 5 = 32$ and $11 + -5 = 6$. Then make 32×10^6 into 3.2×10^7.
(b) One way is to enter $6.4 \times 5 \times 10^{(11-5)}$. Other ways are possible on different calculators.

Exercise 9B

1 9.46×10^{12} km
2 1600 kg

Activity 9C

Offer 1 gets bigger (£2,048) by the end of the year than the total of £1,200 for Offer 2.

Exercise 9C

1 (a)

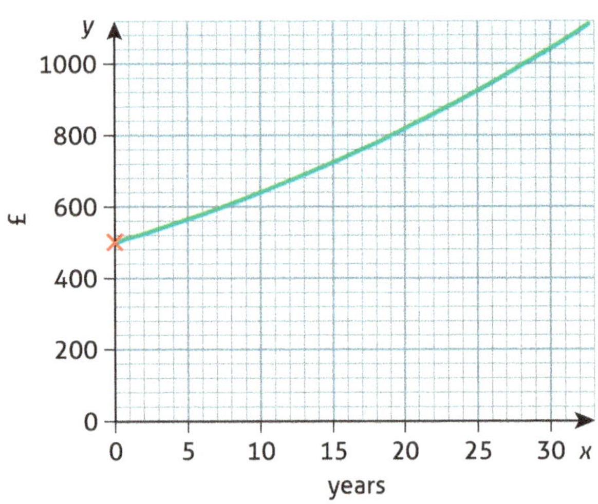

(b) 29 years

2 (a)

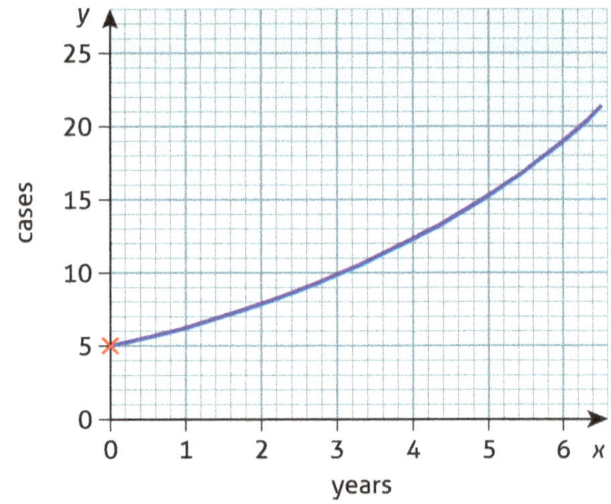

(b) 5 weeks

Activity 9D

(a) Blue (b) Green
(c) Red doubles in about 4 years, green in 7 years; blue is off the scale but would be 14 years.
(d) The red and green curves were easy, the blue one less so.

Exercise 9D

1 (a) After 3 years (b) 7th year (c) =B3*1.035 or =1.035^A4

Activity 9E

Red is $y = 0.9^x$, green is $y = 1.5^{-x}$, blue is $y = 0.5^x$

Exercise 9E

1 (a) A(u)(w) B(p)(t) C(s)(q) D(r)(v)
(b) $y = 0.01^x$
2 (a) 126.5 million (b) 2098
(c) (i) 25.69 % (ii) 25.95 % (iii) 0.997 represents a decrease of 0.3 %.

Activity 9F

(a) (i) 94.5 (ii) 56.8 (iii) 0.35
(b) 2.7×10^{-23}%
(c) (i) 50.0 (ii) Half-life is 12.25 years.
(iii) It is half of its value at 12.25 years.

Exercise 9F

1 $x = 2.55$
2 (a) 28 years
(b) No. Population growth rates are tending to fall in most countries, so it is likely that Kenya will follow this trend and that the rate will get lower. In some countries the growth rate has become negative.
3 4.60 years

Activity 9G

(a) A logarithmic vertical scale with equal intervals for 0.01 to 0.1, 0.1 to 1 and 1 to 10.
(b) The scale should include lower intervals, e.g. 0.0001 to 0.001, 0.001 to 0.01 and 0.01 to 0.1.

Exercise 9G

1 120 dB is 10 000 times as loud as 80 dB, and 120 dB is 3162 times as loud as 85 dB.
120 dB is therefore over 3000 times as loud as the normal safe limit.
2 (a) $2^0 = 1$, so 1000 (b) 2 days (c) 2828
(d) 22 627.
3 (a) =(A1+A2)/2 (b) =SQRT(A1*A2)
(c) Yes, 7.5 and 7.348469
(d) No, they are equal when A1 = A2.

Test yourself 9H

1 (a) 0.000 000 04 (b) 2.68×10^{-22} m^3
2 (a) 4.5×10^9 years (b) 25%
3 (a)

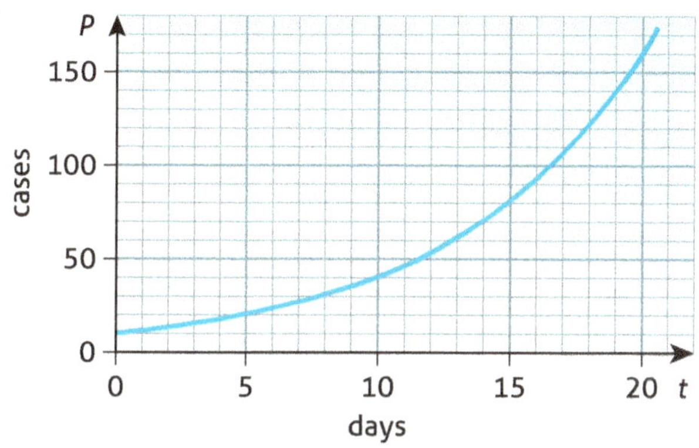

(b) 10 is the initial number of cases.
(c) 0.2 represents $\frac{1}{5}$ and $2^{0.2}$ is the multiplier for the daily increase in cases.
(d) $10 \times 2^{0.2d} = 1\,000\,000$ or $2^{0.2d} = 100\,000$
(e) 83 days to the nearest day.
4 (a) Logarithmic
(b) The business has grown rapidly so the logarithmic scale enables a better representation of the whole graph.
(c) 9 years to the nearest year.
(d) £31.6 million
5 (a) 34.9% (b) during the 7th year (c) =0.9*B3
(e) There are two possible explanations: that the 10% figure was too high, or that the population has bottomed out and started to climb back up.

Practice question
(a) 3.9×10^{13}
(b) 1×10^{-6}
(c) 3.9×10^7 m so 39 000 km. This is less than 400 thousand, so, no, it would not reach the Moon.

Chapter 10

How big is a cubic metre?
- 100 cm
- 10 000 cm^2
- 1 000 000 cm^3
- Yes

Activity 10A

(a) 13.5, 40.5, 450 (b) 1237.7 cm^2 (c) Chunky chips

Exercise 10A

1 490 cm^2
2 225 kg
3 Halving the weight of clay means halving the volume of the statue. The scale factor for the volume is therefore 0.5.
The scale factor for the lengths is $\sqrt[3]{0.5} = 0.794$.
The scale factor for the areas = 0.630.
The cost of the gold is less but not by as much as half.

Activity 10B

(a) Yes
(b) That up to a certain level, wealth and happiness are associated but beyond that level this is no longer so.
(c) There are no units on the scales.

Exercise 10B

1 A 7-ton truck (1017 kg)
2 (a) How far he travels on 1 litre of petrol. (b) 9.25 litres

Activity 10C

It can be on the right-hand wall (as seen in the diagram) or along the right-hand half of the bottom wall.

Exercise 10C

1 (a) 3 m by 2 m (b) 4 m by 2 m (c) 1.125 m^3
(d) 6.25 m^2 (e) 525 g
(f) Approximately 2 m to the furthest part of the vegetable plot.

Activity 10D

Your own drawing.

Exercise 10D

1 (a) Slope is 1 in 18, so yes.
(b) Slope is 1 in 11, so no.
(c) 4.8 m
2 3 staff in the first room and 3 in the second room.
6 altogether.

Activity 10E

(a) (i) 100 m (ii) 530 m (b) Pathways (c) Lakes
(d) (i) 120° (ii) 300°

Exercise 10E

1 (a) 1 kilometre (b) 3.2 km (c) 30 cm
2 (a) 9 cm (b) 1.375 km^2
3 (a) 50.25 m (b) 0.75 m (c) 5.29 metres per minute
(d) 0.0789 metres per minute

Activity 10F (1)

(a) C (b) Side, Plan, Front

Activity 10F (2)

8 but 2 of them are reflections of each other, so 7 is a possible answer.

Exercise 10F

1

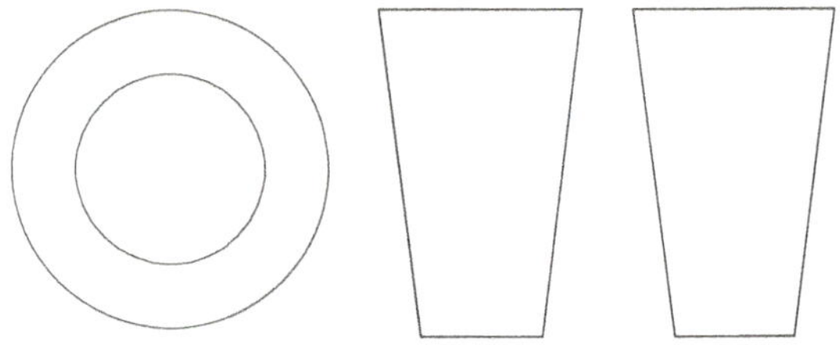

2

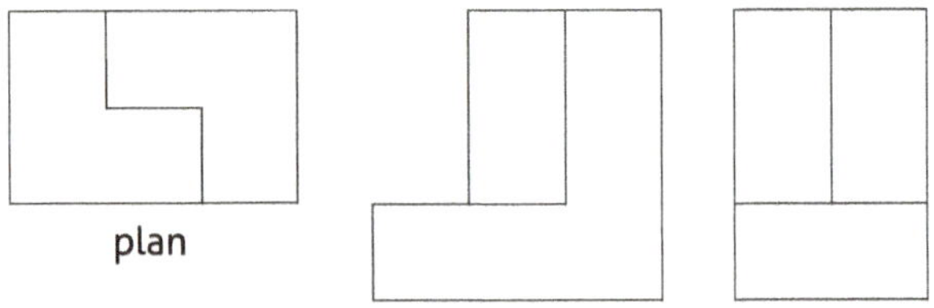

3 (a) 8 cm^3 (b) 32 cm^2

Test yourself 10G

1 (a) 15.408 m (b) 72,567 m^2
2 843,750 tonnes of waste
3 The longest distance of a point in the room to the bath is from the bottom right-hand corner of the room.

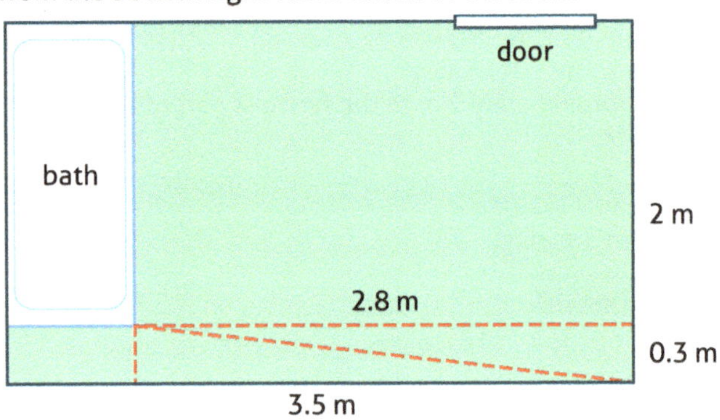

The distance is m = 2.82 m.
It is not possible to have an electric socket.
There is, however, plenty of room for a walk-in shower on one of the 2 walls not occupied by the bath or the door to the room.

4 (a)

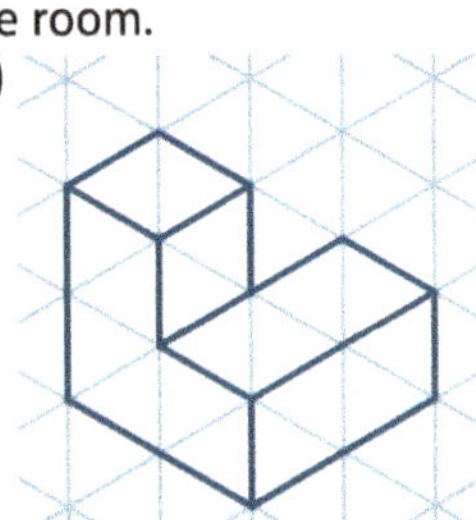

(b) 4 cm^3 (c) 18 cm^2

5 (a)

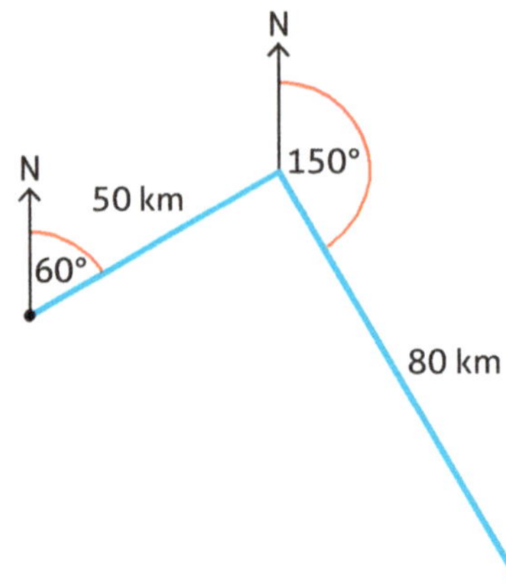

(b) 4 hours 20 minutes (c) 94 km (d) 21.7 km/h
(e) 298° (f) 37.6 km/h (g) 32.8 km/h

Practice question

(a) 7.8 g (b) 62.5 kg
(c) Approximately 2.5 cm (for a width of 0.1 mm on an ant).
(d) About a million times stronger.
(e) Not worry: We are much stronger. They would not be able to move as their legs would not be strong enough.
Worry: There may be many more of them; They may have a poisonous sting.

Chapter 11

Activity 11A

(a)

Height	1.7 m
Weight	70 kg
Heart rate	90 beats/min (or 60)
Length of foot	20 or 30 cm
Width of hand	10 cm
Circumference of head	50 cm

(b) Your opinion with reason.

Exercise 11A

1 Approx. 8000 to 9000
2 Answer in range 2 to 3 hours.

Activity 11B

Description	Number	'Units'
World population	7.8	billion people
UK population	66.3	million people
Population of London	9.3	million people
Capacity of a coach	56	people
Capacity of a train carriage	80	people
UK motorway speed limit	70	mph

Exercise 11B

1 Approx. 13 million
2 Approx. 700 000
3 Answer in range 300 to 2000

Activity 11C

(a) E.g., counting for part of the tray and multiplying up.
(b) How many layers and how to choose a typical section.
(c) Sensible estimate

Exercise 11C

1 (a) Decide what counts as 'internet use' – does it include streaming films? Listening to music? Then work out how many hours each day of the week that would take up.
(b) The number of teenagers in the UK; the number of hours in a known time period that they use the internet, probably one day or one week.
(c) She must assume the number of hours of internet use per person, the number of teenagers in the UK.
(d) 3 or 4 billion hours
2 Answer in range 0.5 million to 2 million (with assumptions and calculation).
3 Answer in range 50 to 70 hours (with assumptions and calculation).

Activity 11D

(a) Answer in range 0 to 0.02%.
(b) Answer in range 50% to 100%.
(c) Answer in range 15% to 70%.
(d) Answer in range 0 to 0.05%.

Exercise 11D

1 $\frac{1}{12}$
2 You are more likely to win money on a scratch card if you buy one.

Activity 11E

(a) kilogram or tonne (b) minute (c) millilitre
(d) metre (e) kilogram (f) gram

Exercise 11E

1 Approx. 16 cm^2
2 (a) Approx. 10 m, plus suitable method.
(b) 15 to 25 kg, plus suitable method.
(c) 5 to 9 mph, plus suitable method.
3 Approx. 80 kg

Activity 11F

(a) 100 (b) 10 000 (c) 10 000 (d) 100
(e) Your description

Exercise 11F

1 (a) Identify how many hours a week he plays games, making estimates about how this might change during holidays and the revision period leading up to the exams.
(b) 100 or 1000 hours
2 C

Activity 11G

(a) 10 million tonnes per year.
(b) Not enough to influence policy as other factors to take into account.

Exercise 11G

1 (a) Approx. 100 million
(b) Mobile phone companies assessing the pricing of their packages.
(c) Yes, probably. Companies will be able to update pricing once they have actual data.
2 (a) Approx. 10 million
(b) Turkey farmers and shops to assess whether they have enough turkeys to supply the demand.
(c) May need to be more accurate than that. Justification: e.g., if the demand turned out to be 12 million then they would not have enough.

Test yourself 11H

1 Answer in range 18 000 to 25 000
2 2.5 to 3 million kg
3 Answer in range 10 000 to 25 000
4 3 to 10 billion
5 Approx. 5000 office blocks
6 1000 medals
7 20 to 100 m^2

Practice question

(a) Assumptions: London is too far, Midlands and most of North, and some of Wales would be within 2 hours; this is approx. one quarter of the UK. 70 million ÷ 4 = 17.5 million, so about 20 million people.
(b) 2% of people would visit every 2 years so 200 000 a year. Approx. 10 years.

Chapter 12

Activity 12A

Probability of 1st class passenger surviving is 0.62.
Probability of surviving passenger being 1st class is 0.42.
Not the same.

Exercise 12A

1 (a) 0.452 (b) 0.129 (c) 0.549 (d) 0.285
(e) Jim is wrong (with reason). (f) 0.555

Activity 12B

(a)

	Identified as faulty	Identified as OK	Total
Faulty	96	4	100
OK	199	19 701	19 900
Total	295	19 705	20 000

(b) See Worked example 12B.
(c) It ensures we only have whole numbers.

Exercise 12B

1 (a) 20, 20, 20, 0, 10, 10 (b) $\frac{2}{3}$
2 (a) 8500, 1500, 8075, 425, 30, 1470 (b) 22.4% (c) 0.4%
(d) E.g., no – it will reduce the spam getting through but also block more real emails.

Activity 12C

E.g., effectiveness and cost of test, how ill people get.

Exercise 12C

1 (a)

	Test positive	Test negative	Total
Has illness	45 000	5000	50 000
Not ill	2500	47 500	50 000
Total	47 500	52 500	100 000

(b) 0.095 (c) 0.947

2 (a) 750, 250, 675, 75, 25, 225 (b) 0.25 (c) 0.036

Activity 12D

No, there are millions of people in the UK.

Exercise 12D

1 (a) 3.65 (b) 2500

2 (a) $\frac{1}{365}$ (b) Likely (with comment).

Activity 12E

No, see Worked example 12E.

Exercise 12E

1 (a) 5000

(b) More than $\frac{1}{5000}$. It would be $\frac{1}{5000}$ if there was no other evidence. The additional evidence increases the probability.

2 (a)

	Identified as lie (positive result)	Identified as true (negative result)	Total
Lie	200	50	250
Truth	225	525	750
Total	425	575	1000

(b) Lawyer is wrong, 47% chance.

(c) 42.5%

Activity 12F

(a) £60,000

(b) More than £60,000. See Worked example 12F.

Exercise 12F

1 (a) Statement (iii)

(b) The player does not have a better chance of winning a prize. Each number is equally likely to come up every time.

Test yourself 12G

1 (a) 0.02 (b) 0.024

2 (a)

	Identified as drug taker (positive result)	Identified as not taking drugs (negative result)	Total
Takes drug	4950	50	5000
Does not take drug	100	4900	5000
Total	5050	4950	10 000

(i) 50.5% (ii) $\frac{100}{5050} \approx 0.02$ (iii) $\frac{50}{4950} \approx 0.01$

(b)

	Identified as drug taker (positive result)	Identified as not taking drugs (negative result)	Total
Takes drug	990	10	1000
Does not take drug	180	8820	9000
Total	1170	8830	10 000

(i) 11.7% (ii) $\frac{180}{1170} \approx 0.154$ (iii) $\frac{10}{8830} \approx 0.001$

(c) Less than 6% of sportspeople at the event took the drug.

3 (a) 198, 2, 980, 8820 (b) $\frac{8820}{8822} \approx 0.9998$

4 (a) For those taking a second test, their probability of having the illness is $\frac{1}{6}$, so a starting frequency divisible by 6 is helpful. It is also useful for the starting frequency to be divisible by 100 as percentages are involved for the second branches.

(b) 98, 2, 25, 475

(c) 0.797

(d) Most of those who have two positive tests do actually have the illness but there is still a probability of about 20% of having a false positive.

5 Calculations show that option A reduces false negatives, option B reduces false positives – the reduction for option B is much greater so that is the better option.

Practice question

(a) 1000, 9000, 960, 40, 90, 8910

(b) 0.086

(c) Interviewing the students is the better option (with explanation).

Chapter 13

Activity 13A

(a) A change from 0.2 to 0.1 is a change of 50%.

(b) First one sounds better.

Exercise 13A

1 (a) 0.3% (b) 4.3% (1 dp) (c) 11.1% (1 dp)

2 (a) Risk changes from 6 in 100 000 to 18 in 100 000.

(b) Risk changes from 25 per 100 000 to about 44 per 100 000.

(c) Risk changes from 125 per 1000 to about 131 per 1000.

Activity 13B

Left-hand chart emphasises the changes in the unemployment rate.
Right-hand chart shows the general trend.

Exercise 13B

1 How many customers each has.
2 (a) A multiple bar chart with percentages starting at zero.
(b) A multiple bar chart with percentages starting at say 15% or 16%.
(c) A multiple bar chart with percentages starting at 13%, 14% or 15%.

Activity 13C

Just eating chocolate seems extremely unlikely to increase the relative number of Nobel prizes. Chocolate is a luxury not easily affordable in poorer countries – richer countries can afford both more chocolates and spend more money on education and research, so are more likely to win more Nobel prizes.

Exercise 13C

1 Little, if any, correlation.
2 (a) Scatter diagrams drawn.
(b) (i) Positive correlation (ii) Positive correlation
(c) (i) Both dropped over time. (ii) Both dropped over time.

Activity 13D

(a) People who bother to take part in polls are self-selected and tend to have extreme views.
A random selection of subjects – perhaps from the voters list (but are all people on voters list?).
(b) People who use landline phones tend not to have up-to-date mobile phones or mobile phones at all. Better to ask people who have mobile phones – in a mobile phone shop perhaps?

Exercise 13D

1 (a) Not many rough sleepers may have bedded down at this time, also in August warmer weather so might be more rough sleepers. To reduce bias, count rough sleepers from say midnight till 2 am and then over a period of time, i.e. summer and winter, weekdays as well as weekends, etc. Establish which areas to survey to include as many as possible.
(b) Biased because it would only 'capture' rough sleepers who had been ill and who were recorded by the hospital as rough sleepers.

Activity 13E

(a) Partly random and partly not
(b) Wholly random
(c) Partly random and partly not
(d) Not at all random

Exercise 13E

(a) Unless there is some underlying medical reason, baby weight is more or less random, so a light baby will tend to be followed by a less light sibling, similarly for heavy babies.
(b) Assuming her scores are partly random and partly not: her mean score for the ten practice runs is 42, after a score of 47 she will tend to regress towards 42 so cannot be certain of scoring 43 or more.

Test yourself 13F

1 (a) Insect bites and car accidents are caused by a third factor. The correlation is only by chance.
(b) Both related to temperature.
2 (a) People want to be seen to be doing the right thing.
(b) When people know they are being observed they may act differently than they normally would.
3 (a) A change of 3.3 to 3.1 represents a drop of 6%.
(b) Not the best way to show the information.
4 Biased because only patients admitted during working hours were surveyed.
5 (a) There could be a third factor accounting for the correlation.
(b) Smoking could be the cause, as many coffee drinkers also smoke.

Practice question

(a) No, there is a random element.
(b) There is weak negative correlation between hours of sunshine and profit but it is difficult to see any corelation between rainfall and profit.
(c) Amber thought the 20% percentage reduction in risk was '20 percentage points'. The new risk will be 0.01%.
(d) The profit axis is not labelled with units, we cannot tell how much the profit is or how much it has increased by – was there no profit in month 1? – origin not labelled.

Chapter 14

Activity 14A

No. of doughnuts	Cheapest way
12	3 packs of 4
13	3 packs of 4 and 1 doughnut
14	2 packs of 5 and a pack of 4
11	2 packs of 5 and a doughnut

Exercise 14A

1 (a) 25% (b) £70 (c) Customers will be tempted to buy an extra item with the offer,
2 (a) £105.58 (after 3 payments of £180)
(b) Cost similar; many reasons for choice.

Activity 14B

No. She would lose money.

Exercise 14B

1 (a) £190 (b) Sell the rest at a discount.
2 £8
3 (a) £24.16 (b) Your own answers.

Activity 14C

(a) Yes
(b) Reduce the price.

Exercise 14C

1 (a) £279.65 (b) £2.12

Activity 14D

Issues include time needed and likely demand.
Other answers will be your own.

Exercise 14D

1 (a) People will not buy it.
(b) Danger of making a loss.
2 Your own answer.

Activity 14E

(a) 19p (b) 20 (c) Sell 95 but no profit. (d) £5.40
(e) Approx. 35p

Exercise 14E

1 £6.50

Activity 14F

Your own answers.

Exercise 14F

1 You expect to lose £10.
2 (a) £4000 (b) 0.2 (c) £1300
(d) £2500 (e) (i) £800 (ii) £500

Activity 14G

Your own answers.

Exercise 14G

(a) This should give a good idea, as it takes into account how their customers behave as well as their situation.
(b) This might give an idea of the trend of the frequency of events that trigger claims.
(c) Only slightly relevant, as it suggests trends in climate change.
(d) This is more reliable than 1 year's data on triggering events for flood claims.
(e) Height above sea level and proximity of rivers likely to flood are relevant.
(f) Not relevant, except that some newer proprties are more likely to be built on floodplains.

Test yourself 14H

1 Lose £3467.50 (or £3469.40)
2 The credit card is better value.
3 (a) Perhaps sell it at £1 per bottle. Give reasons for this.
(b) Only if she sells all her remaining stock at £2.03 a bottle.
4 7 hours or more
5 £6.25 or £6

Practice question
(a) £43.74
(b) £39.25
(c) £46.49

Chapter 15

Activity 15A

The winning driver effectively gets (25 + 1) points as they always pick up the fastest lap point.
A driver can always win the championship if they win 11 races (286 points).

Exercise 15A

1 Same order but Liverpool and Queens Park Rangers tie for first.
2 (a) 42
(b) 3 will divide 78 exactly (= 26), so equal scores may be possible.
One possible set of positions is (1 + 2 + 11 + 12), (3 + 4 + 9 + 10), (5 + 6 + 7 + 8).
3 No. The best the team can finish is (1+10+11+12) giving 34 points. The worst one of the other teams can finish is (9 + 8 + 7 + 6) = 30 points.

Activity 15B

For example, adding the judges' rankings, making lowest the best, gives, in order: E(10); B(12); C(12); F(14); A(18); D(18).

Exercise 15B

1 (a) One third of 180 seats is 60 seats; she will be an MP.
(b) Less than one. Small parties cannot get elected.
2 Your own answers.

Activity 15C

(a) 105 (b) Answer in range 35 208 to 35 381 boys.

Exercise 15C

1 (a) The white region has the larger area.
(b) 5 kg at £8
(c) Oscar
(d) Richard
2 China: one, later two, child policy; more male live births; in adulthood males tend to die younger. Peru: births usual biological gender difference, but with males succumbing to illness/disease more than females, although proportionately less males dying at 66+. UK: begins with the usual biological difference, with males dying earlier.

Activity 15D

Just possible

Exercise 15D

1 About the same (≈ 53 years, 2022 – 1969 = 53, 1969 – 1916 = 53).
2 Matilda born in about (2020 – 40 – 96) = 1884. Karl Marx only died in 1863, so someone could have met both Karl Marx and Matilda.
3 Assuming 5 people per m^2, need 1.6 billion m^2 but only have 380 million m^2, not possible.

Test yourself 15E

1 (a) 90 voters
(b) Total votes for each of 1st, 2nd and 3rd choice should all be 90.
(c) 1st: 32; 2nd: 44; 3rd: 14
(d) Cherner

2 Yes, it is reasonable.
3 (a) True (b) False (c) True (d) False; weighs 20.83 kg.
4 The time taken by the whales is about 8 times that taken by the walker. If the distance swam looks less than 8 times the distance from Lands End to John O'Groats, the walker has the greater average speed.
5 Your own answers.

Practice question

(a) (i) =C4/B4 (ii) =B4/(B4+25000) (iii) =D4*E4
(b) (D3=) 8.10 (D4=) 8.50 (E4=) 0.98 (F4=) 8.34
(c) *Gladiator* *Platoon* *Hamilton*
(d) This ranking is different.

Chapter 16

Activity 16A

E.g., they might have got better anyway.

Exercise 16A

1 (a) Group C
(b) People who like coffee might have different sleep patterns to other people.
2 (a) A likely conclusion is that smiling results in a larger amount spent.
(b) The café owner might smile at people he knows are likely to spend more.
(c) In favour of friend's design: It would stop any decisions about who to smile at.
In favour of original design: Maybe the spending would be different on different days anyway.

Activity 16B (1)

(a) E.g., toss a coin for each person; heads means treatment.
(b) E.g., prevents bias.

Exercise 16B

1 (a)

Methods	Will it guarantee 130 treated?	Equally likely?
Assign either 0 or 1, at random, to each of the 260 schools. Schools with 1 are in the treatment group.	No	Yes
Number the schools 1 to 260. Choose 130 different random numbers from 1 to 260. The schools with those numbers go into the treatment group.	Yes	Yes
Use a spreadsheet to give a random number to each school. Sort the schools in order of size by random number. The top 130 go into the treatment group.	Yes	Yes
Put the schools in alphabetical order of name. The top 130 go into the treatment group.	Yes	No

(b) The final method is not random. Could introduce bias as schools from one particular area may be similar to each other and have similar names, so end up in the same group.
(c) E.g., it is easier to ensure all year 9s in a school do the same thing than to make sure each student does what has been assigned.

Activity 16C

E.g., they might think the herbal tea would help and so do better.

Exercise 16C

1 (a) B
(b) The bracelets made from the different metal.
(c) The researchers should not know who is getting a copper bracelet and who is getting a bracelet made from the other metal.

Activity 16D

Trial 2 (see Worked example 16D).

Exercise 16D

1 (a) A possible reason in favour of a larger sample could be:
- The study would provide stronger evidence.
- It would allow for some to drop out of the study.

(b) One possible reason against a larger sample could be:
- It would cost more.
- It would be unethical to try a new medication on too many people.

(c) No, it would no longer be a randomised trial and comparisons with the control group would not be valid.

Test yourself 16E

1 (a) Participants have an equal chance of being allocated to each group.
(b) There is a control group.
2 (a) Participants don't know whether they are in the treatment group or control group.
(b) Neither participants nor those running the trial know who is in the control group.
3

Feature	Reason for use
Randomised	Ensures factors not being tested do not cause a difference between groups
Control group	Allows comparison
Blind	Prevents distortion of results due to participants knowing what is expected
Placebo	Helps to ensure participants do not know whether they are in the treatment or control group

4 (a) B (b) True, False, True, False (justifications will vary)
5 (a) (i) B, C (ii) C (iii) D, E, G
(b) They will be able to record results in a standard way.
(c) No. This trial is double-blinded and does not include a placebo.

Practice question

(a) E.g., it could be that Aisha expects the tea to help and sleeps better because of that.

(b) E.g., toss a coin for each person and those who get heads are allocated to chamomile tea.

(c) The group with hot water is the control group. They are needed to compare whether the chamomile tea has an effect.

(d) On average, each person with chamomile tea got 3 more minutes per night. Not convincing evidence.

Chapter 17

Activity 17A

(a) Normal distribution

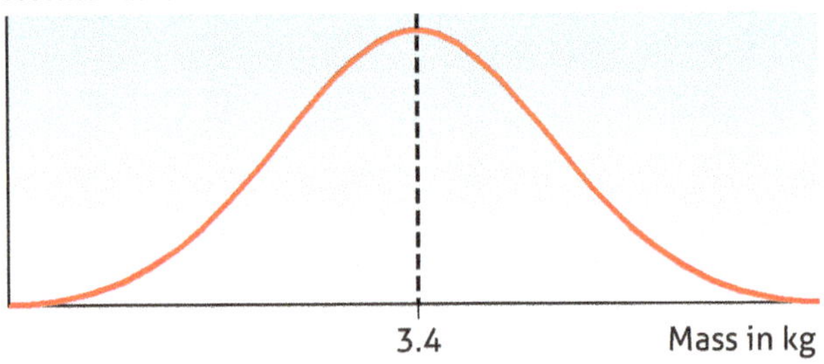

(b) Approximately Normal

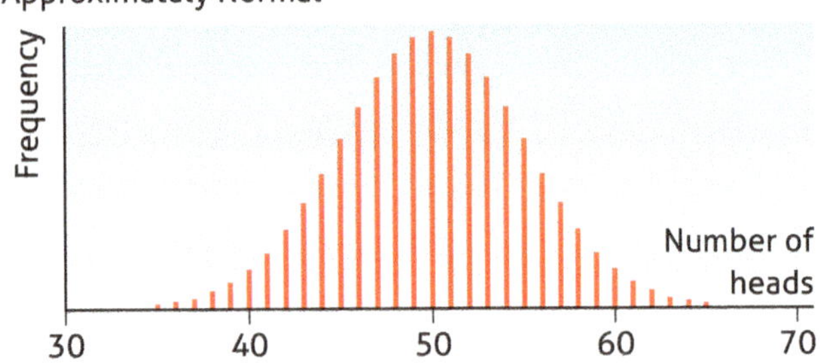

(c) Normal distribution

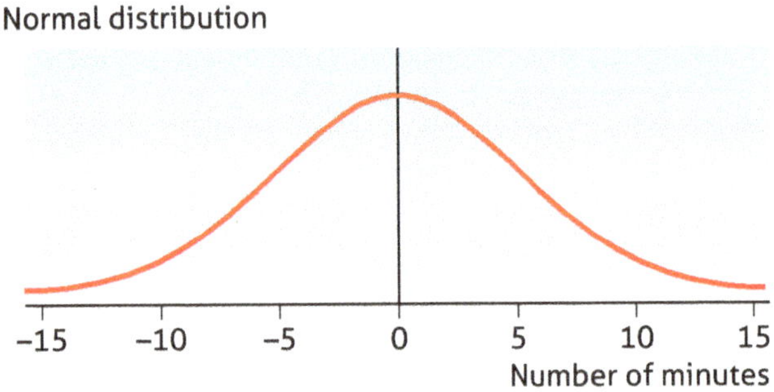

(d) Not Normal (uniform distribution)

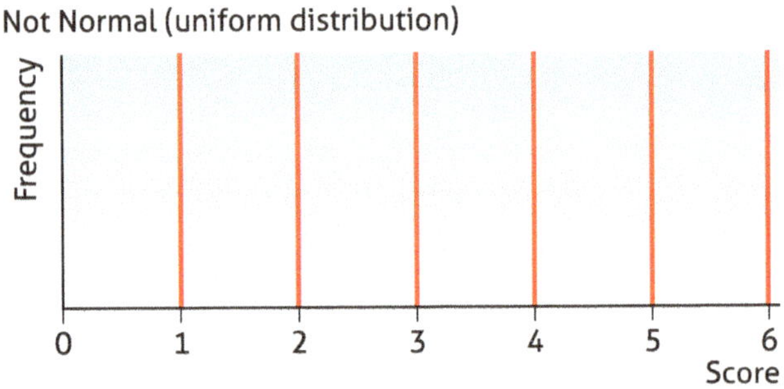

(e) Approximately Normal

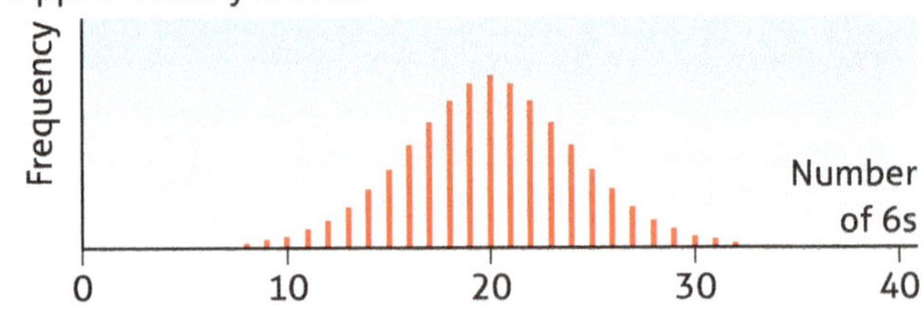

(f) Not Normal

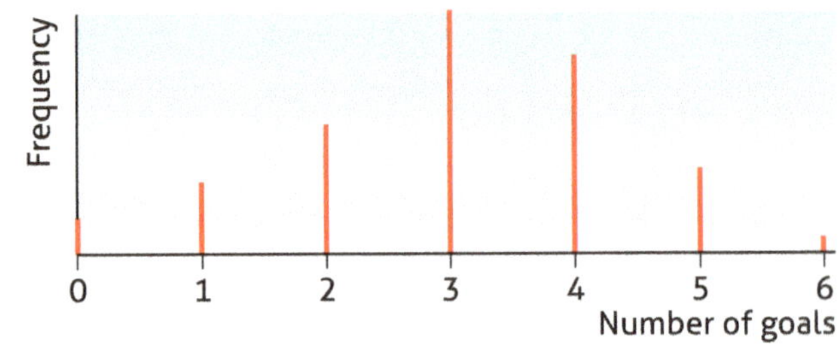

Exercise 17A

1 A 47.5% B 13.5% C 47.5% D 81.5% E 16% F 16%

2 (a) (i) 16% or 0.16 (ii) 2.5% or 0.025 (iii) 68% or 0.68

(b) Sam: $z = 0.4$; Amy: $z = -1.4$; Leon: $z = -0.4$; Amy is most unusual.

Activity 17B

(a) 10 of each score.

(b) Your own work.

(c) The distribution will be roughly uniform and the more rolls that are carried out the closer the distribution is to being completely uniform.

Exercise 17B

1 (a) 50 (b) 300

2 No, but it is highly likely to be between these two values.

3 (a) 420 (b) No

Activity 17C

Approximately Normal distribution for both dice.

Exercise 17C

1 (a) 2.77 (3 sf)

(b) Yes, a high *z*-score, so evidence that the die is biased.

2 $z = 1.43$; the *z*-score is not unusually high, so Marina does not have enough evidence.

Activity 17D

(a) Method A

(b) Increase the number of spins.

Exercise 17D

1 (a) 200

(b) $z = 2.5$; there is enough evidence to suggest that the coin might be biased. However, a high *z*-score is not definitive proof.

2 (a) 200 (b) 10

(c) (160), 170, 180, 190, 200, 210, 220, 230

(d) $z = 3.9$; 239 is nearly 4 standard deviations from the mean, so the evidence suggests that there is some bias.

3 (a) 450 (b) 15

(c) $z = -3.53$; 397 is more than 3 standard deviations below the mean, so it is likely shoppers are more likely to turn right than left.

Activity 17E

Method: calculate the z-score and then estimate the area under the Normal distribution curve.

Exercise 17E

1 (a) 69.15% (b) 77.34% (c) 96.41% (d) 99.18%
(e) 67.58%
2 4938 bags

Activity 17F

Step 1: calculate the z-score.
Step 2: use symmetry to find the area under the Normal distribution curve.

Exercise 17F

1 (a) 34.46% (b) 65.54% (c) 11.51% (d) 88.49%
(e) 54.03% (f) 22.95%
2 27.43%

Test yourself 17G

1 (a) −2.67
(b) Yes, it is unlikely that such an extreme z-score would occur by chance.
2 46.81%
3 0.0668
4 58.9%
5 (a) 5000 (b) 50
(c) $z = -92.76$; Yes, it is unlikely that such an extreme z-score would occur by chance, which suggests the company is probably exaggerating its claim.

Practice question
(a) Symmetrical about the centre, bell shaped.
(b) Your own answer.
(c) $z = 3.75$; it is unlikely that such an extreme z-score would occur.

Chapter 18

Activity 18A

(a) Suitable (b) Not suitable (c) Suitable

Exercise 18A

1 **Assumptions:** Word length, sentence length, choice of words vary between authors.
Data collection: Decide on the best author fingerprint. If necessary run a small pilot. Decide on an adequate size of sample text.
Processing: Use samples of the three authors' work to find their fingerprints. Then take a sample of the unknown author's text and see which it matches best.
Check whether the text files, etc. be analysed using appropriate software. (There is some suitable free software available on the internet.)
Interpretation: Decide whether the sample of the unknown author's text clearly matches one of the three authors' fingerprints but not the other two. Is more data needed?

2 **Assumptions:** Summer moisture can be measured by the total rainfall over June, July and August (summer) and the summer temperature by the mean temperature over the three months. The 'hardness' of the subsequent winter can be measured by the mean temperature over the three winter months (December, January, February). The 13th-century saying implies a negative association between mean summer rainfall and mean winter temperature, and a positive association between mean summer temperature and mean winter temperature.
Data collection: All the weather data for particular weather stations can be found on the Met Office website, some going back over 100 years.
Processing: Enter the data into a spreadsheet and use it to draw both the relevant scatter diagrams. An alternative approach is to find the proportion of years when each part of the saying is true.
Interpretation: If there is a weak association it may be helpful to collect data from more Met Office sites.

Activity 18B

These are just some of many possible answers.
(a) Take some of the cod caught by fishermen in a number of different places.
(b) Go to a number of towns in different parts of the country and select people sleeping on the street.
(c) Choose, say, every 100th mobile phone leaving the factory.
(d) Select people from the town's electoral role.
(e) Choose a number of universities and then a number of psychology students from each of them.

Exercise 18B

1 In some of these situations the choice of problem solving procedure may depend on the available budget; in such cases there can be realistic alternatives to these answers.

(a) Cluster	(b) Opportunity	(c) Stratified
(d) Cluster	(e) Quota	(f) Systematic
(g) Self-selected	(h) Cluster	(i) Stratified
(j) Simple random	(k) Self-selected	

Activity 18C

(a) Online are self-selecting. Face-to-face in the street is opportunity sampling, usually with quotas.
(b) Possible bias due to positioning, e.g. outside a baby-ware shop, butchers, large mall, etc.

Exercise 18C

1 (a) Leading question
(b) Some people may answer incorrectly to avoid incriminating themselves.
(c) Difficult to understand, so respondents may not answer the intended question.
(d) Biased against other possible answers (e.g. the hare).

2 It is not a truly stratified sample. The numbers sampled from each part of the town should be proportional to its population size rather than its area.
3 Many people use mobiles and no longer have landline telephones. So, the sample may well be unrepresentative.

Activity 18D

(a) 350
(b) In E9 median is larger than upper quartile. In C14 the total is less than one of the items.

Exercise 18D

1 (a) Your chart (b) Your spreadsheet
(c) As ranked by your spreadsheet.
(d) IQR = 318 - 290 = 28; 190 is an outlier
(e) Mean = 300.0 (to 1 dp), standard deviation = 30.4
2 (b) (10, 1205) (c) (165 to 175) microns
(d) No outliers by either method.

Activity 18E

Observations based on:

	Mean	Median	IQR	sd
PM	57.26	57.00	10.00	7.20
President	54.59	55.00	9.00	7.24

They are remarkably similar.

Exercise 18E

1 Your own answer but it should include information like Mean 820.9, sd 7.4, range 34, Median 822, Lower quartile 819, Upper quartile 824, IQR 5. It may also consider whether 797 is an outlier.
2 Your own answer, but it should include information such as

	Mean	Median	SD	IQR
Rest	89	86	11.5	14.8
Walk	100	99	11.5	12.0
Jog	112.8	114	10.8	11.0

Activity 18F

(a) The mean is 299 792 450 + the mean of 7, 5, 9, 6, 3.
(b) The standard deviation is that of 7, 5, 9, 6, 3.

Exercise 18F

1 (a) S = total score, H = number of hits
(b) Mean = 29, sd = 7.2
2 (a) Mean = 5.1, sd = 3.4 (1 dp)
(b) Mean = 2551, sd = 34

Activity 18G

(a) Shape, approximate location of mean, mode and median
(b) Use of a Normal probability plot.

Exercise 18G

1 A: Neither B: Uniform C: Normal D: Neither
2 (a) Approximately Normal
(b) Probably not, assuming mean time per km of 4 m 25 s (which is optimistic).
3 (a) A Normal, B Uniform
(b) A Approximately 15% to 20%; B Approximately 30%; A is the safer.

Activity 18H

Good: It states the question clearly.
Needing improvement:
It does not explain how the data answer the question.
It does not summarise the data.
It does not give the source of the data.
The stages of the cycle are present but not stated explicitly.

Exercise 18H

Your own answers. They may include the following ideas.
1 **Accept:** Seems obvious choice but it does not take account of variation, although A has more call-outs over the weekend, B has larger of either – 72 on Friday – but this might be chance.
Ask for more data: So far only for one week. Are there some significant differences in populations between the areas?
Employ different method: Probably the best option – need more data and need to take into consideration differences in demand over time, e.g. end of week and weekends – giving B a far greater spread in demand than A.
2 **Seek more information:** Jasper is not quite right with the 30% over 70 mph it is 'only' 24%.
These are speeding figures; there is nothing about accident rates. All 50 cyclists may have never had an accident or most may have had an accident. There is no way to tell from the data.

Test yourself 18I

1 It would be unethical to carry out an experiment dropping cats from high places, so only secondary data can be used. In fact all the original data concerned cats who were actually taken to the vets after they had survived their falls. So the data were biased.
2 (a) Between 900 and 910
(b) People visiting the town hall may be unrepresentative of the population.
(c)

Age range	18–24	25–44	45–64	65–84	85+
Voters	76	56	36	28	4

(d) A Biased – many voters may just want the sports centre in their locality.
B is a poor opportunity sample and likely to be unrepresentative.
D Self-selected sample – only people who listen in and who probably have strong views so make the effort to phone.
C Is the best – reasonable size, represents the whole population of the town in proportion to age distribution (but does miss out the under-17s and those not registered to vote).

3 (a)

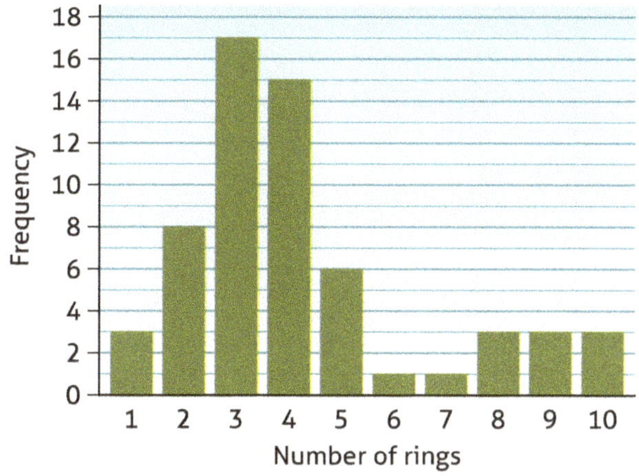

(b) Median = 4; IQR = 2

(c) Jan is wrong, it is 28 out of 60 = 46.7%.

(d) After 6 or more rings possibly but almost certainly after 8 or more. The distribution at these numbers of rings is flat, suggesting a machine setting for switching. For human waiting times a roughly bell-shaped curve is more likely.

4 (a) A grouped frequency chart of the data shows the distribution is not Normal. Its mode is at low wave heights and it is positively skewed.
The distribution is not Normal as can be seen from this grouped frequency chart, which is positively skewed. In addition, the mean (6.46 m) and median (6.30) are different, although the difference looks quite small. The modal group is not symmetrical with the distribution. Stronger evidence could be gained by using the normal probability plot.

(b) (i) Mean: True is 7.46 (increased from 6.47).

(ii) Median: True is 1 metre more than the measured value.

(iii) Standard deviation: No difference.

5 (a) Not bell-shaped but positively skewed. Mean (6.75) and median (6.50) different.

(b) Mean = 6.75, standard deviation = 2.56

(c) No outliers ('non-outlier' range is $1.6 \leq x \leq 11.9$)

(d) Median = 6.5, Q1 = 5.05, Q3 = 8.2

(e) No outliers ('non-outlier' range is $0.3 \leq x \leq 12.9$)

(f) The data are very spread out suggesting the need for improved measurement. More observations will be helpful if they are more accurate.

Practice question

(a)

Time (m)	60–69	70–79	80–89	90–99	100–109	110–119	120–129	130–139
Frequency	2	2	2	6	6	2	4	1

(b)

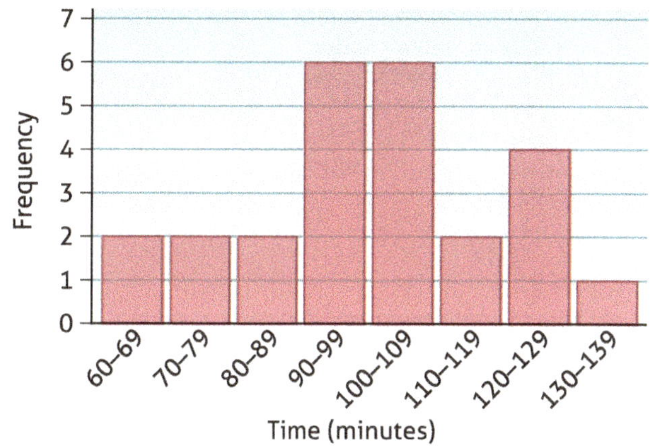

(c) Approximately bell-shaped, mean = median = 100

(d) (i) 84% (ii) 16%

(e) There are 2 × 20 = 40 km of bank but 1 km has been done so 39 km remain.

(f) 25 sections per km so 975 sections to be done.
120 volunteers doing 2 sections is 240 per day.
975 ÷ 240 = 4.0625, so a 5th day is needed.
120 volunteers doing 3 sections is 360 per day.
975 ÷ 360 = 2.708, so 3 days are enough.

(g) On average, yes. 6 hours is much more than 3 times the mean.

Chapter 19

Activity 19A

It may be polluted.

Exercise 19A

1 (a) (i) Life expectancy of women in the town

(ii) Records of deaths

(iii) H_0: Life expectancy of women in the town is unchanged from that in the past; H_1: Life expectancy is different from that in the past.

(b) (i) The number of earthquakes of magnitude at least 3 each year.

(ii) Geological records (the British Geological Survey)

(iii) H_0: The number of 3+ earthquakes per year now is the same as it was in the past.
H_1: The number of 3+ earthquakes per year now is different from that in the past.

(c) (i) The numbers of questions students get right.
(ii) The students' marks.
(iii) H_0: The mean number is 25; H_1: The mean number is not 25.

2 (a) H_1: $\mu \neq 100$
(b) Normal distribution with mean 100 and standard deviation 7.07.
(c) $z = \dfrac{89 - 100}{7.07}$
(d) ± 1.645
(e) There is not enough evidence to reject the null hypothesis. Alice's theory that singers are different is not supported.

Activity 19B

The distribution is Normal, with mean 300 and standard deviation 10.
That would make 280 to 320 the mean ±2 standard deviations, so account for 95% (the vast majority).

Exercise 19B

1 (a)

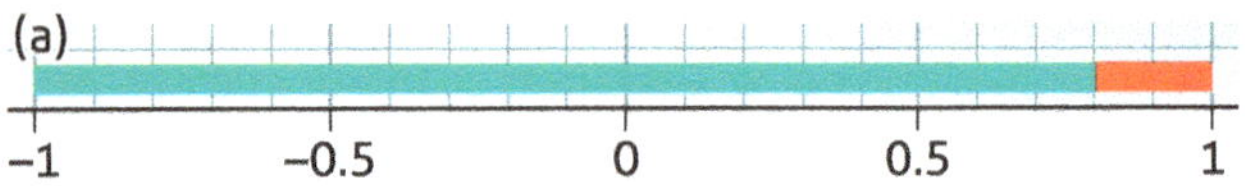

(b)

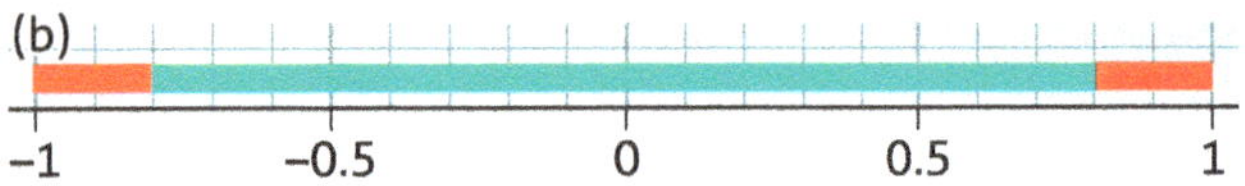

2 (a) (i) 1-tail
(ii) H_0: Average size = 2.4; H_1: Average size > 2.4
(iii) A sample of household sizes in the town.
(b) (i) 1-tail
(ii) H_0: The number of rough sleepers is unchanged. H_1: The number of rough sleepers has increased.
(iii) Collect a (cluster) sample of numbers in several towns.
(c) (i) 2-tail
(ii) H_0: The probability that an interviewee is male is $\frac{1}{2}$. H_1: The probability that an interviewee is male is not $\frac{1}{2}$.
(iii) The numbers of males and females interviewed by particular people.

Activity 19C

Observed frequency, f_o	Robins	Blackbirds	Sparrows	Total
Park A	8	14	10	32
Park B	12	18	18	48
Total	20	32	28	80

The distributions are similar.

Exercise 19C

1 Correct: b, f; Wrong: a, c, d, e

2

Frequency	Foxes	Badgers	Deer	Total
Wood P	16	24	40	80
Wood Q	24	36	60	120
Total	40	60	100	200

The distributions are similar.

Activity 19D

(a) 2 rows and 3 columns
(b) 2 parks and 3 types of bird
(c) It will be 2 × 3.

Exercise 19D

1 (a)

f_o	A	S	B	D	Total
Area X	16	78	18	8	120
Area Y	8	32	15	25	80
Area Z	35	40	20	5	100
Total	59	150	53	38	300

(b) 3 × 4
(c)

f_1	A	S	B	D	Total
Area X	23.6	60	21.2	15.2	120
Area Y	15.73...	40	14.13...	10.13...	80
Area Z	19.66...	50	17.66...	12.66...	100
Total	59	150	53	38	300

(d) H_0: The proportions of people with different drinking habits are independent of the area.
H_1: The proportions of people with different drinking habits are not independent of the area.
(e) $X^2 = 57.91$
(f) Cell (Y,D) or Cell (Z,A)

Activity 19E

(a) No.
Bolton's goals + Preston's goals = Total goals
(b) 2

Exercise 19E

1 2 × 8 or 8 × 2
2 (a) 3 × 4 (b) 6 (c) 16.81

Activity 19F

H_0 is accepted. The evidence suggests the distribution of the different types of birds is independent of the park.

H_0 is rejected. The evidence suggests the distribution of the different types of birds is not independent of the park.

Exercise 19F

The solutions above for Exercises 19D (Q 1) and 19E (Q 2) continue:

The test statistic is $X^2 = 57.91$ and the critical value is 16.81.

Since 57.91 > 16.81, the evidence does not support the null hypothesis at the 1% significance level. The evidence suggests that the proportions of people with different drinking habits do depend on the area they live in.

Test yourself 19G

1 (a) The test is 1-tail (b) $N(50,5^2)$ (c) $z = \dfrac{58 - 50}{5} = 1.6$
(d) ± 1.96 are the critical values for a 2-tail test.
(e) 'The evidence is not sufficient to support Simon's theory that girls prefer tall men.'

2 (a) Significant for a 1-tail test at the 5% significance level. Not for a 2-tail test.
(b) The software has used the test statistic when finding the p-value.

3 (a)

Expected frequency, f_e	Before 16	16/17	18/19	After 19	Total
Town A	5.12	9.28	16	9.6	**40**
Town B	7.68	13.92	24	14.4	**60**
Town C	3.2	5.8	10	6	**25**
Total	**16**	**29**	**50**	**30**	**125**

(b) For Town C before 16, $f_e < 5$.
Combine 'Before 16' and '16/17' into one category or combine Towns B and C.

(c) Combine 'Before 16' and '16/17' into 'Before 18'.
The complete test is as follows.

This test is based on the proportions of people with different ages of finishing formal education in 3 towns.

H_0: The proportions finishing education at different ages are independent of where they live.

H_1: The proportions are not independent of where people live.

Significance level: 5%
Test: χ^2 test on a contingency table, 1-tail
Assuming H_0 is true, the expected frequencies are as follows:

Expected frequency, f_e	Before 18	18/19	After 19	Total
Town A	14.4	16	9.6	**40**
Town B	21.6	24	14.4	**60**
Town C	9	10	6	**25**
Total	**45**	**50**	**30**	**125**

$X^2 = 22.4$ (3 sf), $\nu = (3 - 1) \times (3 - 1) = 4$, the critical value is 9.488.

$22.4 > 9.488$, the null hypothesis is rejected in favour of the alternative hypothesis.

The evidence suggests that the proportions of people finishing education at different ages are not independent of where they live.

4 (a) Blue is the acceptance region and red is the critical (rejection) region.

(b) 10%

(c) There is no limit to the possible size of the test statistic X^2.

5 (a) Your own answer.

(b) Ignoring the Total cells, you can fill 6 of the cells in the body of the table leaving no choice for numbers in the remaining cells.

(c) So, there are 6 free variables in the system and 6 degrees of freedom in the table. $(3 - 1) \times (4 - 1) = 6$

Practice question

(a) H_0: The proportions of the different types of animals are independent of the date.
H_1: The proportions are not independent of the date.
The test is 1-tail.

(b) $\nu = (7 - 1) \times (2 - 1) = 6$; Critical value = 12.59

(c)

	Observed frequency, f_o			Expected frequency, f_e		
Hedgehogs	12	2	**14**	7.583 ...	6.416 ...	**14**
Foxes	6	6	**12**	6.5	5.5	**12**

(d) $X^2 = 34.20$ (to 2 decimal places)

(e) Since $34.20 > 12.59$; H_0 is rejected in favour of H_1.
The evidence suggests that the proportions are not independent of date.

(f) Combining the red and grey squirrels gives $X^2 = 7.47$ and $\nu = 5$.
The critical value is now 11.07.
Since $7.47 < 11.07$, H_0 is accepted. The evidence now suggests that the proportions of the different animals are indeed independent of the date.

Chapter 20

Activity 20A

The line does not appear to fit the points particularly well but this is exaggerated by the scale used for the vertical axis.

Exercise 20A

1 (a) Correlation does not imply causation.
(i) Clear water does not necessarily attract fish.
(ii) Fish do not make the water clear.
(b) Absence of pollution.

2 (a)

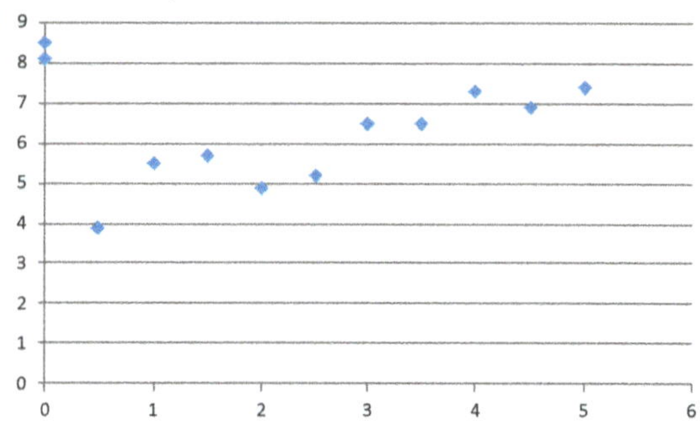

(b) The two who received placebos.

(c) Positive association/correlation

(d) The medication improves sleep by 25 minutes on average.

3

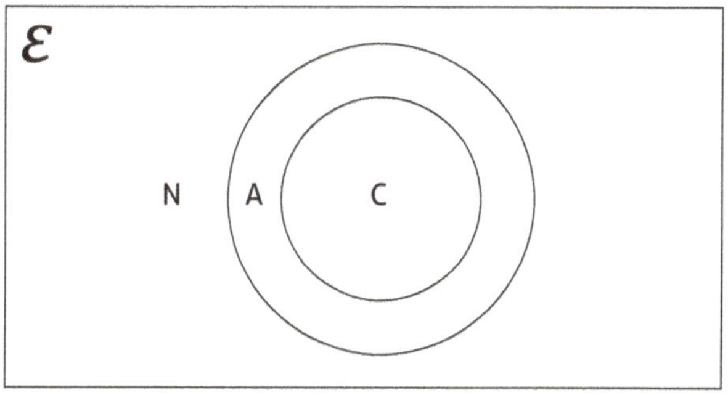

Activity 20B

(a) No

(b) Comparing the differences in their scores with the general spread.

Exercise 20B

1 (a)

Book	A	B	C	D	E	F	G	H
Length rank	8	5	2	4	1	7	3	6
Score rank	8	5	7	3	1	4	2	6

(b)

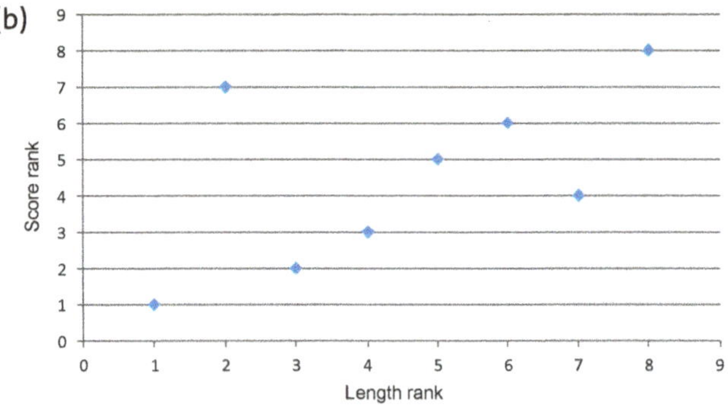

(c) Some positive association despite two outliers.
(d) 0.571 (e) No

2 (a) −0.833, they disagree
(b) (i) 0.738 (ii) −0.857
(c) Robin and Amalie; Amir, Mohan and Wilfred

Activity 20C

(a) You must rank their times.

Person	C time	C rank, x	S time	S rank, y	$d = x - y$	d^2
Sally	22	1	31	4	−3	9
Bella	45	6	30	3	3	9
Kangwa	38	5	38	7	−2	4
Louis	46	7	36	5	2	4
Sam	35	4	22	1	3	9
Mat	23	2	24	2	0	0
Ollie	27	3	38	7	−4	16
Hector	62	8	38	7	1	1
				Σ	0	52

(b) $r_s = 1 - \dfrac{6 \times 52}{8 \times (8^2 - 1)} = 0.381$ (3 dp)

(c) Interchanging x and y does not change the value of d^2.

Exercise 20C

1 (a) 2-tail
(b) H_0: There is no association between the amount of fertiliser and the yield.
H_1: There is an association.
(c) 0.682
(d) The test is significant. The evidence supports the alternative hypothesis.

2 The value of r_s is 0.943 but correlation does not imply causation.
Fine weather may have influenced both variables.

Activity 20D

(a) (i) By plotting the points on a scatter diagram
(ii) Calculating $r_s = -0.8095...$
(b) It gives the same answer of −0.8095...

Exercise 20D

1 (a) Positive correlation

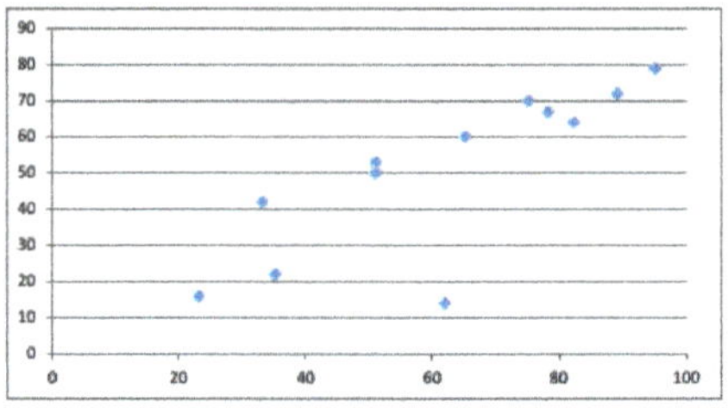

(b) (62, 14)
(c) 0.807; confirms positive correlation.

2 (a) Negative correlation
(b) (i) The correlation is strong.
(ii) Countries with higher life expectancy tend to have lower birth rates.
(c) Correlation does not imply causation. An underlying cause for both could be low income.

Test yourself 20E

1 (a)

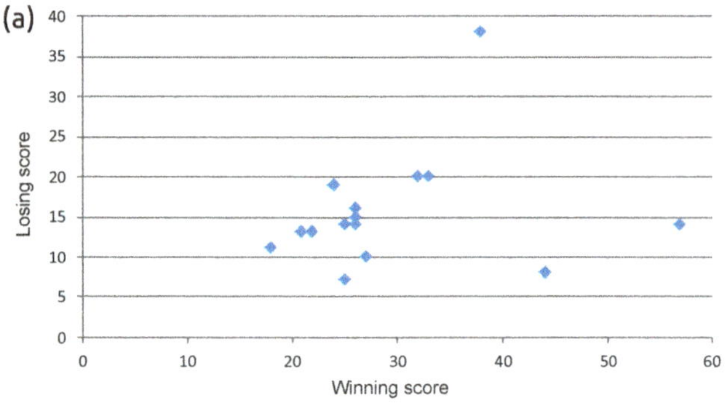

(b) Neither Owen nor Fiona is supported by the scatter diagram; Hari's statement is true.
(c) Take a larger sample, including different types of matches not just the Six Nations.

2 (a)

House	A	B	C	D	E	F	G	H	I	J
Price (£1000)	8	10	1	7	3	5	9	4	6	2
Bedrooms	7.5	10	1	6	2	4	9	4	7.5	4
Land area (ha)	10	7	1	8	4	9	2	6	5	3

(b) (i) 0.942 to 3sf (ii) 0.479 (iii) 0.342
(c) Your own answer.

3 (a) H_0: There is no association between the temperature and the weight of the fish caught.
H_1: There is an association.
(b) 0.6833 (c) 0.700 (d) The data do not support his theory.

4 (a) Self-selected. The sample is not random and likely to be unrepresentative of all people.
(b) H_0: There is no association between consumption of the drink and increase in weight.
H_1: There is a positive association.
(c) $r_s = 0.7255$. The critical value is 0.5035, so the result is significant.
(d) You can never prove anything with statistics.
The sample was not random.
Correlation does not imply causation, etc.

5 (a) 0.8095...
(b) (i) The ranks of the numbers in column A.
(ii) (The number being ranked, the numbers among which it is being ranked, 0 for the largest to be ranked 1, or 1 for the smallest to be ranked 1.)
(c) 0.8095... Spearman's coefficient of rank correlation.
(d) 0.7929..., the pmcc of the original data.

Practice question

(a)

Person	A	B	C	D	E	F	G	H
Time	21	15	18	14	22	2	16	29
Errors	9	8	6	11	5	10	5	2

(b) The outlier is F. A time of 2 minutes is impossible so there is an error.
(c) H_0: There is no association between time taken and errors made. H_1: There is a negative association.
(d)

	Time	Time rank, x	Errors	Errors rank, y	$d = x - y$	d^2
A	21	3	9	2	1	1
B	15	6	8	3	3	9
C	18	4	6	4	0	0
D	14	7	11	1	6	36
E	22	2	5	5.5	−3.5	12.25
G	16	5	5	5.5	−0.5	0.25
H	29	1	2	7	−6	36
				Σ	0	94.5

$r_s = -0.6875$; critical value is -0.7413. Insufficient evidence to reject the null hypothesis.
(e) Given the size of this sample, there is now enough evidence to support the tutor's theory.

Index

P

Q

R

S

T

U

V

W

Z

Photo credits

Cover, p1, p3, p4 © Eric Isselée/stock.adobe.com; p6, p9 © Franzi draws/stock.adobe.com; p7 © Matthias Schulz/stock.adobe.com; p10, p127–129, p141 © Google Earth; p11 *l* © Pixel Embargo/stock.adobe.com; p11 *r* © Andrew_rybalko/stock.adobe.com; p16 © DG-Studio/Shutterstock.com; p19 © Wachiwit/stock.adobe.com; p30 © Mandia/Shutterstock.com; p32 © Kanjana/stock.adobe.com; p33 © Nigel Cooke/Action Plus Sports Images/Alamy Stock Photo; p35 © Maxx-Studio/Shutterstock.com; p37 © Jessica J Jimenez/Shutterstock.com; p38 © Fiphoto/Shutterstock.com; p41 © Lakov Kalinin/stock.adobe.com; p42 © Jarek Kilian/Shutterstock.com; p45 © Bogac/Shutterstock.com; p46 © Alisonhancock/stock.adobe.com; p47 © Jon Le-Bon/stock.adobe.com; p52 © Bankoo/stock.adobe.com; p54 © Nito/stock.adobe.com; p57 © ECLIPSE PRODUCTION/Shutterstock.com; p58 © Banphote/stock.adobe.com; p60 © DGLimages/Shutterstock.com; p63 © Richard Carey/stock.adobe.com; p64 © Jerry Sharp/Shutterstock.com; p68 © Angela Ksen/stock.adobe.com; p69 *t* © Pavlo S/Shutterstock.com; p69 *b* © Photo12/Ann Ronan Picture Library/Alamy Stock Photo; p72 © Diego Barbieri/Shutterstock.com; p77 William Wesen/https://commons.wikimedia.org/wiki/File:Vupperfalls.jpg; p85 © Mikadun/Shutterstock.com; p98 *t* © Guillem Gaber/Shutterstock.com; p98 *b* © William87/stock.adobe.com; p100 © Barrie Bird/Shutterstock.com; p102 © BillionPhotos.com/stock.adobe.com; p110 © Zhengzaishanchu/stock.adobe.com; p111 *t* © Barry Barnes/stock.adobe.com; p111 *b* © Marcel/stock.adobe.com; p121 *t* © KerrysWorld/Shutterstock.com; p121 *b* © Chinch/Shutterstock.com; p126 © Xiaorui/Shutterstock.com; p133 © Lareen/Shutterstock.com; p135 © DenisMArt/stock.adobe.com; p142, p249 © Rawpixel.com/stock.adobe.com; p144 © BGStock72/stock.adobe.com; p145 © Zephyr_p/stock.adobe.com; p147 © SMSka/stock.adobe.com; p148 © Sikov/stock.adobe.com; p154 *t* © PBaishev/stock.adobe.com; p154 *b* © MisterStock/Shutterstock.com; p157 © Paul Maguire/stock.adobe.com; p159 © Anton/stock.adobe.com; p161 © Rawpixel.com/Shutterstock.com; p162 © Graham oakes/Shutterstock.com; p164 © Luke Schmidt/Shutterstock.com; p166 *t* © Daniel Ernst/stock.adobe.com; p166 b, p205 © Monkey Business/stock.adobe.com; p178 © MDBPIXS/stock.adobe.com; p179 © WoodysPhotos/Shutterstock.com; p181 *t* © BasPhoto/Shutterstock.com; p181 *b* © Artmim/stock.adobe.com; p184 *t* © Guteksk7/stock.adobe.com; p184 *m* © Cristina.A/Shutterstock.com; p187 © Arcurs Corp/peopleimages.com/stock.adobe.com; p191, p246 © Fizkes/stock.adobe.com; p197 © CoolR/Shutterstock.com; p199 *b* Microsoft Excel; p199 *r* © Popova Olga/stock.adobe.com; p201 © Upsidedowncake/stock.adobe.com; p208 © Morten Normann Almeland/Shutterstock.com; p211 © Blane13/Shutterstock.com; p213 © Xiaoliangge/stock.adobe.com; p214 © Vladdeep/stock.adobe.com; p219 © Sergei Mishchenko/Shutterstock.com; p228 © Andrey Popov/stock.adobe.com; p230 *t* © Vlad61_61/stock.adobe.com; p230 bl © Henk Wallays/Wirestock Creators/stock.adobe.com; p230 br © Tim's insects/stock.adobe.com; p232 © Milan/stock.adobe.com; p233 © Spirenko/stock.adobe.com; p235 © Ray Park Stock Photo/stock.adobe.com; p236 © Ysbrandcosijn/stock.adobe.com; p238 © Tupungato/Shutterstock.com; p240 © FJAH/Shutterstock.com; p244 © IrynaV/stock.adobe.com; p252 © Sripfoto/stock.adobe.com; p254 © Piotr Wawrzyniuk/stock.adobe.com.